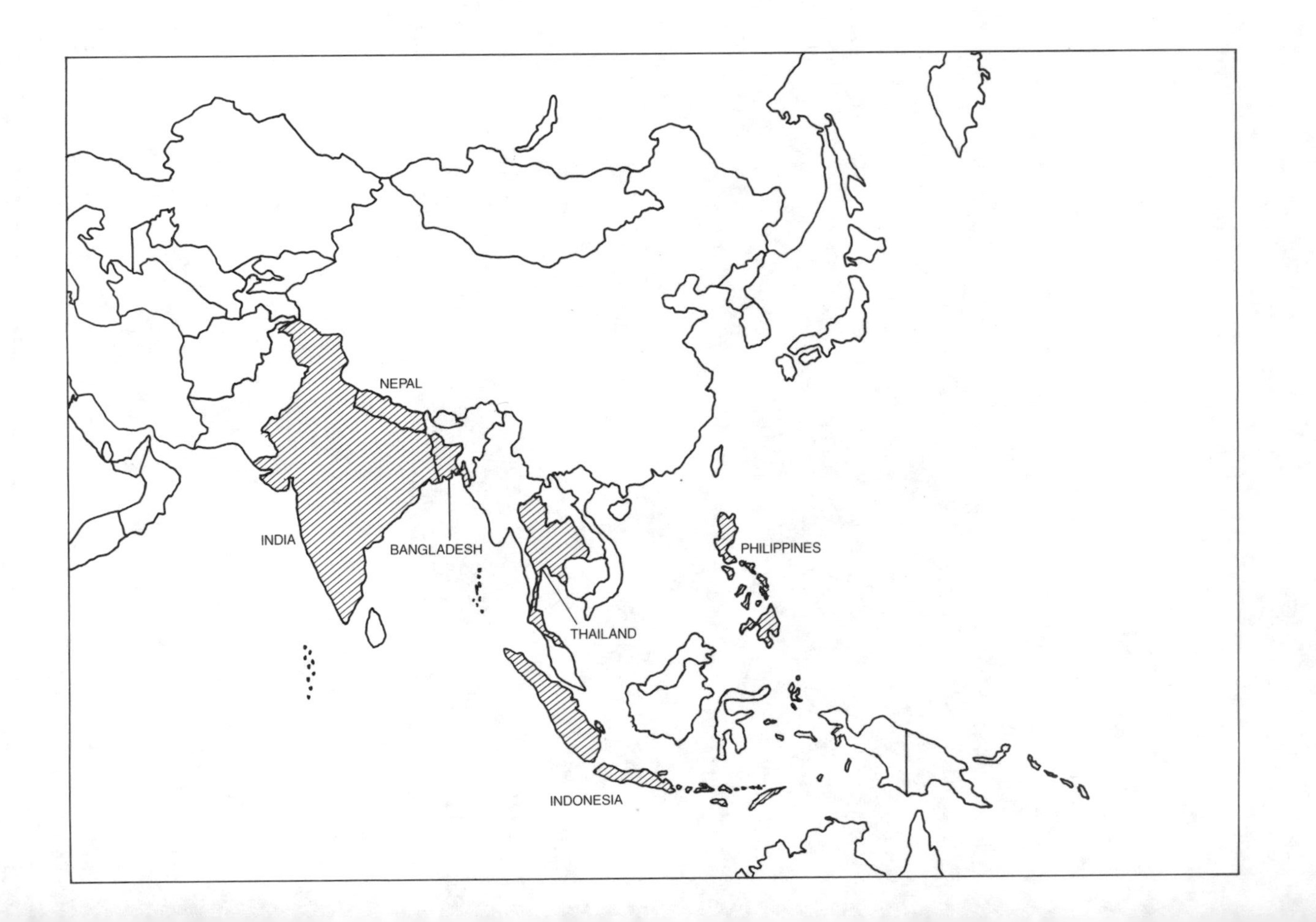
NEPAL
INDIA
BANGLADESH
PHILIPPINES
THAILAND
INDONESIA

NON-GOVERNMENTAL ORGANIZATIONS AND THE STATE IN ASIA

This book is in the Non-Governmental Organizations Series, which comprises three regional studies and a synthetic overview volume. Each of the titles presents detailed empirical insights into the work of Non-Governmental Organizations in agriculture. The case material is set within the context of NGOs' relations with the state and their contribution to democratization and the consolidation of rural civil society.

The books are written at a time of massively increased funding for NGOs, of increasing fiscal stringency in the public sector attributable to structural adjustment, and of growing pressures on NGOs from governments and donors to rethink their roles.

Against the background of a broad review of institutional activity at the grassroots, each book explores specific questions concerning the work of NGOs in agricultural development:

- How good/bad are NGOs at promoting technological innovation and addressing constraints to change in peasant culture?
- How effective are NGOs at strengthening grassroots/local organizations?
- How do/will donor pressures influence NGOs and their links to the state?

Eschewing populist acclaim for NGOs, these books draw on a large volume of new empirical material to provide a comprehensive review and critique of the performance and capabilities of these organizations.

These books will find a wide readership among students, academics and practitioners in the field.

Non-Governmental Organizations and the State in Asia is edited by **John Farrington** and **David J. Lewis**, both at the Overseas Development Institute, with **S. Satish** of the Administrative Staff College of India and **Aurea Miclat-Teves** of the International Institute for Rural Reconstruction (Philippines).

NON-GOVERNMENTAL ORGANIZATIONS SERIES

Co-ordinated by the **Overseas Development Institute**

The titles in the series are:

RELUCTANT PARTNERS?
Non-Governmental Organizations, the State and Sustainable Agricultural Development
John Farrington and Anthony Bebbington with Kate Wellard and David J. Lewis

NON-GOVERNMENTAL ORGANIZATIONS AND THE STATE IN LATIN AMERICA
Rethinking Roles in Sustainable Agricultural Development
Anthony Bebbington and Graham Thiele with Penelope Davies, Martin Prager and Hernando Riveros

NON-GOVERNMENTAL ORGANIZATIONS AND THE STATE IN AFRICA
Rethinking Roles in Sustainable Agricultural Development
Edited by Kate Wellard and James G. Copestake

NON-GOVERNMENTAL ORGANIZATIONS AND THE STATE IN ASIA
Rethinking Roles in Sustainable Agricultural Development
Edited by John Farrington and David J. Lewis with S. Satish and Aurea Miclat-Teves

NON-GOVERNMENTAL ORGANIZATIONS AND THE STATE IN ASIA

Rethinking roles in sustainable agricultural development

Edited by John Farrington and David J. Lewis
with
S. Satish and Aurea Miclat-Teves

London and New York

First published 1993
by Routledge
11 New Fetter Lane, London EC4P 4EE

Simultaneously published in the USA and Canada
by Routledge
29 West 35th Street, New York, NY 10001

Typeset in Garamond by
Witwell Limited, Southport

Printed and bound in Great Britain by
Biddles Ltd, Guildford and King's Lynn

British Library Cataloguing in Publication Data

A catalogue record for this book is available from the British Library

Library of Congress Cataloging in Publication Data
Non-governmental organisations and the state in Asia : rethinking roles in sustainable agricultural development / edited by John Farrington and David J. Lewis with S. Satish and Aurea Miclat-Teves.
p. cm.
Includes bibliographical references and index.
ISBN 0-415-08847-X. — ISBN 0-415-08848-8 (pbk.)
1. Rural development—Asia. 2. Non-governmental organisations—Asia. I. Farrington, John.
HN655.2.C6N66 1993
307.1'412'095—dc20 92-45820
CIP

CONTENTS

FIGURES

TABLES

BOXES

CONTRIBUTORS

K V Bhat is Project Officer at the State Watershed Development Cell, Third Floor Podium Block, Bangalore–5600 001, India.

Bryan Bruns is based at 39/1 Ban Daun Ngeun, A Pong Phayo, 56140 Thailand.

Jerry Buckland was an agricultural officer at the Mennonite Central Committee, Box No. 13, Feni–3900, Bangladesh.

J Carew-Reid works in the Environment Planning Program of the International Union for the Conservation of Nature, PO Box 3923, Kathmandu, Nepal.

Lapu-Lapu Cerna is President of the Mag-uugmad Foundation, PO Box 286, Cebu City 6000, Philippines.

S Chakraborty is Director of the Ramakrishna Mission, Lokasiksa Parishad, PO Narendrapur, Dist 24 Parganas (South), West Bengal, India 743508.

C Das works in the Farming Systems Programme of the Ramakrishna Mission Ashram Narendrapur-740508, South 24-Paraganas, W. Bengal, India.

Jeanette Denholm is a member of the research staff at ICIMOD, P O Box 3226, Kathmandu, Nepal.

John Farrington is a Research Fellow at the Overseas Development Institute, Regent's College, Inner Circle, Regent's Park, London NW1 4NS, UK.

Carlos Fernandez is Under Secretary of the Department of Agriculture, Eliptical Road, Diliman, Quezon City, Philippines.

Aloysius P Fernandez is Director of MYRADA, NO.2 Service Road, Domlur Layout, Bangalore, 560 071 India.

Edward Giordano is Coordinator of Auroville Greenwork Resource Centre, ISAI Ambalam, Auroville 605 101, Tamil Nadu, India.

Julian Gonsalves is Director of the Sustainable Agriculture Programme at

IIRR, Rm. 38, Elena Apartments, 512 Romero Salas St., Ermita, Manila, Philippines.

Peter Graham is now Agricultural Administrator, Mennonite Central Committee, PO Box 2904, Mbeya, Tanzania and formerly occupied a similar position at MCC, Bangladesh.

Aroon Hwai-Kham is a researcher on the staff of the Appropriate Technology Association, 143/171–172 Pinkloa Nakornchaisri Road, Bangkoknoi, Bangkok, Thailand.

Lanthong Jonjuabsong is a Research Officer at the Ubon Farming Systems Research Office, Ubon Province 10900, Thailand.

Delfin Ganapin Jr is Under Secretary, Department of Environment & Natural Resources (DENR), Visayas Ave., Quezon City, Philippines.

Kamal Kar is Training Organiser and Programme Coordinator, Seva Bharati Krishi Vigyan Kendra, PO Kapgari, Dist Mindapore, West Bengal, India.

Mafruza Khan is a researcher on the staff of Proshika, 5/2 Iqbal Road, Mohammadpur, Dhaka-1207, Bangladesh.

N Prem Kumar is a Member of Faculty of the Administrative Staff College, Bella Vista, Hyderabad 500–49, India.

Warlito Laquihon works in the agricultural programme of the Mindanao Baptist Rural Life Centre, PO Box 94, 8000 Davao City, Philippines.

Wattana Leelapatra is a Research Officer at the Ubon Farming Systems Research Office, Ubon Province 10900, Thailand.

David J Lewis is a Research Assistant of the Overseas Development Institute, Regent's College, Inner Circle, Regent's Park, London NW1 4NS.

B Mandal is Farming Systems Research Adviser at the Ramakrishna Mission Ashram, Narendrapur-740508, South 24-Paraganas, W. Bengal, India.

P M Mane works in the Aga Khan Rural Support Programme, Navarangpura, Ahmedabad-380009, India.

James Mascarenhas has worked extensively on Rapid Rural Appraisal at MYRADA, No.2 Service Road, Domlur Layout, Bangalore, 5660 071 India.

Aurea Miclat-Teves worked as an Associate of the International Institute of Rural Reconstruction, Km 39, Silang, Cavite, Philippines.

Shams Mustafa is Research Economist in the Research and Evaluation Division of the Bangladesh Rural Advancement Committee, 66, Mohakhal G/A, Dhaka-1212, Bangladesh.

Ms Fahmeena Nahas is a staff member of Friends in Village Development, Bangladesh (FIVDB), PO Box 70, Khadimnagar Sylhet, Bangladesh.

Krishna Oil is Senior Programme Officer of the Environment Planning Program of the International Union for the Conservation of Nature, PO Box 3923, Kathmandu, Nepal.

Bishnu Hari Pandit is Executive Director of the Nepal Agroforestry Foundation (NAF), PO Box 916, Bisalnagar, Kathmandu, Nepal.

Sanzidur Rahman is Research Economist in the Research Evaluation Division, Bangladesh Rural Advancement Committee, 66, Mohakhal G/A, Dhaka-1212, Bangladesh.

Min Bahadur Rayachhetry is a consultant to Winrock, PO Box 1312, Kathmandu, Nepal.

Gregorio D Reyes is Chief Science Research Specialist at the Ecosystems Research and Development Bureau (ERDB) College, Laguna 4031, Philippines.

Mark Robinson, formerly a Research Fellow at ODI, is now at the Institute of Development Studies, University of Sussex, Brighton, Sussex BN1 9RE, UK.

Tess del Rosario is a staff member of the Department of Agriculture, Eliptical Road, Diliman, Quezon City, Philippines.

Asgar Ali Sabri is a staff member of Proshika, 5/2 Iqbal Road, Mohammadpur, Dhaka-1207, Bangladesh.

Dipankar Saha is Training Organiser at the R K Ashram Krishi Vigyan Kendra, PO Nimpith-Ashram Dist South 24 Parganas, West Bengal, India – 743338.

S Satish is a Member of Faculty of the Administrative Staff College of India, Bella Vista, Hyderabad 500–49, India.

Ghulam Sattar is Research Economist, Research and Evaluation Division of the Bangladesh Rural Advancement Committee, 66, Mohakhal G/A, Dhaka-1212, Bangladesh.

Armin Sethna has worked as an associate of the Aga Khan Rural Support Programme, Navarangpura, Ahmedabad-380009, India.

Parmesh Shah is Director of Agriculture at the Aga Khan Rural Support Programme, Navarangpura, Ahmedabad-380009, India.

Anil C Shah is Director of the Aga Khan Rural Support Programme, Navarangpura, Ahmedabad-380009, India.

Mohammed Shahabuddin is Principal Programme Coordinator at Proshika, 5/2 Iqbal Road, Mohammadpur, Dhaka-1207, Bangladesh.

Narayan Kaaji Shrestha has worked as a Consultant for Winrock, Nepal, PO Box 1312, Kathmandu, Nepal.

Irchmani Soelaiman is a staff member of LP3ES, JL-S, Parman 81 Slipi, Jakarta Barat, Kotali Pos 493 JKT, Indonesia.

John Sollows formerly worked as Adviser to the Ubon Farming Systems Research Office, Thailand, and is now with the International Development Research Centre, Tanglin PO Box 101, Singapore 9124.

P Swaminathan is Deputy Director (Extension) at the UPASI Krishi Vigyan Kendra, Coonoor – 643 101, Niliris DT – Tamil Nadu, India.

Duman Singh Thapa is Project Director, NCRP, PO Box 126, Kathmandu, Nepal.

Niran Thongpan is a Research Officer at the Ubon Farming System Research Office, Ubon Province 10900, Thailand.

Carlos Tomboc is Director, Ecosystems Research & Development Bureau (ERDB), College, Laguna 4031, Philippines.

T J P S Vardhan is Projects Officer at Action for World Solidarity 12-13-304, St No.9 Tarnaka, Secunderabad-500 017 (AP), India.

M P Vasimalai is Programme Director of PRADAN, 45B, TB Road, Arasaradi, Madurai – 625 010, Tamil Nadu, India.

Rev. Harold Watson is Director of the Mindanao Baptist Rural Life Centre, PO Box 94, 8000 Davao City, Philippines.

FOREWORD

The green revolution in Asia was highly successful in many respects. Food production per capita has risen by over 20 per cent since the 1960s. This has been achieved by a top-down approach, made possible because the revolution focused on a limited set of new technologies – high-yielding varieties together with fertilizer and pesticide packages – and on the most favoured agricultural regions of Asia – well irrigated lowlands with relatively large farms.

Today, thirty years on, the challenges are different. The sustainability of the green revolution gains are being questioned. Falling water tables, salinization, untoward effects of heavy pesticide and fertilizer use, and the first signs of yeild plateaux, are raising doubts as to whether the high levels of productivity can be maintained. But equally important is a shift of attention to the needs of those millions (probably well over half a billion) who live on Asia's less well-favoured lands. These lands are characterized by their great variety and the predominance of conditons adverse to productive agriculture – floods or droughts, deficient and infertile soils, susceptibility to erosion and inaccessibility. Unlike the larger farmers on the green revolution lands, the people who live there count farming as only one of the ways by which they achieve a livelihood. For a variety of reasons the top-down, transfer of technology approach is not appropriate to their needs.

This is not to say that the potential for development is absent; in many respects the returns to investment may now well be higher than in the green revolution lands. Nor is it impossible to produce appropriate, yet highly productive, technologies. The point is that the sheer heterogeneity, both of circumstance and need, will defy any attempt to identify homogenous solutions or organize research and development on a centralized model. Where the untapped potential lies is in the skills, capabilities and ingenuity of the rural people themselves, and in the energies and commitment of the numerous non-governmental organizations (NGOs), both large and small, that have sprung up in the countries of Asia in recent years.

This volume attests to the high level of potential that exists, providing many examples of successful partnerships between rural people, NGOs and government organizations (GOs). But the objective of the book, and the detailed studies on which it is based, is less to convey this message, which is in any case rapidly gaining acceptance, and more to provide a sober analysis of the complexity of the challenges involved.

NGOs can play many roles – as supporters, lobbyists, catalysts and innovators. Often they are small and flexible, but sometimes they are as large and powerful as GOs. This diversity poses problems for the people the NGOs are trying to help and for the GOs with whom they interact. As this book demonstrates, there are no blue-prints for success, but experience so far suggests that NGOs are at their most useful as facilitators, encouraging the emergence of local initiative, on the one hand, and enhancing the responsiveness of government services, on the other.

Gordon Conway
The University of Sussex

PREFACE

John Farrington, Overseas Development Institute

Interest in participation by the rural poor in the design and implementation of changes affecting their livelihoods, until recently a concern only of lobbyists on the fringe, is now moving to occupy centre stage in development debates. But practical experience with participatory approaches has so far been small-scale and localized. What organizational and institutional conditions need to be met if participatory approaches are to be implemented on a large scale? This book is one of a set of four that address this question.[1]

The book's focus is on agricultural change, particularly on how the types of technology and management practice necessary for sustainable improvement in agricultural productivity among small-scale, low-income farmers might best be developed. Institutionally, its focus is primarily on the work of non-governmental organizations (NGOs) in this sphere, but also on that of government research and extension services and, importantly, on the scope for closer interaction between the two.

These twin foci – on agriculture and on NGOs – are central to a number of policy concerns: the rural poor in many countries continue to rely for much of their livelihood on agriculture, yet renewable natural resources are coming under increasing population pressure, government research institutes have found agricultural technologies harder to design for difficult than for well-endowed areas, and not all of the changes that have been adopted have proven technically or institutionally sustainable. At the same time, assistance to developing countries provided by and through NGOs is increasing rapidly and now amounts to almost one-fifth of net bilateral flows. Growing interest in NGOs is driven by perceptions of their role in democratic pluralism, by the hope that they might share some of the costs of providing development services and, significantly in the present context, by their perceived ability to reach the rural poor. This book is, above all, an empirical study: by focusing on their work in agriculture it attempts to move the debate about NGOs' potentials and limitations from the rhetorical to the concrete.

The book and its companion volumes bring together the findings of a research study initiated by the Agricultural Research and Extension Network of the Overseas Development Institute (ODI) in 1989. The study grew out of a background paper,[2] given at a workshop on farmer participation in agricultural research at the Institute of Development Studies, the University of Sussex.[3] The paper reviewed around 100 experiences of working with farmers in research and extension, many of them received from Network members.

During 1988 and 1989 we discussed our interest in the institutional prerequisites for large-scale implementation of these methods with practitioners in the field, many of whom expressed willingness to write up their experience in what has hitherto been a sparsely documented field. By 1991 documentation was in progress in eighteen countries covering Africa, Asia and South America, and this book is one of three continent-wide compilations of experiences. In addition to these books, efforts to make available the results of the Asia component of the study to policy-makers at national and international levels have been made through:

- the publication of case studies in the ODI Network
- thematically focused papers presented at international seminars (e.g. Farrington and Bebbington 1991; Bebbington and Farrington 1992)
- sponsorship of national workshops in Bangladesh (Hassanullah, forthcoming) and the Philippines (Gonsalves and Miclat-Teves 1991)
- an Asia regional workshop held in Hyderabad, India, 16–20 September 1991.

The workshops in particular highlighted the extent to which entrenched attitudes, often perpetuated by lack of information, continue to act as a barrier to effective linkages between NGOs and the public sector: barriers that we hope this book and the broader story of which it is part have begun to break down.

NOTES

1 The other three are *Reluctant Partners? Non-Governmental Organizations, the State and Sustainable Agricultural Development* by John Farrington and Anthony Bebbington with David J. Lewis and Kate Wellard; *Non-Governmental Organizations and the State in Africa: Rethinking Roles in Sustainable Agricultural Development*, edited by Kate Wellard and James G. Copestake; *Non-Governmental Organizations and the State in Latin America: Rethinking Roles in Sustainable Agricultural Development* by Anthony Bebbington and Graham Thiele with Penelope Davies, Martin Prager and Hernando Riveros.

2 Farrington and Martin (1987).

3 'Farmers and Agricultural Research: Complementary Methods' workshop co-ordinated by Robert Chambers and held at the Institute of Development Studies, University of Sussex, in July 1987. For the edited proceedings, see Chambers *et al.* (1989).

ACKNOWLEDGEMENTS

This book is the culmination of a joint enterprise over several years among many individuals and organizations. The names of those who took responsibility for the final form in which ideas are presented are found at the beginning of each section, but many more contributed to the development of those ideas as the study from which this book derives took shape. All the sections not attributed were written by the editors.

We owe an intellectual debt to Frances Korten and Norman Uphoff for their seminal work on local organizations. Early work on participatory methods by Robert Chambers and participants at the 1987 'Farmers and Complementary Methods' workshop, together with that of Stephen Biggs and Adrienne Martin, first led us to ask how such methods might be implemented on a wide scale. Funds from the Nuffield Foundation allowed a round of initial visits to countries in the region which proved catalytic. Gordon Conway and Anthony Bottrall at the Delhi office of the Ford Foundation provided both financial support and intellectual stimulus during an extended period of fieldwork in India. They also helped us to link with the Administrative Staff College (ASCI) in Hyderabad which provided major support by allowing Dr Suryanarayan Satish to work with us for eighteen months, and by hosting the Asia regional workshop 'NGOs, Natural Resources Management and Links with the Public Sector' in September 1991, for which IDRC (Delhi) also provide support. Dr K V Raman, director of the National Academy for Agricultural Research Management (NAARM), generously provided intellectual and material inputs into the workshop and into the study as a whole. Without ASCI and NAARM the workshop would not have happened, nor would the continuing interaction among many of the eighty participants, and the study itself would not have benefited from the wealth of presentations and discussions. We are indebted to the principal and staff of ASCI, to the director of NAARM, to participants at the workshop and especially to Dr Satish for his skills and commitment in arranging the workshop and his intellectual inputs into the India case studies.

In Bangladesh, Bruce Currey of Winrock, Richard Holloway and Aroma Goon of PRIP/PACT, and Asmeen Khan and subsequently Ray Offenheiser of the Ford Foundation were generous with both financial and intellectual support. Dr M Hassanullah, with support from Winrock, was instrumental in bringing NGOs and government together for a national level workshop on our theme.

In the Philippines, Julian Gonsalves, Scott Killough and Aurea Miclat-Teves of the International Institute for Rural Reconstruction (IIRR) worked with us to identify illuminating case studies, provided support in writing them, and brought them to a national level workshop. John Graham of the Singapore office of the International Development Research Council (IDRC), and Frances Korten at the Ford Foundation (Manila) provided financial support for both the IIRR/ODI and ASCI/ODI workshops and were a source of constructive critique.

In Indonesia, Larry Fisher (World Neighbors) and David Winder (Ford Foundation) generously gave their time to discuss ideas, and Ford provided financial support.

In Thailand, Virayut Sujirakulkit of the Appropriate Technology Association (ATA), and John Sollows of the Ubon Farming Systems Research Institute provided invaluable insights into NGO–government interaction and helped to commission case studies. In Nepal, Gerry Gill of Winrock provided financial support, and both he and Kaaji Shrestha were a source of ideas and critique over many months.

The Ford Foundation (New York) provided generous support during the writing-up of all four volumes of the study, and Walt Coward, John Gerhard and Peter Geithner took time to argue through their ideas and discuss our own. Part of John Farrington's time was covered by a core grant from ODA.

Closer to home, Chuck Antholt of the World Bank was a strong source of support during his regular visits to ODI, and John Howell and Mark Robinson at ODI commented on proposals and drafts. Alison Saxby typed and organized the manuscript with consummate ease, and Kate Cumberland very capably handled an enormous volume of typing and organizational work in the earlier stages of the study.

We owe a particular debt to the rest of the ODI 'team' working on this study: Tony Bebbington, James Copestake and Kate Wellard shared ideas, enthusiasm and enormous personal commitment throughout the study.

Finally, the views expressed here are the responsibility of the editors and authors, and do not necessarily reflect those of any of the above individuals or organizations.

ABBREVIATIONS

ACAP	Annapurna Conservation Area Project (Nepal)
ADAB	Association of Development Agencies in Bangladesh
ADC	Area Development Centre
ADMAS	All-India Co-ordinated Research Project, Monitoring and Surveillance of Animal Diseases
ADO	Agricultural Development Officer
ADP	Agricultural Development Programme
AERDD	Agricultural Extension and Rural Development Department, University of Reading
AGRC	Auroville Greenwork Resource Centre (India)
AI	Artificial Insemination
AICRP	All-India Co-ordinated Research Project
AKRSP(I)	Aga Khan Rural Support Project (India)
ALP	Agri-Livestock Programme (Philippines)
AME	Agriculture, Man and Ecology
ANGOC	Asian Non-Governmental Organizations Coalition for Agrarian Reform and Development (Philippines)
APROSC	Agricultural Project Services Centre (Nepal)
APTT	Appropriate Technology for Tibetans
ARPP	Agricultural Research and Production Project
ATA	Appropriate Technology Association (Thailand)
ATA	Agricultural Technical Assistant
ATD	Agricultural Technology Development
ATS	Agricultural Technology System
AVA	Association of Voluntary Agencies (Indonesia)
AVI	Auroville International
AVRDC	Asian Vegetable Research and Development Centre (Thailand)
AWS	Action for World Solidarity (India)
BADC	Bangladesh Agricultural Development Corporation
BAIF	Bharatiya Agro-Industries Foundation (India)
BARC	Bangladesh Agricultural Research Council
BARD	Bangladesh Academy for Rural Development

BARI	Bangladesh Agricultural Research Institute
BAU	Bangladesh Agricultural University
BBP	Bauddha Bahunipati Project (Nepal)
BCRSP	Bangladesh Co-ordinated Soybean Research Project
BCSIR	Bangladesh Council for Scientific and Industrial Research
BFD	Bureau of Forest Development (Philippines)
BIRC	BAIF Information Resource Centre (India)
BKB	Bangladesh Krishi Bank
BNP	Bangladesh Nationalist Party
BONGO	Business-oriented NGO
BRAC	Bangladesh Rural Advancement Committee
BRIAH	Bharatiya Research Institute for Animal Health (India)
BRRI	Bangladesh Rice Research Institute
CARP	Comprehensive Agrarian Reform Programme Philippines
CBCP	Catholic Bishops' Conference of the Philippines
CC	Cluster Club
CDO	Chief District Officer
CDP	Crop Diversification Programme
CDR	Complex, diverse, risk-prone
CEC	Commission of European Communities
CECI	Canadian Center for International Studies and Cooperation
CENRO	Community Environment and Natural Resources Officer (Philippines)
CHEC	Commonwealth Human Ecology Concern
CHIRAG	Central Himalayan Rural Action Group
CHT	Chittagong Hill Tracts (Bangladesh)
CICFRI	Central Inland Capture Fisheries Research Institute
CIDA	Canadian International Development Agency
CIMMYT	International Maize and Wheat Improvement Centre
CMC	Central Management Committee
CODE-NGO	Caucus of Development NGO Networks (Philippines)
CPAR	Congress for People's Agrarian Reform (Philippines)
CPC	Chief Programme Co-ordinator
CPR	Common property resource
CRRI	Central Rice Research Institute (India)
CRS	Central Research Station
CSIRO	Central Scientific and Industrial Research Organization (Australia)
CSSRI	Central Soil Salinity Research Institute (India)
CSWCP	Cebu Soil and Water Conservation Programme (Philippines)
CTR&TI	Central Tasar Research and Training Institute (India)
CU	Calcutta University
CUSO	Canadian Universities Service Overseas

DANIDA	Department of International Development and Co-operation, Denmark
DAR	Department of Agrarian Reform (Philippines)
DDS	Deccan Development Society (India)
DENR	Department of Environment and Natural Resources (Philippines)
DFL	Disease-Free Laying (of silkworm eggs)
DFPR	Department of Forestry and Plant Research (Nepal)
DHP	Dihydroxypyridine
DLDB	District Livestock Development Board (India)
DLG	Department of Local Government (Philippines)
DLYC	District Level Youth Council (India)
DMMMSU	Don Mariano Marcos Memorial State University (Philippines)
DNES	Department of Non-Conventional Energy Sources (India)
DoA	Department of Agriculture
DoAE	Department of Agricultural Extension (Thailand)
DoE	Department of Environment
DoF	Department of Fisheries
DoH	Department of Health
DoL	Department of Livestock
DoRR	Department of Relief and Rehabilitation (Bangladesh)
DRDA	District Rural Development Agency (India)
DRDS	District Rural Development Society (India)
DSCWM	Department of Soil Conservation and Water Management (Nepal)
DST	Department of Science and Technology
EC	European Community
ED	Executive Director
EIG	Employment and Income Generation Programmes (Bangladesh)
ERDB	Ecosystems Research and Development Bureau (Philippines)
FAO	Food and Agriculture Organization (United Nations)
FCA	Forest (Conservation) Act (India)
FEVORD-K	Federation for Voluntary Organizations for Rural Development in Karnataka (India)
FFF	Federation of Free Farmers (Philippines)
FFW	Federation of Free Workers (Philippines)
FIVDB	Friends in Village Development Bangladesh
FMB	Forest Management Bureau (Philippines)
FMD	Foot and Mouth Disease
FPAN	Family Planning Association of Nepal
FPR	Farmer Participatory Research
FSR	Farming Systems Research
FSRI	Farming Systems Research Institute (Thailand)

FWE	Foundation for World Education
GDP	Gross Domestic Product
GK	Gonoshosthya Kendra
GNP	Gross National Product
GO	Governmental Organization
GoB	Government of Bangladesh
GoI	Government of India
GONGO	Government-sponsored NGO
GoWB	Government of West Bengal (India)
GRID	Grass Roots Integrated Development (Thailand)
GRINGO	Government-formed NGO
GVM	Village Development Association (Gram Vikas Mandal) (India)
HF	Haribon Foundation (Philippines)
HMG	His Majesty's Government
HMGN	His Majesty's Government of Nepal
HRD	Human Resources Development
HUDCO	Housing and Urban Development Corporation
HYV	High-Yielding Varieties
ICAR	Indian Council for Agricultural Research
ICC	Interim Consultative Council (Philippines)
ICCO	Inter Church Co-ordination Committee for Development Projects (Netherlands)
ICDP	Intensive Cattle Development Programme (India)
ICIMoD	International Centre for Integrated Mountain Development
ICRISAT	International Crops Research Institute for the Semi-Arid Tropics
IDARA	Information Development and Resource Agency
IDE	International Development Enterprises
IDRC	International Development Research Centre
IDS	Integrated Development Systems (Nepal)
IFG	Intensive Feed Garden (Philippines)
IGVGD	Income Generation for Vulnerable Group Development (India)
IIED	International Institute for Environment and Development
IIRR	International Institute for Rural Reconstruction (Philippines)
IISc	Indian Institute of Science
IITA	International Institute for Tropical Agriculture
ILO	International Labour Organization
INGI	International NGO Forum on Indonesia
INGO	International NGO
INRA	Institut National de Recherche Agronomique (France)
INSAN	Institute for Sustainable Agriculture Nepal
INTSOY	International Soybean Programme of the University of Illinois
IPCL	Indian Petro-Chemicals Limited

IRDP	Integrated Rural Development Programme
IRRI	International Rice Research Institute
ISFP	Integrated Social Forestry Programme (Philippines)
ISNAR	International Service for National Agricultural Research
IT	Intermediate Technology
IUCN	International Union for the Conservation of Nature
IVRI	Indian Veterinary Research Institute
IVS	International Voluntary Services
JPC	Joint Project Committee
KAF	Konrad Adenauer Foundation
KAU	Kerala Agricultural University (India)
KDP	Karnataka Development Progress (India)
KMP	Peasant Movement of the Philippines
KR	Key Rearer
KVIB	Khadi and Village Industries Board (India)
KVK	Krishi Vignan Kendra (Farm Science Centre) (India)
LDCs	Less developed countries
LDO	Local Development Officer
LIFT	Local Initiatives for Farmers' Training (Bangladesh)
LLP	Lab-to-Land Programme (India)
LP3ES	Institute for Social and Economic Research, Education and Information (Indonesia)
LSP	Loka Siksha Parishad (India)
MBRLC	Mindanao Baptist Rural Life Centre (Philippines)
MCC	Mennonite Central Committee (Bangladesh)
MCRC	Murugappa Chettiar Research Centre (India)
MFI	Mag-uugmad Foundation (Philippines)
MoHRD	Ministry of Human Resources Development
MoU	Memorandum of Understanding
MYRADA	Mysore Relief and Development Agency (India)
NABARD	National Bank for Agricultural and Rural Development (India)
NAF	Nepal Agroforestry Foundation
NASSA	National Secretariat for Social Action (Philippines)
NCL	National Chemical Laboratory (India)
NCRP	Nepal Coppice Reforestation Project
NCS	National Conservation Strategy
NDP	National Development Programme
NDRI	National Dairy Research Institute (India)
NERAD	North-east Rainfed Agriculture Development Project (Thailand)
NET	Foundation for Self-Reliance in North-east Thailand
NFESC	Non-formal Education Service Center
NGO	Non-governmental organization
NIA	National Irrigation Administration (Philippines)

NILG	National Institute of Local Government
NPA	New People's Army (Philippines)
NPC	National Planning Commission
NWDB	National Wastelands Development Board
NWDPRA	National Watershed Development Project for Rainfed Areas (India)
OD	Outreach Desk (Philippines)
oda	Official development assistance
ODA	Overseas Development Administration (UK)
ODI	Overseas Development Institute
OFCOR	On-Farm Client-Oriented Research
ORNOP	Organisasi Non-Pemerintah (Indonesia)
ORP	On-farm Research Programme (India)
PACT	Private Agencies Collaborating Together (USA-based)
PAD	Participatory Approach to Development (Philippines)
PAU	Punjab Agricultural University (India)
PBSP	Philippine Business for Social Progress
PC	President's Council
PEDO	People's Education Development Organisation (India)
PF	Progressive Farmer
PHILDHRRA	Philippines' Partnership for the Development of Human Resources in Rural Areas
PIDOW	Participative Integrated Development of Watersheds (India)
PMC	Project Management Committee
PO	People's Organization
PRA	Participatory Rural Appraisal
PRADAN	Professional Assistance for Development Action (India)
PRDO	Private Research and Development Organization
PRIA	Society for Participatory Research in Asia
PRRM	Philippine Rural Reconstruction Movement
PTP	Progeny Testing Programme
PW	Poultry Workers
R&D	Research and Development
RD	Rural Development
RDD	Rural Development Department (Nepal)
RDO	Rural Development Officer
RDP	Rural Development Programme
RDRS	Rangpur Dinajpur Rural Service (Bangladesh)
REFORM	Resource and Ecology Foundation for the Regeneration of Mindanao (Philippines)
RHC	Red-headed Hairy Caterpillar
RKM	Ramakrishna Mission (India)
RLF	Revolving Loan Fund
RPC	Regional Programme Co-ordinator

RTC	Regional Theological Centre
SALT	Sloping Agricultural Land Technology
SAP	South Asia Partnership (Nepal)
SAU	State Agricultural University (India)
SCF	Save the Children Fund
SDC	Swiss Development Co-operation
SIDA	Swedish International Development Agency
SLCC	State-Level Committee for Consultation
SMAP	South Mindanao Agricultural Programme (Philippines)
SMS	Subject Matter Specialist
SNC	Standing Committee
SPO	Special Projects Office (Philippines)
SSNCC	Social Services National Co-ordination Council (Nepal)
SWDC	State Watershed Development Cell (India)
TN	Tamil Nadu (India)
TNAU	Tamil Nadu Agricultural University (India)
ToT	The 'Transfer of Technology' approach to agricultural research and dissemination
TPC	Tree Planting Campaign
TRI	Tea Research Institute (India)
TVS	Tasar Vikas Samity (India)
UMN	United Mission to Nepal
UNCHS	United Nations Centre for Human Sciences
UNDP	United Nations Development Programme
UNEP	United Nations Environment Programme
UNESCO	United Nations Educational, Scientific and Cultural Organization
UNO	United Nations Organization
UP	Uttar Pradesh (India)
UPASI	United Planters' Association of Southern India
UPLB	University of the Philippines at Los Baños
USAID	United States Agency for International Development
USC	Unitarian Service Committee of Canada
VANI	Voluntary Action Network India
VDA	Village Development Animator (Thailand)
VHSS	Voluntary Health Services Society (Bangladesh)
VIDCO	Village Development Committee
VLW	Village-Level Worker
VRO	Voluntary Resource Organization
VSO	Voluntary Service Overseas
VSWC	Vivekananda Social Welfare Centre (India)
WCS	World Conservation Strategy
WDP	Watershed Development Programme (India)
WFP	World Food Programme

WN	World Neighbors
WUA	Water Users' Association (Indonesia)
WWF	World Wildlife Fund
XISS	Xavier Institute of Social Service (India)
YC	Youth Club
YFA	Youth for Action (India)
ZP	Zilla Parishad [District Council] (India)

Chapter 1
INTRODUCTION

BACKGROUND

> Most research systems in developing countries were organised to serve commercial farmers operating in more favourable and homogeneous agroecological conditions than those in resource-poor farming contexts.
> (Merrill-Sands and Kaimowitz 1990: 1)

The focus of this book, drawing on evidence from Bangladesh, India, Nepal, Indonesia, Thailand and the Philippines, is on the institutional arrangements necessary to promote agricultural technology development (ATD) for the enhancement of livelihoods among the rural poor. The focus of practically all public sector ATD institutions has been on male farmers having secure access to land. Parts of this book share the same focus, but much of our concern is with those men and women whose livelihoods are based on agriculture-related activities, but who do not have secure access to land, and with women, including those who, for cultural reasons, do not normally engage in field-work. As we shall discuss, almost 50 per cent of those seeking livelihoods in rural areas of Bangladesh have insecure access to land, as do substantial proportions of the rural poor in India and the Philippines. To focus exclusively on male farmers is, therefore, to neglect large groups of the rural poor.

The underlying motivation for assembling and analysing evidence in relation to these concerns derives from three broad trends:

- limited public sector success in meeting the needs of the rural poor
- the recent establishment of large numbers of non-governmental organizations (NGOs) which claim advantages over the public sector in reaching the rural poor
- the increasing weight attached to views that the prospects of successful change are enhanced if the poor participate in its design.

The first two of these trends relate specifically to institutions, the third to wider changes which have institutional implications.

LIMITED PUBLIC SECTOR SUCCESS IN MEETING THE NEEDS OF THE RURAL POOR

The absolute numbers of rural dwellers in Asia is high, and will continue to grow in absolute terms for some decades to come. As the data in Table 1.1 indicate, for instance, it is only in Indonesia that employment in agriculture has already begun to level off. It is expected to do so in Thailand before the turn of the century, but in the South Asia countries and in the Philippines the agricultural labour force is expected to rise for at least a further two decades, implying a continuing increase in population pressure on renewable natural resources. Estimates of the incidence of rural poverty are provided in the final column of the same table. Given the problems of defining and measuring rural poverty, these figures should be interpreted with caution. They suggest high levels of poverty in South Asia (especially in Bangladesh) and in the Philippines, but somewhat lower levels in Indonesia and Thailand.

The task facing those promoting change in agriculture is therefore particularly difficult: in most of the countries discussed here, efforts to improve the conditions of the large numbers of rural poor will, over at least two more decades, be faced by increasing rural populations on, in many areas, a deteriorating resource base. However, much could still be done to improve the effectiveness of such efforts. For instance, two major recent studies of agricultural research in less-developed countries (ldcs) from which the opening quotation derives, indicate that public sector institutions mandated to develop agricultural technology have, over several decades, given priority to a clientele of well-resourced farmers operating in homogeneous areas.[1] In Chapter 2 our review of the successes and failures of the Green Revolution allows closer examination of this trend. For the present, it is sufficient to note that those seeking livelihoods in difficult farming areas, characterized by some combination of low and unreliable rainfall, poor soils and hilly topography – the complex, diverse and risk-prone (CDR) areas of Chambers *et al.* (1989) – pursue different kinds of agricultural enterprises from well-resourced farmers, face different sets of opportunities and constraints, and so require different kinds of technology. On the whole, research into such technologies has been allocated low priority, as have the support services necessary to ensure full realization of the potential of technological change, ranging from input supply and marketing, to the wider requirements of communications and educational infrastructure. Yet, as a number of recent studies have argued (Richards 1985; Chambers *et al.* 1989) livelihoods in CDR areas *can* be enhanced through improvements in agricultural technology, providing that it is developed in response to local needs, and in ways which build upon indigenous knowledge and local capacities for experimentation.

Table 1.1 Demographic characteristics of the case-study countries, and the incidence of poverty

	Number of rural dwellers[a] *(m)*	*Period of anticipated levelling-off of labour force in agriculture*[b] *(years)*	*Economically active population relying on agriculture*[c] *(%)*	*Rural population in poverty*[a] *(%)*
Bangladesh	97	2010–2020	72	86
India	604	2020–2025	68	51
Indonesia	128	1985–1990	53	44
Nepal	12	beyond 2025	92	61
Philippines	36	2020–2025	49	64
Thailand	43	1995–2000	68	34

Notes: a Based on *World Bank Social Indicators of Development 1990*, Washington, DC: World Bank and Johns Hopkins University Press

b Based on *World Demographic Estimates and Projections 1950–2025*, New York: United Nations, 1988

c Based on *Key Indicators of Developing Asian and Pacific Countries*, Manila: Asian Development Bank; July 1990

THE EMERGENCE OF NGOs

A second context underlying our concern with appropriate institutional arrangements for ATD in difficult areas is the recent increase in numbers of institutions outside the public sector concerned with livelihood enhancement in difficult areas. While small-scale private commercial sector activity has long existed in these areas in the form of small traders and moneylenders, there is a continuing lack of provision of new agricultural technologies such as seeds, equipment and agrochemicals by commercial *companies*, largely because of the diverse agro-ecological conditions of these areas and the wide range of technologies that would be required to meet them. To the private commercial sector, these constitute highly fragmented markets, often distant from the main commercial centres and, because of poor infrastructure, difficult and costly to reach (Pray and Echeverria 1989). By contrast, the voluntary sector has perceived substantial opportunities for enhancing the livelihoods of the rural poor and has greatly increased its presence in difficult areas over the last decade. On a global scale, international funds channelled through non-governmental organizations (NGOs) reached approximately US$7 billion in 1990 – the equivalent of 16 per cent of total bilateral aid flows – as against only US$3.6 billion in 1983 (Williams 1990; Clark 1991).

Subsequent sections of this chapter define the characteristics of the voluntary sector more closely, and elaborate the specific questions central to this study to which they give rise regarding the organization and management of ATD. Our present concern is with the wider perceptions of their characteristics which have led to such substantial increases in funding, and with the questions that these wider perceptions leave unanswered.

Three broad perceptions can be identified: first, the notion that NGOs represent a force towards democratic and pluralist civil society, second, a view that NGOs have particular strengths in poverty alleviation and sustainable development, and third, that they offer the prospect of enhancing the efficiency of public sector service delivery.

- *NGOs as a force for democracy* NGOs have increasingly become associated with grassroots development (Carroll 1992), with development and democracy (Lehmann 1990) and with alternative development and empowerment (Friedman 1992). The concern in these and many other writings has been less with the establishment of multi-party democracy in the western liberal sense, and more with the establishment of checks and balances on the use and abuse of power, with the struggle for liberty to express views at odds with those of established interests, and with increasing representation of the views of the poor (Healey and Robinson 1992).
- *NGOs as poverty alleviators and sustainable developers* A perspective rapidly gaining ground is that NGOs' commitment to poverty alleviation underpins a strong presence in rural areas, that their respect for self-determination encourages them to support the establishment of mechanisms and grassroots organizations through which the rural poor can express views on their needs, and that their small scale and flexibility allow rapid response to these needs (Korten 1987; Clark 1991).
- *NGOs as efficiency-enhancers* A third set of writings has been concerned with the potential that NGOs offer for enhancing the efficiency of service delivery in general, and of government services in particular. Broadly, the arguments are, first, that strong presence in rural areas and detailed knowledge of the needs of the poor allow NGOs to deliver more appropriate services to the poor more cost-effectively than the public sector could (Farnworth 1991; World Bank 1991a; 1991b). This view, in particular, has led to substantial increases in funding allocations to NGOs. A second argument is that the innovations – whether technological, methodological or institutional – developed by NGOs would enhance the efficiency of the public sector if it were to adopt them and apply them on a wider scale (Morgan 1990; Hulme and Edwards 1992). A specific facet of this view is that many NGOs are concerned with technologies that are more environmentally sustainable than those relying on high inputs of agrochemicals and mechanical power, and so offer prospects of efficiency enhancement in the long run (Haverkort *et al.* 1991). A third argument is that NGOs can influence the agenda of public sector organizations informally through personal contacts, and more formally through representation on advisory bodies. This 'demand-pull' can be sustained in the long term by a gradual

take-over of NGOs' responsibilities by the grassroots organizations that they seek to support (Abed 1991; Carroll 1992).

Two difficulties are inherent in these general arguments in support of NGOs: first, they contain a number of inconsistencies. The most important of these is rooted in perceptions of long-term roles. Widely pervasive democracy is seen by many to involve shifts in the agenda of government in order to reverse 'urban bias'. If the rural poor are to play a fuller role in democratic processes, more resources must be channelled into those areas to bring levels of infrastructure, and health and education, closer to those prevailing elsewhere. Clearly, decisions on broad patterns of resource allocation of this kind lie firmly in the mandate of government, as does the wider responsibility for monitoring their impact. Wide-scale involvement of NGOs in the provision of rural services runs the risk that governments will see themselves absolved from these obligations, and may make it difficult for NGOs to implement their widely stated intentions to withdraw from particular activities once sustainable interaction between government and grassroots organizations has been established. Problems of this kind have been noted in Bangladesh by Sanyal (1991) and more widely in Asia by Holloway (1989).

The second major difficulty is that the majority of pro-NGO statements discussed above are pitched at a high level of generality. In particular, they give the impression that relations between NGOs and government organizations (GOs) will be relatively simply arranged on a functional basis according to their respective comparative advantage. Given the wide range of NGO types, the diversity of biophysical and socio-economic conditions in which they operate, and the prospect of widely divergent views between NGOs and government on both the futures of the rural poor and the means of achieving those futures, it seems implausible to suggest that NGO–GO relations would be anything other than complex, so that the achievement of adequate working relations would require much effort. The long history of tension in the wider relations between NGOs and the state is further evidence of the difficulties likely to be encountered (Tandon 1989).

Some recent work has begun to challenge the generalized nature of the arguments outlined above. Fowler (1991), for instance, argues that for Africa the democratizing potential of NGOs is likely to be more limited than is widely held. Riddell and Robinson (forthcoming), reporting on a study of the impact of NGOs' rural income-generating activities in Africa and Asia, doubt whether they have been able to reach the lowest 10–20 per cent of the rural poor, and Carroll's (1992) impact assessment study in Latin America suggests that NGOs have been unable to reach the poorest 20–30 per cent. At the very least, these recent impact assessments suggest the need for greater agnosticism regarding the impact and cost-effectiveness of NGOs' work, for more detailed case-by-case assessments of their impact, and for stronger internal mechanisms of performance assessment. They also raise the question of whether strategies oriented towards welfare and income support would not be more appropriate than income-generation to meet the needs of the poorest.

NGOs AS FACILITATORS OF PARTICIPATORY PROCESSES

The third context underlying this study is the increasing recognition that if the poor do not participate in decisions on the types of change necessary to enhance their livelihoods, then the impact of such change is likely to be less than desired. Furthermore, local organizations need to be established that will serve as fora for the discussion of proposals for change. In the dynamic context, these organizations might develop local ideas and capacities for change and interact with government, drawing down services and attempting to shift government's agenda through feedback on previous attempted changes or in response to newly emerging needs. Although aspects of these arguments have been touched on in the two previous sections, they are given separate treatment here since the strengthening of interaction between GOs and the rural poor need not, in principle, involve NGOs in an intermediary role.

In the context of ATD, these views on the need for involvement of local people both in decisions on specific changes and in longer-term interaction with GOs has become known as 'farmer participatory research (FPR)'. Its proponents argue that farmers have considerable curiosity and experimental capacity of their own; from direct and inherited experience they have detailed knowledge of agroclimatic conditions, soils and the types of crops and crop combinations that meet their requirements for food security and cash sales. Furthermore, farmers are capable of providing assessments of formal trials that are useful in guiding scientists' future work (Ashby 1987). ATD work by the public sector that seeks to draw upon and build up local knowledge and experimental capacities therefore have a much better chance of generating adoptable technologies than those which do not (Rhoades and Booth 1982; Farrington and Martin 1987; Chambers *et al.* 1989).

OBJECTIVES OF THE STUDY

The study was initiated with the objectives of exploring what institutional arrangements might be made for the wider implementation of FPR and what the scope for closer NGO–GO interaction might be in such arrangements. Historically, the study evolved from a review conducted by the Agricultural Research and Extension Network of the Overseas Development Institute of FPR methods (Farrington and Martin 1987) as a background paper for the 1987 conference 'Farmers and agricultural research: complementary methods' from which the book *Farmer First* emerged (Chambers *et al.* 1989). Over 100 experiences – largely unpublished – of FPR were reviewed, most of which derived from the work of NGOs, others being conducted by university teams or by special donor-funded projects. Very little FPR experience appeared to have been gained by government research organizations. These early observations suggested that GOs may not have the capacity – even if they had the inclination – to establish on anything more than a pilot scale the close contacts with local communities necessary for FPR methods to succeed.

The central question that emerged from these observations was, therefore, whether scope existed for closer interaction between NGOs and government research and extension services in the implementation of participatory methods and, if so, what form such interaction might take. This can conveniently be subdivided into a number of interrelated issues.

First, NGOs' experience in developing and implementing participatory methods might, if adequately documented, serve as material from which government services could learn in developing participatory methods of their own or, in the case of many public sector services whose mandate was not yet broad enough to embrace notions of participation, in the development of more orthodox farming systems approaches.

Second, from the evidence reviewed, NGOs appeared to have developed a strong array of participatory methods in problem diagnosis and, to a lesser degree, in the evaluation and dissemination of technologies, but had fewer methods to offer in the screening, testing and adaptation of technologies and resource management practices. This, taken together with the complexity of agricultural problems and systems interactions in CDR areas, and with NGOs' lack of specialist research skills and facilities,[2] suggested that NGOs may have to rely on outside sources – among them, public sector research and extension services – for technologies to test and adapt to local circumstances.

Third, NGOs' awareness of local conditions and their familiarity with farmers' requirements places them in a strong position to articulate needs and opportunities into public sector research and extension services. The broad multi-sectoral approach (e.g. in agriculture, health and education) taken by many – though potentially a source of over-stretching and therefore of weakness – enhances their capacity to identify what is feasible in the context of local opportunities and constraints. The intention, clearly expressed in the rhetoric of many NGOs – though not yet widely evident in their praxis – to facilitate the emergence of local self-sustaining membership organizations capable of taking over many of their functions offers some prospect that any 'demand-pull' of this kind on government services which NGOs initiate might be sustained in the longer term.

Three types of potential interactions can be distilled from these postulates: in the first, GOs adopt and 'scale up' innovations developed by NGOs, whether in technologies, research methods or institutional arrangements. In the second, NGOs and GOs work together, the strengths of one compensating for the weaknesses of the other in performing ATD functions. Specifically, GOs would conduct research; NGOs would field-test, disseminate and provide the feedback necessary to influence subsequent research agenda. In the third, over a longer period, NGOs would support the emergence of grassroots organizations capable of taking over many of their functions, including interaction with GOs. Posited in economic terms, these interactions can largely be explained in terms of comparative advantage and of the consequent 'gains from trade' that might be made.

As the study evolved, it quickly became apparent that purely function-based analysis of the scope for NGO–GO interaction would generate misleading prescriptions. The importance of the wider socio-political context in which NGO–GO interaction takes place has been alluded to briefly above. To disregard this would certainly lead to overstatement of the extent to which apparently successful interaction might be replicated.

It was therefore suggested (Farrington and Biggs 1990)[3] that a set of wider conditions would have to be met if NGO–GO interaction were to have any prospect of success.

First, overall relations between NGOs and the state would have to be at least neutral and, if possible, favourable to the presence of NGOs. In the context of the countries studied here, conditions have been less than favourable in Indonesia for many years, so that the presence of NGOs is weak and their activities limited to non-confrontational issues. In Thailand, macropolitical conditions facing NGOs have fluctuated widely so that the poor prospects of NGO–GO collaboration during the 1980s appeared to improve in 1991, but then suddenly worsened in the wake of government suppression of student unrest in 1992 (see Chapter 7).

Second, the prospects for successful interaction are enhanced where NGOs and government share similar visions for the future of the rural poor. In many countries this requires a reversal of 'urban bias' and a commitment to invest in the physical and human infrastructure necessary to improve the conditions of the rural poor.

Third, differences in the development models pursued by each side may make positive interaction difficult. For instance, a broadly 'modernizing' model pursued by GOs may see as desirable: the buying-out of small farmers by large, the consolidation of land with larger units, the introduction of mechanization and the conversion of previously independent small farmers into a labour force on which large farms can draw. While some NGOs may share this view, and may, for instance, gear their actions to the relief of hardship during the perceived transition of the rural poor into this new role, others may be aiming at the establishment of a self-supporting class of small farmers and its consolidation over the long-term through income-enhancing activities. For some NGOs, such activities may be smaller scale versions of those pursued by a modernizing state (e.g. high-yielding seeds and agrochemicals, but power tillers instead of tractors). For others, they may be rooted in 'alternative' types of agriculture relying on 'low external inputs' (Haverkort *et al.* 1991).

Finally, even if these three sets of conditions are met, there may be a good deal of variation in the contextual factors influencing whether NGOs and GOs will work together. Some NGOs, for instance, may see some governments, or government departments, as a threat to their independence. In other cases, NGOs may feel that their credibility with their own clientele is at risk if they associate with a government which has a long history of corruption and/or of failed development projects.

In all, the analysis of NGO–GO relations is more complex than that of relations among government departments for two principal reasons:

- A wider range of diversity exists among the philosophies, objectives and modes of operation of NGOs than GOs.
- There is no statutory obligation for NGOs and GOs to work with each other, in the same way as, for instance, government research departments are required to collaborate with extension departments. If the advantages perceived by one side of working with the other are outweighed by perceived disadvantages, then the one can ignore the other and continue to work independently within its own mandate.

Five objectives for the study on which this book is based can be derived by drawing together the various strands of argument in the above discussion. These objectives are summarized in Box 1.1.

Box 1.1 Study objectives

1. To stimulate and support the documentation of NGOs' experience in ATD, particularly that involving interaction with GOs.
2. To obtain and document GOs' views on interaction with NGOs, where possible, on the specific experience described in (1) by NGOs.
3. To document the contextual factors relevant to the experiences and views obtained in (1) and (2).
4. To analyse these views and experiences against their contexts, seeking to identify why certain types of interactions were chosen, to what extent they succeeded, and why.
5. To identify the advantages that might be gained from particular types of interactions in likely future contexts, and to identify the particular initiatives that might be taken by NGOs, GOs and funding agencies to realize such advantages.

STUDY METHODOLOGY

The basic methodological framework was conceived in the early stages of the study. However, certain elements were modified in the light of experience as the study evolved, and others were influenced by the various constraints that arose.

With extensive prior knowledge of the type and location of NGOs' experience in ATD and of the experiences of government agencies and NGOs in trying to influence each other and/or work together, it would, in principle, have been possible to devise a procedure for stratified sampling to reflect e.g. particular kinds of socio-political contexts, or particular types of links between GOs and NGOs. In practice, this information was not available in advance, largely for the very reason that so little NGO experience in this area had previously been documented. In retrospect, our view is that the diversity of NGO types and operational contexts is so great that in practice it would have been virtually impossible to pursue this procedure.

An alternative would have been to concentrate resources in a small number of countries, and to conduct in-depth studies of both the wider context of state–NGO relations and of the range of observable NGO–GO interactions. Such an approach was rejected,[4] again partly because of lack of prior detailed knowledge of the sampling population, but also because initial indications were that the range of experience was extremely diverse and so unlikely to be captured in a small number of in-depth studies.

Two overriding principles influenced the decision to adopt a more flexible methodology than would be permitted with a predetermined sampling scheme. The first was that the inadequate initial information on NGOs' activities could best be overcome through a series of reconnaissances involving discussions with the staff of both NGOs and GOs in the countries where NGOs were known to be active. These discussions were intended to allow local knowledge and opinion to influence the design of the project. The second was that the initiators of the study had no preconceived notions over the extent to which – given the potentially wide diversity of NGO social histories, structures and roles – it would be possible to draw out valid generalizations.

The first step in a methodology appropriate to these broad principles was, therefore, to make preliminary visits to countries where NGOs were known to be promoting livelihood generation through agriculture-related activities. Hypotheses were discussed and refined through these visits and, in most cases, agreement reached on the respective roles of NGO and GO staff, study initiators and, in some cases, local consultants, in preparing case studies.

Given the importance identified in the study's conceptual development of locating information on functions and roles into their political and economic context, case studies were required to include adequate reference to the wider context. But, in addition, it was decided to prepare a number of country overview papers which would document the evolution and current status of wider NGO–state relations.

Factors taken into account in the identification of case studies included the following:

- *NGOs' role in agricultural change* the attempt was made to capture as wide a range of diversity as possible, covering both on-farm (crops, animals, trees) and off-farm (trees, pasture) resource management, and both production and processing activities.
- *NGOs' interaction with government* again, the emphasis was on documenting as wide a range of diversity as possible, embracing the range from collaborative to conflictive interaction.
- *NGOs' interaction with each other* during reconnaissance visits, it became apparent that the absence or poor quality of interaction among NGOs had diminished the quality of their relations with their clients and with government, and efforts were made to ensure that, where relevant, the quality of NGOs' interaction with each other was documented in the case studies.
- *NGO and GO learning processes* the types of activities and roles undertaken by some of the longer-established NGOs had evolved in the light of experience. In some cases, GOs' experience of working with NGOs had also been modified over time. Where relevant, these processes were described in the case studies.

In all of the above, the emphasis was on understanding the factors that led NGOs to successful identification of where their comparative advantage lay in agricultural technology development and of productive modes of interaction with GOs. More of the case studies discussed here are therefore likely to reflect 'success' than occurs in reality. However, several cases of NGO failure – in agricultural technology, in relations with GOs and with each other – and failure to learn from experience, are documented, and even where case studies record some degree of success, numerous obstacles had to be addressed and are described.

As a final note, it should be stressed that aspects of the methodology went beyond what is normally required of a research study. As the study progressed, the research co-ordinators found themselves supporting not merely the documentation of experience but a wider process of reflection on roles among some of the NGOs and GOs. These, in turn, requested support for workshops at national (Bangladesh;[5] Philippines[6]) and at regional[7] level at which views and experiences could be exchanged. To a modest extent, therefore, those involved in the study also became part of a process of familiarization and the broadening of perspectives which are a prerequisite of closer interaction between the two sides.

THE SEARCH FOR A CONCEPTUAL FRAMEWORK[8]

The search for a framework for analysis of the large body of empirical material documented in this study requires some analysis of the frameworks used in related studies. A first point of reference is provided by the ISNAR studies on Research-Extension Linkages and on On-Farm Client-Oriented Research (Merrill-Sands and Kaimowitz 1990) which developed a number of arguments and concepts relevant to the present study. They were based on the premises that ATD was more likely to be relevant to small farmers' requirements, first, if clients were clearly identified and included in the ATD processes, particularly in on-farm research, and second, if strong two-way flows of information existed between researchers, technology transfer agents and farmers. Enhancing farmer participation and promoting linkages were seen mainly as management issues, but three contextual factors outside the control of managers were recognized as important: external pressure, resources and agro-ecological diversity. The authors saw a strong positive correlation between external pressure and the degree of responsiveness to farmers' needs. Ideally, this should be exercised by farmers themselves but it was recognized that small farmers were unlikely to be able to do so. NGOs were therefore seen as potentially useful interpreters and articulators of small farmers' requirements. NGOs were also seen as useful additional sources of funds to permit FPR, especially where agro-ecological diversity placed particularly severe demands on GOs.

These studies did not explore how such interactions might be realized, nor whether there were qualities that NGOs might bring to the relationship beyond the three contextual areas defined. There appears to have been, for instance, no consideration that innovations generated by NGOs – whether in technologies, methodologies or institutional arrangements – might be adopted and 'scaled up' by the public sector.

The principal concept around which the ISNAR studies were organized was that of the agricultural technology system (ATS), defined as all the individuals or groups working on the development, diffusion and use of new and existing technologies, the actions in which they engage, and the relations between them. The ATS comprises sub-systems composed of individuals and organizations engaged in particular activities with specific goals. Thus, the basic research sub-system develops new knowledge; the applied research sub-system uses this knowledge to develop technologies to address specific issues; the adaptive research sub-system adapts these to local agro-ecological and socio-economic conditions. These are disseminated through a technology transfer sub-system.

For the sub-systems making up the ATS to operate in a mutually reinforcing fashion, organizational procedures (i.e. linkage mechanisms) are needed to facilitate interaction. These may be operational, i.e. providing links for specific ATD activities, or structural.

Some of the concepts are of value to the present study. The distinction between, for instance, NGOs' efforts to set up operational links for specific field trials, and their efforts to achieve formal representation on GO advisory or decision-taking bodies in a structural context is useful in the analysis of some of the empirical material that follows.

There are, however, four difficulties in applying the ATS framework deriving from the ISNAR studies directly to analysis of NGOs' work and of linkages between them and GOs:

- NGOs cannot neatly be grouped into a single sub-system such as technology transfer. While none in the present study is involved in basic research, some fall into the applied, some into the adaptive and others into the technology transfer sub-systems. Criteria would therefore be needed against which interaction between NGOs and their sub-group peers can be explained, as well as interaction among sub-systems. Such explanations are likely to be as varied as the individual NGOs themselves.
- Many NGOs engage in ATD on an 'issue' basis, conducting research where necessary, followed by testing, adaptation and dissemination. A single activity conducted by a single NGO may therefore straddle several sub-systems, so that any need to link with other organizations is less systematic than the 'technology pipeline' concept underlying the ATS implies.
- Some NGOs conduct different, and only loosely related ATD activities, perhaps engaging in applied research to meet the needs of some clients, and in stimulating feedback to GOs on behalf of others. In these cases, links from one 'box' to another are, contrary to the ISNAR premise, no longer exclusively *external* (i.e. between different organizations).
- Many NGOs have external links, usually with other NGOs, and, even if primarily concerned with technology transfer, bring in ideas from outside in ways which would be inconceivable for a technology transfer (or even adaptive research) arm of government. There is thus considerable cross-national 'leakage' to and from sub-systems.

Perhaps the strongest reason for making only sparing use of the ATS framework, however, is that the pressures to collaborate are radically different among GOs than among NGOs, or between NGOs and GOs. Government departments are mandated to work together to ensure a smooth flow of technologies down the 'pipeline', and, though in reality perhaps in too few cases, a substantial flow of feedback. A successful outcome of the activities of one GO depends to a high degree on the quality of interaction with others occupying positions higher or lower in the pipeline.

By contrast, NGOs collaborate among themselves only when the perceived advantages outweigh disadvantages. They apply similar criteria to interaction with government. We postulate here that NGO–GO interaction may, at the least positive extreme, be conflictive; at a more indifferent position they may ignore each other; more positively, one may act in ways intended to inform the other, or, more specifically, have the other incorporate lessons that it has generated. At the positive end of the spectrum, links may be fully collaborative.

These gradations of interaction provide at least a rudimentary scale against which the empirical material is classified towards the end of this book. Although supplementing the concepts we draw from ISNAR's ATS framework, this classification remains unlikely to capture the full range of interaction. Nor will it provide more than the beginnings of an explanation of why each side interacts with the other in the ways described – the contexts are too complex to permit accurate reduction to simple explanations – nor any clear indication of how each side is likely to react to changing circumstances. In so far as such explanations are possible, they have to be sought through combinations of this classification with others that seek to explain *inter alia* the different roles adopted by NGOs and GOs when interacting on ATD, the predispositions of each towards certain roles and the perceptions held by each of the potential advantages and disadvantages of working together. These classifications are developed in the final part of this book. We now turn to the definition of terms and concepts used throughout the book.

DEFINITIONS

The terms interaction, link and collaboration are used throughout the text, but are not interchangeable. As Figure 1.1 indicates, interaction is the broadest term, not only embracing both link and collaboration, but also allowing the possibility of conflictive interaction. Collaboration, at the opposite end of the spectrum, implies a formalized dependence of one partner on another for at least part of the success of its activities, as when, for instance, GOs might contract NGOs to deliver inputs based on the technologies that they (GOs) have developed. Link occupies an intermediate position, implying either formal or informal positive interaction, but of a less mutually dependent kind than that designated by collaboration.

Agriculture is defined broadly as the sustainable exploitation of renewable natural resources and includes annual and perennial cropping, agro-forestry and livestock as well as the conservation measures needed for long-term maintenance of the resource. A narrower definition of agriculture had been adopted for the early part of the study, but this was broadened to embrace the important complementarities between on- and off-farm resource use in difficult areas, and to take into account the fact that innovative interactions between NGOs and GOs existed in (often, the newer) government departments concerned with Environment or Natural Resources, for example, from which agriculturalists could learn.

Technology is used here to mean both hardware (equipment and inputs) and increments in knowledge required for improved resource management. The term 'technology generation and transfer' is widely used in preference to 'research and extension'. The term 'technology' focuses attention on the applied and adaptive end of the spectrum of research activities, which is where the majority of inter-institutional links of the kind discussed are to be found. As Merrill-Sands and Kaimowitz (1990) argue, technology generation and transfer allows due attention to the role of inputs and services in the analysis of technology development and delivery; it also allows private commercial and voluntary activity into the debate, whereas extension has come to be widely associated with only the public sector. It should be noted that 'transfer' implies a two-way flow of technical information between researchers, transfer agents

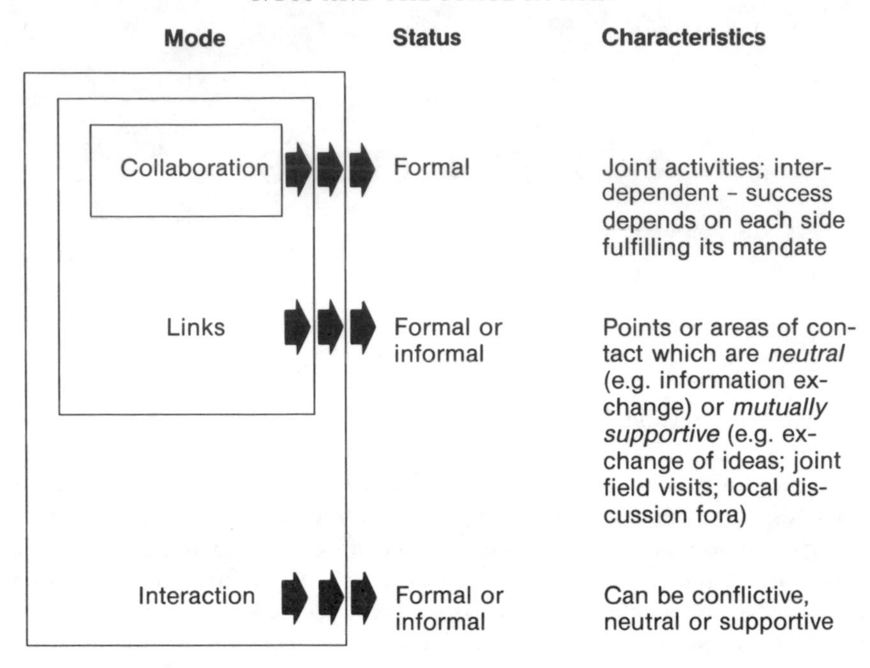

Figure 1.1 The differing characteristics of forms of interaction

and end-users.

NGOs are notoriously difficult to define. At its broadest, the definition of NGOs embraces everything outside the public and private commercial sectors. However, this definition is too broad to be operationally useful. What we are concerned with here is the livelihood generating (i.e. not relief) and environmental work of small to medium-scale private, independent, non-profit organizations. Major philanthropic foundations are excluded. Figure 1.2 provides a 'path' to indicate the types of NGOs of principal interest to the study. Essentially, these are south-based, although a number of north-based organizations are also considered (e.g. Mennonite Central Committee in Bangladesh), as are the southern branches of a number of north-based NGOs that have varying degrees of independence from their head offices (e.g. ActionAid in The Gambia).

In terms of scale, the second criterion on the vertical axis of Figure 1.2, the study is more concerned with supra-community level organizations that provide a service to grassroots organizations. There seems greater likelihood that these higher-level organizations may be able to provide a bridge between grassroots organizations and government services, or influence the latter on behalf of the former. Many of these are non-membership organizations which

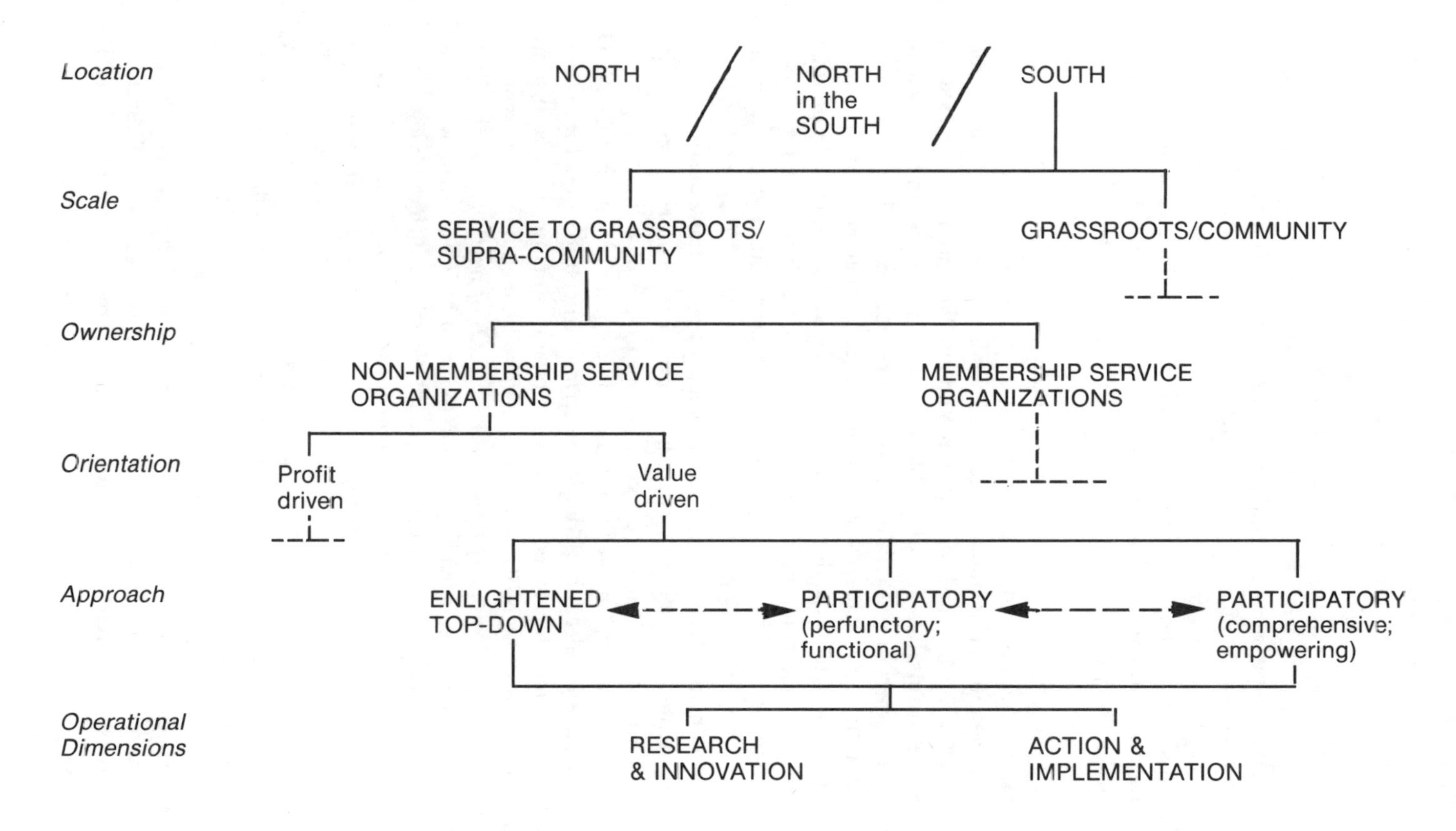

Figure 1.2 NGOs: diversity in the crowd

have a primarily value-driven orientation, although in some countries, some NGOs now offer their staff a structured career with promotion prospects and salaries in excess of their counterparts in government. In some large NGOs, graduate staff may well consider a choice between banking or NGO work.

Other NGOs continue to work with staff whose primary motivation is religious, humanitarian or political. Our distinction between membership and non-membership organizations corresponds with that made earlier by Carroll (1992) and by Fowler (1991).

The NGO work which has been documented has exhibited a wide spectrum of approaches, ranging from the enlightened 'top-down' through the functionally participatory to the more fundamentally empowering. NGOs in this last category seek not only to facilitate the emergence of self-sustaining local organizations, but also to stimulate institutional change in public sector bureaucracies intended to make them more responsive to the rural poor. Finally, as indicated at the foot of Figure 1.2, while the majority of case studies documented are concerned with action and implementation, a substantial number see themselves in roles that are more reflective and innovative, whether in respect of research methods, or inter-institutional links.

Substitution occurs when NGOs undertake tasks that government would normally be expected to undertake. They may do so independently, or when government 'contracts in' the services of NGOs to perform a task for which it lacks resources. NGOs are increasingly being drawn into this role under conditions of privatization in many countries, and in this way may be instruments in an externally imposed schedule of structural adjustment.

Complementarity involves a shared, differentiated contribution to joint objectives. An example might be a government programme in which government provides inputs, an NGO provides staff training and another NGO participates in site selection. Perhaps the term partnership offers the clearest statement of intent for an equal working relationship in which a distinct contribution is made by NGOs and government, although it is a term which is increasingly being distorted by those (whether GOs or northern-based NGOs) having a predetermined agenda for involving local NGOs in their work.

It is sometimes the case that collaborating NGOs have different views from the government agency of the desired outcome of the partnership. An NGO may take on a temporary substitution role combined with a training input which is intended to lead to a situation in which government takes over the role after a finite period. This is BRAC's approach in Bangladesh. However, the government may require more time and may become unwilling to end the relationship, turning it into a longer-term substitution of roles.

The notion of 'empowerment', from the ideas of Paulo Freire (1972), the Brazilian educationist, has generated much discussion. Different NGOs have adapted the concept and talk instead of 'conscientization' or 'mobilization'. Broadly, these ideas refer to the creation of an environment of inquiry in which people question and challenge the structural reasons for their poverty through learning and action. 'Critical consciousness' is seen by Freire as a prerequisite for cultural emancipation and is the major objective of many NGOs' interventions. As F H Abed, founder of the Bangladeshi NGO BRAC, points out: 'BRAC and many other NGOs firmly believe that adoption of innovations, which is the goal of any extension programme, cannot take place unless the capacity of the receivers is properly developed' (Abed 1991: 4).

The success of this approach has been difficult to evaluate. In Bangladesh, there are documented cases in which landless group members have secured better employment conditions and wage rates from employers, women have challenged dowry customs and sexual harassment, and have obtained access to public resources such as land and waterbodies (Kramsjo and Wood 1992). However, 'empowerment' is inevitably a slow process, which makes its impact difficult to evaluate. Some observers have pointed to the 'patronage' element existing in some NGOs' activities and suggested that this is at odds with the concept of empowerment, since groups may remain dependent upon outsiders' motivation and authority (e.g. Nebelung 1987). Others have shown how the language of conscientization can all too easily turn to rhetoric (Hashemi 1989) and be used as 'buzz-words' by some groups to achieve their objectives.

The concept of participation is perhaps in even greater danger of being devalued than that of partnership. Thus:

> We are working in a participatory mode – these farmers always let me know as soon as they feel sick.
> (Remark made by an entomologist to one of the editors, referring to the spraying of insecticide by farmers, wearing minimal protective clothing, high into the canopy of cocoa trees)

At a more general level, the rhetoric of participation is frequently used by planners to justify – through reference to frequently perfunctory consultation – decisions already taken.

Two broad types of participation can be distinguished for the purposes of this volume: at a functional level, farmers might be consulted on the types of technological changes that might meet their needs, and be invited to test and comment on them. A second type of participation leans more towards wider-ranging empowerment. It implies a commitment by an external agent to allow the pace and character of change to be determined by local communities, to support (in ways that they specify) their capacity to analyse the constraints and opportunities that they face and to facilitate (again, in response to their requests) access to the resources needed for improvement of their conditions.

The scope of this involvement is, therefore, much wider than that of functional forms of participation: it implies a willingness to be led by the wishes of a community not merely to a wide range of possible changes within agriculture, but also beyond, into, for example, the realms of education, health and small-scale industry. It also implies a much longer-term social organizing and grassroots institution building involvement than most GOs can countenance.

Biggs (1989a), as a contribution to the OFCOR study, distinguishes four categories of participation:

- *contractual* in which researchers merely hire inputs (land, labour) from farmers, but make little effort to seek their opinions
- *consultative* in which the farmers' opinions are actively sought
- *collaborative* in which the farmers are in control of the particular trial or experiment
- *collegiate* in which farmers and researchers interact as equals, the opinions of the former being taken on board by the latter when future research agenda are designed.

The first three categories defined by Biggs relate broadly to functional participation; the fourth has elements of a wider-ranging empowerment.

Agricultural researchers have been criticized, sometimes rightly, for practising excessively perfunctory forms of participation. However, it is equally clear that this wider, empowering, form lies well beyond their mandate. In government it may lie within the mandate of, for instance, a community development department. However, it is more than likely that no single GO has the mandate to take on its full implications. NGOs' freedom of action gives them a distinct advantage in this context.

STRUCTURE OF THE BOOK

Taking forward the arguments outlined above concerning the importance of context, Chapter 2 examines the diversity of political and economic contexts in the case-study countries, examining their political and economic conditions, NGOs' relations with the state, and differing perceptions on the nature of NGOs. It concludes by examining the comparative advantage of NGOs and government in ATD, drawing out the principal lessons from the Green Revolution.

Chapters 3 to 6 examine empirical evidence on NGOs' work in ATD and their interaction with GOs in the four main case-study countries – Bangladesh, India, Nepal and the Philippines. Each opens with an overview of macro-economic and political conditions, of government initiatives in rural poverty alleviation and of the extent that they converge with NGO perspectives. Chapter 7 follows a similar format in presenting briefer empirical material from Thailand and Indonesia.

A synthesis of the case-study material against concepts developed in this introduction is attempted in the final two chapters: Chapter 8 identifies the principal roles adopted by each side in NGO–GO interaction, and examines why the outcome of interaction frequently falls below expectations. Chapter 9 attempts to identify what advantages may be gained by closer NGO–GO links in the future, and what policy and strategy initiatives will be necessary on both sides, and on the part of donors, if these advantages are to be realized. The empirical material in Chapters 3 to 7 demonstrates a great deal of diversity in NGO types, in their ATD activities and in relations with government. The dangers of drawing general conclusions inadequately borne out by the empirical material are therefore particularly acute, and this final chapter attempts to distinguish what is generalizable from what is not, and to identify issues that remain unresolved.

NOTES

1 The studies on On-Farm Client-Oriented Research and on Research-Extension Linkages conducted from the International Service for National Agricultural

Research (ISNAR) (Merrill-Sands and Kaimowitz 1990).

2 Though it should be noted that a small number of NGOs have exceptionally strong agricultural research capabilities – see, for instance, the experience of the Bharatiya Agro-Industries Foundation in India and that of the Mennonite Central Committee in Bangladesh, recounted in this volume.

3 This broader analytical process subsequently benefited from extensive discussions with A J Bebbington and D J Lewis.

4 This type of approach has, however, been adopted by Kaimowitz for certain Central American countries in order to generate results of interest to an organization having a Regional mandate.

5 Held in August 1991, BARC, Dhaka (Hassanullah, forthcoming).

6 Held in July 1991, IIRR, Silang, Cavite (Gonsalves and Miclat-Teves 1991).

7 The Asia Regional Workshop on NGOs, natural resources management and linkages with the public sector, Hyderabad (India), 16–20 September 1991.

8 This section draws on a more detailed analysis by Bebbington and Thiele (1993).

Chapter 2

THE POLITICAL ECONOMY OF STATE-NGO RELATIONS

THE DIVERSITY OF THE POLITICAL AND ECONOMIC CONTEXTS

ECONOMIC AND POLITICAL CONDITIONS

Economic and political contexts vary widely among the six countries reviewed here. Contrasting colonial histories among, for example, Indonesia, the Philippines and South Asian countries have had specific cultural and administrative legacies. Others (Thailand, Nepal) have been touched only lightly by colonialism. Violent political upheaval (e.g. marking the independence of Bangladesh) gave birth to political movements whose members went on to establish NGOs, and some were eventually constituted as political parties. In other countries, NGOs arose from long traditions of social welfare provision by local groups. Both elected and military governments are to be found in the region, but while many NGOs find the scope of their activities circumscribed by military government in Indonesia, the 1991 military coup in Thailand ironically appeared to offer, in its early months, a more favourable environment for NGOs than had the previous civilian government.[1] Inefficiency, bureaucratic procedures and, in some cases, corruption are longstanding points of friction in NGO–state relations in South Asia.

Economic and social conditions also vary widely. Per capita GNP is a highly imperfect measure of income, but differences in Table 2.1 are wide enough to illustrate the gap between, at the one extreme, Thailand, and at the other, some of the South Asian countries. Differences in GNP growth rates and in life expectancy are equally noticeable. Nepal is particularly heavily dependent on agricultural production, but all six countries have a high proportion of the work-force engaged in agriculture. Severe distributional problems among ethnic groupings, geographical location, or both, occur in the majority of countries. Hill farmers in Nepal, Thailand and the Philippines have particularly been the subject of unsuccessful government development efforts or have been neglected altogether; as have farmers in the outer islands of Indonesia, those operating under a range of difficult conditions in India, and the rural landless in Bangladesh.

Table 2.1 Overview of economic and social indicators – Asia case-study countries

	Per capita GNP (1990) in 1987 US$	*Average annual GDP growth (%) 1990*	*Agriculture as % of GDP 1989*	*Agricultural labour force (%) 1985–8*	*Life expectancy at birth 1989*
Bangladesh	170	5	44	57	51
India	370	5	30	63	58
Indonesia	530	6	23	54	61
Nepal	170	-2	58	93	51
Philippines	680	6	24	43	64
Thailand	1,170	10	15	72	66

PERCEPTIONS OF NGOs IN ASIA

The development and the perceptions of the NGO sectors in different Asian countries have been formed through diverse experiences. This section draws upon discussions at the Asia Workshop in Hyderabad, where participants discussed and compared this diversity.[2] An almost bewildering range of typologies proposed by participants illustrated the ways in which perceptions of the distinguishing features of NGOs have formed along different paths in Asia, reflecting local conditions. In Indonesia for example the translation of the politically neutral term non-governmental organization from English into Organisasi Non-Pemerintah (ORNOP) carries an anti-government overtone, leading some in the NGO sector to prefer Lembaga Pengembangan Swadaya Masyarakat (People's Self-Reliant Development Groups) in order to emphasise their development role (Holloway 1989).

The NGO sectors in different countries have developed widely differing characteristics according to the history of the state, the amount of development assistance received and the gaps perceived in public service provision. Charitable and community organizations have a very long history in most Asian countries and the roots of present-day NGOs are traceable in many cases to the traditions which have guided these types of community work.

For example, there are a multiplicity of NGOs active in India, and these can be categorized in complex ways according to membership, the degree of voluntarism and party political affiliations, for example. In a recent study Robinson (1991) provides a concise summary of existing categories, which serves as a general guide to NGO typologies in Asia:

1 large national NGOs working in several states in different parts of the country and sometimes acting as intermediaries channelling funds from donors to smaller local NGOs
2 large national NGOs working in most districts of one state
3 medium-sized national NGOs working in a large number of villages in one or two districts of one state

4 small national NGOs working in a group of villages in one locality
5 large international NGOs with in-country representation providing funding and support to national NGOs
6 small international NGOs working directly in one or two localities.

Category 4 is the most common type of NGO in India.

The term NGO therefore has different meanings in different contexts. In the Latin American context, the role of membership organizations is more pronounced than in Asia, where most of the NGOs with which we are concerned in the present study are non-membership. The relative importance of and the relations between international and national NGOs involved in rural development also varies from country to country. Foreign NGOs may be active in implementation or 'partner' roles with local organizations (e.g. Oxfam). All the countries discussed in this volume are host to international private voluntary agencies such as World Neighbors, either with an operational presence, or providing technical assistance and funds to other organizations.

In the Philippines the term NGO refers to private, non-profit organizations aiming at addressing the needs of disadvantaged sectors of society by development activities. There are two main sub-categories: first, national and local NGOs, which are organizations with full-time staff working in community-organizing, education, sectoral development, human rights, disaster relief or humanitarian and charity work (e.g. PHILDHRRA); and second, people's organizations which are membership associations of people such as farmers, forest dwellers or fisherfolk organized at the grassroots by the people themselves and engaged in community organization, sectoral development, community education and co-operatives (e.g. KMP, CPAR).

NGOs in the Philippines context are therefore distinguished clearly from people's organizations (POs). NGOs seek to provide an umbrella structure to service and support these POs, while POs are said to exist in their own right and may be serviced by several NGOs. In Bangladesh, such a distinction is not in common use. An NGO is usually taken to include the groups which it forms, and most of these groups have very little independent life outside the overall framework of NGO activities (although the NGO may consider this to be a desirable long-term outcome). However, differences in perceptions may be wider than those in reality: for instance it was sometimes found during the collection of these case studies that POs in the Philippines, as in the Bangladeshi context, were little more than extensions of a particular NGO.

Another distinguishing feature of NGOs in the Philippines is the extensive network of coalitions and umbrella groupings based around issues such as health, organization and land reform. By contrast, the proliferation of NGOs in Bangladesh has not led to effective formal mechanisms for NGO–NGO co-operation, and the Association of Development Agencies in Bangladesh (ADAB) has remained relatively weak. However, there is evidence that considerable 'informal' networking between NGOs takes place, particularly at the sectoral level. In Bangladesh, while such groupings of NGOs are less common and less effective, the activities of large, individual NGOs, including BRAC and Proshika, case studies of which are presented in Chapter 3, have received a high profile both nationally and internationally.

Bangladesh probably contains more NGOs than any other country of similar size. The prevailing NGO approaches centre on two main strategies: first, mobilization work, where NGOs concentrate their efforts on organization of the rural poor (e.g. Nijera Kori), and second, rural livelihood creation, where NGOs are concerned primarily with income-genera ion activities, frequently centring their activities around credit provision (e.g. Grameen Bank).[3] Many of the better known NGOs follow a combination of both types of approaches which are combined to varying degrees (BRAC, Proshika, FIVDB).

The coverage and size of the NGOs is also an important factor with local NGOs centred upon a small number of districts in the country (including FIVDB and Saptagram in Bangladesh) and national NGOs working in a larger number of districts (BRAC and Proshika). NGOs in Bangladesh are also strongly differentiated by their focus: some are 'single-issue' NGOs (e.g. Women for Women) whereas others are multidimensional and work with men and women in urban and rural areas, in such sectors as agriculture, apiculture, aquaculture, forestry, and livestock, and in a range of activities, such as consciousness-raising, legal awareness and credit provision (Proshika, BRAC).

Nepal's voluntary sector has until recently been subject to patronage and control by the royal family. Some regard the role of the king in creating the Social Services National Co-ordination Council (SSNCC) in 1977 as adding to NGO visibility and credibility in Nepal (PACT 1989), while others (see Chapter 5) have regarded the SSNCC as a stultifying force. In the often remote communities of Nepal, a long tradition of voluntarism motivated local development and welfare activities by community and religious groups. These local traditions have helped to shape the structure and activities of local NGOs. Informal grassroots associations exist today all over the country in the form of *guthi* (religious groups often based on land endowments), *parma* (labour-sharing societies) and *dhikuri* (credit associations). A women's NGO formed in 1917 with the aim of encouraging women's participation in the handloom industry provides an early example in Nepal of an NGO seeking to improve the conditions of particular disadvantaged groups. However, what is equally illuminating is that the attention it focused on poor rural women was seen as a threat by government, which subsequently forced its dissolution.

A growing number of NGOs in Nepal are beginning to work as catalysts in local decision-making and popular action, identifying local resources and seeking to create sustainable, external linkages. In addition to the 140 NGOs formally registered in 1991 (Shrestha 1991), eighty-two of which are active outside the Kathmandu valley, a number of newly formed people's associations or non-formal groups are beginning to register with the government's

Community Development Office. The group includes both traditional and modern groups, such as youth and women's associations.

In Thailand, distinctive cultural and religious traditions also have shaped the NGO sector. Groups of Buddhist monks seeking to combine spiritual and community development activities, have, for instance, been active in reforestation.

Despite the fact that popular organizations have been active in Indonesia for generations, development-oriented (as opposed to welfare) organizations started to become more visible only during the 1960s. The political context of a repressive state still makes Indonesia a dangerous place for many NGO activities, especially those which can be construed as being political or concerned with rural organization. Nevertheless, Indonesia (as well as Thailand) now contains groupings of NGOs similar to those in other countries discussed. For instance, national NGOs provide services and development inputs, some work in up to three or four provinces of the country at a time, in both urban and rural areas (e.g. LP3ES); local NGOs work at province or district level, with a small staff and office, usually informally organized, delivering rural development services to local communities (e.g. Mag-uugmad in the Philippines); and people's organizations, which are self-funded membership organizations for rural agricultural development activities, fisheries and cottage industries (e.g. water users' associations).

This section has shown the diversity of NGO sectors in the countries under discussion. The case studies in this volume are grouped by country so that the special conditions operating in each country can be appreciated. In Chapters 8 and 9 more general conclusions are drawn from the case study material on the types of roles adopted by each side in NGO–GO interaction, and on the scope for closer links.

NGOs AND THE STATE

NGOs in different countries and at various times have faced constraints placed upon their activities by the political climate created by government regimes. Three broad categories of political space and 'room for manoeuvre' for NGOs in different political contexts in Asia can be identified:[4]

- *NGOs working in environments of political repression* (e.g. the Philippines under Marcos, Indonesia today): NGOs find ways of working in 'safe', localized sectors without compromising their overall political and community-based ideals. Such strategies may be regarded by NGOs as temporary ones, building people's organizations so that in the event of political change (which they may themselves contribute to) both NGOs and people's organizations can assume a more direct, policy-centred role.
- *NGOs working with relatively non-antagonistic but bureaucratic government agencies under stable but non-democratic conditions* (e.g. Thailand,

pre-1990 Bangladesh): work can be grassroots and poverty-focused, along with broad-based non-party political 'mobilization' strategies or 'consciousness raising'. But NGO–GO relations are often characterized by mutual suspicion and concerted attempts by government to monitor NGO activities for purposes of control rather than co-operation.

- *NGOs working under conditions of relative democracy*, which may be either long established (India) or recently won (Nepal, Bangladesh, Philippines): in both, the challenge is to build new levels of accountability into government activities and improve the responsiveness of government agencies to the demands of poor people, encouraging those people to articulate their demands. Mechanisms for regulating NGO–GO partnerships through networking, communication and co-ordination are often lacking, but are needed if the potential for partnership is to be explored fully.

These categories of political space are not static, as recent events in Thailand demonstrate (Chapter 7): political changes brought as each regime succeeds the other – or, at times, even within regimes – offer new opportunities and constraints for NGO action. The historical experience of such changes is a major determinant of current NGO–state relations, and will have a bearing on the prospects of success of any functional NGO–GO that might be sought. Political and historical contexts also vary widely among countries. It is within these historical and spatial contexts that NGOs' future room for manoeuvre and their relations with the state will be defined. In Chapters 3 to 7 we turn to an analysis of NGOs' recent agriculture-related activities in the context of the political, economic and social conditions prevailing in each country. But first, we discuss six brief illustrations of areas of agricultural technology development in which NGOs and government research and extension services have developed a comparative advantage.

GOVERNMENT AND NGO COMPARATIVE ADVANTAGES:

Illustrations and lessons

SIX ILLUSTRATIONS OF COMPARATIVE ADVANTAGE

Our review above of government strategies and resource allocations for rural development indicate a bias towards the more favoured areas: the irrigated lowland areas of India, Nepal and the Philippines, and the central lowlands of Thailand. Government clearly has a comparative advantage in developing and transferring agricultural technology for these: farmers' conditions can easily be replicated on experimental stations and improvement in a single input (e.g. in a variety of rice or wheat) when complemented by agrochemicals, reliable water supply and efficient marketing, can easily have wide impact. Furthermore, government institutes' close contact with international research centres can be instrumental in importing techniques and information conducive to the major advances characteristic of the Green Revolution. Also, governments have prioritized these areas deliberately for a number of reasons: in many countries they contain the most articulate and politically powerful farmers, and often generate food surpluses for politically important urban electorates.

Box 2.1, taking India as an example, serves the purposes of, first, illustrating the impact made by government research and extension services in areas such as these, and second, outlining the areas that have been neglected.

Government agencies' lack of success in developing and introducing new technologies for the more difficult areas is rooted partly in inappropriate research methods: conditions in complex, diverse and risk-prone areas can rarely be replicated adequately on-station. Extensive field-work is required, and must incorporate local knowledge and the testing of technologies by farmers themselves if it is to succeed. Part of the problem also, however, lies in the top-down approach taken by many government agencies. The features of this include research and extension agenda predetermined by specialists without reference to farmers' needs; inadequate monitoring of the uptake of technology, and inadequate analysis of low adoption levels; weak mechanisms for obtaining farmers' views on the technologies introduced. Furthermore, government organizational culture has often contributed to inefficiency – through low salaries, rigid hierarchies and poor incentive systems.

Many public sector programmes have encouraged dependency and made little lasting impact after government support is withdrawn. Although public policy increasingly specifies that 'beneficiaries' should participate in project planning and implementation, the reality is that few programmes have been able to achieve success in this respect. Nor are NGOs entirely free from criticisms that they have created dependency and relations of patronage, but NGOs have proved more successful than government agencies in utilizing participatory methods and have made a significant impact, albeit largely on a local scale.

NGOs have therefore moved into 'niches' configured by their skills in certain types of methodology, by the needs of their resource-poor 'clients', and by the space left by government efforts. The 'top-down' approaches favoured by government not only lead to inefficient structures for delivering inputs (which may well be diverted by more powerful rural elites), but also are often based on conceptions about what rural people want and need, and a value system in which judgements are made without consultation, since 'beneficiaries' are widely assumed unable to express their needs.

NGOs have challenged this model with a 'target group' approach, but it is a model not without its problems, since the notion of 'targets' among the poor all too easily becomes an imposed model in which the rights of poor rural people in determining the 'development process', though rhetorically acknowledged, are given insufficient emphasis (Wood 1985). Further brief examples in this section (Boxes 2.2 to 2.6) illustrate a set of typical NGO approaches to the areas in which government services have either disregarded the needs of the poor or have responded to them inadequately. These areas include:

- technologies and management practices adapted to difficult areas (Box 2.2)
- technologies to meet the needs of the rural landless (Box 2.3)
- technologies to meet the specific needs of women (Box 2.4)
- approaches that 'de-mystify' complex technologies and make them suitable for neglected groups (Box 2.5)
- approaches helping local groups to form which then carry forward the technology in a sustainable fashion, linking in with input suppliers and markets (Box 2.6).

A final example (Box 2.7) illustrates innovative efforts by funding agencies to enhance government's responsiveness to the needs of the rural poor, and to open up government to the possibility of collaboration with NGOs.

CONCLUSIONS

The chapter has demonstrated the diverse economic and political conditions of

Box 2.1 Indian agricultural research

India's public system of agricultural research is exceptionally large, comprising 44 central research institutes, 25 national centres and 127 zonal research stations (mainly financed centrally) plus 26 state agricultural universities (financed by a combination of central and state-level funds). Over 20,000 scientists work in the system.

Its most notable successes since 1960 have been in the introduction of irrigated short-strawed fertilizer responsive varieties of wheat and rice, increasing yields of rice from 863 kg/ha in 1966–7 to 1568 kg/ha in 1985–6 and of wheat from 887 kg/ha to 2,032 kg/ha over the same period. By contrast, rainfed sorghum yields rose only from 511 kg/ha to 641 kg/ha and oilseeds from 420 to 591 kg/ha over the period (Randhawa 1988).

The decision to focus on rice and wheat grown under reliable, high-input conditions was undoubtedly correct at the time in the sense that crop production (and, therefore, food security) were greatly enhanced and the returns to research investment high (Mukhopadhyaya 1988). The homogeneous conditions under which rice and wheat were grown could easily be replicated on-station and, once discovered, any innovation was likely to be replicable over a wide area.

Successes in rice and wheat research were, however, achieved only at the expense of disproportionately large allocations of research funds to the areas where these crops are predominantly grown. Thus, average expenditure under the Sixth (1980–5) Five-Year Plan by Central and State governments in the north-western states of Haryana, Himachal Pradesh and Punjab averaged Rs25 per capita of rural population, as against a national average of Rs6.9 and an average in the western (mainly rainfed) states of Bihar and Orissa of Rs3.5 and Rs2.8 respectively (Clay, personal communication).

Low productivity gains in the less favoured states, while clearly attributable in large measure to exogenous factors such as the difficult farming conditions prevailing there, have been compounded by two further factors:

1 the comparative neglect of these areas as measured by per capita research expenditures
2 the inappropriateness of conventional on-station research methods and 'top-down' dissemination systems to the complex, diverse and risk-prone conditions prevailing in these areas.

While the reduction of the first of these biases envisaged in the Seventh and Eighth Plans, and the gradual introduction of participatory methods in parts of the research system, will go some way towards redressing these imbalances, it is clear that the complexities are such that even a doubling of government research resources to these areas would be unlikely to produce widely distributed production gains. Approaches to research have to be developed in which local organizations participate strongly in the adaptation and dissemination of new technologies and management practices, bringing the twin advantages of a reduced burden on the public purse, and better adaptation of technologies to local conditions.

Sources: Randhawa (1988); Raman *et al.* (1988)

Box 2.2 Auroville's environmental work

Auroville is an NGO based upon an experimental community founded in 1968 on the east coast of India near Pondicherry. The 1,000 hectare site was originally severely degraded, but has now been planted with over 2 million trees as part of a continuously evolving programme of environmental regeneration undertaken by the NGO. Land management approaches integrate the use of trees, animals and annual crops over individual watersheds. As the trees grew, new micro-climates were formed and many animal species returned. The Auroville Greenwork Resources Centre was established in 1988 to test the principles and practice of sustainable resource management gained by the NGO against the opportunities and constraints (agro-ecological and socio-economic) prevailing in local communities.

Auroville's achievements include the establishment of a tree seedling nursery, containing varieties proven successful in local areas, offering seedlings for sale; an *index seminum* which lists the details of seeds of over 300 tree species and offers them for exchange throughout and beyond India, and an active biological agriculture programme in which local communities are involved.

See Chapter 4 for further details.

Box 2.3 Proshika's landless irrigation programme

In rural Bangladesh, landlessness has reached massive proportions as the population has increased, lands have been fragmented by the multiple inheritance system and marginal farmers have been dispossessed through mortgage and debt. The numbers of families owning no land at all or less than half an acre reached 50 per cent by the early 1980s. Few new rural employment opportunities have been forthcoming. There had been little productive investment in agriculture, since wealthy landowners obtain higher returns from non-productive economic activities such as renting and moneylending than from investing in the land.

In 1980, the government's Second Five-Year Plan, based on advice by the World Bank to privatize input distribution and remove subsidies, attempted to increase investment in agricultural production primarily through free market distribution of minor irrigation equipment and fertilizer supply. These inputs had previously been supplied through subsidies to highly inefficient government co-operatives and the result had been low levels of equipment use due to inefficient implementation and over generous subsidy.

The Bangladeshi NGO Proshika feared that the programme offered little to benefit the growing numbers of landless households. Subsequent discussions between NGO staff, irrigation specialists and government personnel led to the idea that groups of landless people could own and operate irrigation equipment themselves and sell the water to local farmers. Members of Proshika's landless groups had themselves raised this possibility with local Proshika staff. The Ford Foundation, which supported the Planning Cell of the Ministry of Agriculture, helped to facilitate these discussions. The government, through its agricultural co-operative system, decided, with some scepticism, to experiment with landless irrigation and Proshika was authorized to conduct a parallel programme. The value of the venture, aside from its employment potential, was that it would bring the landless groups into continuous contact with landowners, commercial classes and officials.

Proshika therefore initiated the approach using low-lift pumps and shallow tube-wells during the 1980–1 season with twenty-six groups, before the technology was monopolized by richer entrepreneurial households. These groups sell water each season to local farmers in exchange for a share of the crop. The income of the landless groups helps to create demand for the produce. Over 150 small groups of landless water sellers have since been established. This venture has aimed to empower the poor through facilitating access to a crucial rural resource and is therefore both creating an income generating activity and an innovative approach to the issue of agrarian reform.

Since Proshika's efforts to establish the system, other similar ventures have followed, such as BRAC (see Chapter 3), the CARE Landless Operated Tubewell User's System, the Grameen Bank and the Bangladesh Rural Development Board's Rural Poor Programme.

Source: Wood and Palmer-Jones (1990)

Box 2.4 Friends in Village Development Bangladesh (FIVDB) working with rural women

FIVDB is a local NGO established in 1980 which works with local landless people, particularly women, in building local organizations and generating income. The group-based integrated approach combines livestock and agricultural extension work, functional education, credit and savings, health and children's education.

One of its most successful income-generation programmes has been duck-rearing, an activity largely ignored by the government's research and extension efforts. The appeal of duck-rearing to the rural landless, and to women in particular, lies not only in its economic viability. Its specific appeal is twofold: first, it does not require access to land – exploiting instead the large areas of water to which the poor have access; second, it is essentially a 'backyard' activity which women who, for cultural reasons, do not engage in field-work, can practise within the household compound. FIVDB has therefore promoted improved duck-rearing practices with its women's groups in the Sylhet region of north-east Bangladesh. The NGO has introduced new breeds and developed its own hatchery based on a Chinese model. Some informal links have been established for facilitating research, such as that with Bangladesh Agricultural University. Training has been provided to female group members and other government and NGO extension staff.

For example Mayarun Nessa, a landless FIVDB group member, has learnt how to hatch and rear improved 'Cherry Valley 2000' ducks and received credit to build a flock of thirty-four birds. She has also been trained in vaccination techniques, and so earns extra income by providing this service to others in the community, so enhancing the productivity of their flocks. Her husband sometimes sells their eggs directly in the market; at other times she sells to local traders.

FIVDB's duck-rearing work has therefore directly addressed the needs of landless women by providing a source of increased income, greater respect within the household built on women's successful enterprise, and greater mobility within the community for women undertaking para-veterinary work.

See Chapter 3 for further details.

Box 2.5 PRADAN's work in 'de-mystifying' complex technologies

Professional Assistance for Development Action (PRADAN) is an Indian NGO established in 1983. PRADAN takes the view that many technologies developed by government are costly, complex and risk-prone and therefore of little value to the rural poor.

PRADAN therefore adapts the technologies and provides support to local organizations in technical and management aspects of project design and implementation, and in institutional development.

For instance, the technologies recommended by public sector research institutes for mushroom production include the use of electrically powered humidifiers, sterilized conditions for spawn production, and the use of external heat sources to sterilize the straw used as a substrate in mushroom growing. PRADAN has simplified these and scaled them down by devising simple methods of conserving humidity and by sterilizing straw through anaerobic fermentation.

See Chapter 4 for further details.

Box 2.6 Mag-uugmad Foundation (MFI) and farmer groups

MFI is a small Filipino NGO working with upland farmers on the island of Cebu in the Central Visayas region. Cebu is a predominantly hilly area with 70 per cent of its slopes greater than 18 per cent. There is severe deforestation, and severe population pressure on the remaining land. Originally established as local counterpart to the Soil and Water Conservation Programme of the US NGO World Neighbors, MFI is now an autonomous NGO with a Board of Directors which includes three farmers and one fisherman.

Through needs-based discussions with local farmers, MFI has developed a farmer-based extension strategy for its soil and water conservation programmes in three areas of the island based on the need for a technology which has economic viability, environmental sustainability and social and cultural acceptability to the people who implement it. These technologies include the contruction of contour channels with hedgerows and drainage, soil traps and gullies for soil stabilization, and bench terracing with contour ploughing.

These technologies require considerable labour inputs and MFI seeks to create motivated farmer groups and ensure the development of autonomous, self-sustaining local people's organizations (POs) to continue the work after MFI's support is withdrawn. MFI has therefore built on the local system of *alayon* work groups, based on kinship or residence. These groups have helped to facilitate the reciprocal sharing of agricultural labour tasks for generations by rotating the work and draught animals each week from one farm to another. Now the groups have adapted themselves to the urgent priority of regenerating the uplands.

See Chapter 6 for further details.

Box 2.7 The Ford Foundation and forestry in India: broadening government perceptions

The Ford Foundation has supported programmes in three areas of South Asia where poor land management is causing environmental degradation. Support has specifically been provided to government agencies attempting to improve land management through partnerships with local groups.

As the population of India approaches 1 billion, sensitive management of the country's natural resources is essential for enhanced social and economic welfare, yet land, forest and water resources are being rapidly depleted. Over half of India's forest land is seriously degraded. About 1.5 million hectares of forest land are denuded annually, and demands for firewood and fodder are expected to triple over the next decade. The attempts of senior planners to address these issues in reforestation and watershed management programmes in recent years have made little headway in improving public forest land management, mainly due to a lack of attention to tenure issues, and conflicts over the usufruct and protection rights of rural forest users.

One of the key problems is that although local people are primary users of forest resources, they have no formal role in management. The government has taken over the role of 'policing' forest resources, but it tends to lack the institutional capacity to control access. Farmers have found that the Forest Department has proved far more sympathetic to timber contractors than to the small-scale demands of local forest users.

Throughout South Asia there is therefore an urgent need to develop locally based management systems for common properties and state forest land. Ford has supported activities in the tribal regions of central India, semi-arid western India and the middle hills of the Himalayas. In each region, government forest agencies engaged in the process of establishing joint management programmes were identified. Working relationships have been established with the West Bengal Forest Department, Haryana Forest Department and the Nepal Institute of Forestry, with support for training and research from NGOs, consultants and university-based researchers.

Based on the success of the Arabari experiment in West Bengal in the 1970s, the West Bengal Forest Department has extended the idea to the sal (*Shorea robusta*) forests of SW Bengal. It was found that if villagers protected the natural sal forests from fuelwood cutting, the trees could rapidly regenerate, due to their hardy rootstock. In return for protecting the forests, the Forest Department agreed to give the villagers 25 per cent of all revenue generated from the sale of firewood and timber. By the end of 1989, it was estimated that community-based forest protection committees were protecting 152,000 hectares of forest land.

The Foundation's efforts to facilitate innovation by government departments in this way has attracted more acclaim than criticism. However, a number of problems remain (Hobley, pers. comm.). These include the lack of wide agreement on what should be managed by villagers under joint forest management programmes, what the role of NGOs should be, and what charges Forest Departments should levy for e.g. arranging timber auctions (in West Bengal these charges are currently so high that villagers are left with rewards insufficient to recompense their efforts). Furthermore, external pressures of this kind run ahead of government's ability to sustain institutional change and local-level institutional capacity through e.g. re-training, and have been poorly co-ordinated with donor-supported projects.

Source: Poffenberger (1990)

countries in the region, and the diverse origins, structures and objectives of the state, of government organizations and of NGOs. These general conditions define the extent to which views among NGOs and GOs will converge over their respective roles, and define the roles that each might perform. We have concluded that government research and extension departments have comparative advantages in certain areas but have neglected others, and that NGOs have achieved some success in several of the areas neglected by government. However, we have to be careful to recognize the lack of *hard* data on NGO achievement and acknowledge NGOs' own weaknesses in monitoring and documentation.

NGOs' success in demonstrating how livelihoods might be enhanced among the poor leads, as Korten (1987) points out, to the question of how to assist the large geographical areas still untouched by their efforts. The donor trend towards provision of increased funding for NGOs, if it continues, might imply strategies of encouraging NGOs to substitute for what might normally be expected to be supplied by the state. In our view, this should be no more than one component of a balanced strategy. While there *are* NGOs willing to take on an enhanced service delivery role (see e.g. Aguirre and Namdar-Irani 1992), and numerous new 'opportunistic' NGOs seem likely to be established to take on this function over the next decade, many others exhibit qualities which illustrate how they and GOs might best complement each other, and funding agencies need to devise strategies to bring out these complementarities.

In conclusion, we stress that the actions of, or support for, NGOs in disregard for government, or vice versa, implies an agenda which is unlikely to be resource efficient, and may have wider negative consequences. We suggest that various forms of interactions exist between NGOs and government, along a spectrum from, at one extreme, close partnership, to, at the other, efforts by one to influence the agenda of the other, and that lessons can be learned from these. The central task – for NGOs, government and donors – is to identify types and levels of interactions compatible with the character of the organizations concerned, with the task in hand, and with wider political and economic conditions. In Chapters 3 to 7 we now turn to the discussion of a wide range of case-study material from six Asian countries in an attempt to learn from existing types of interactions.

NOTES

1 However, events of April 1992 illustrate the fragility of the space afforded to NGOs and wider movements for democracy.

2 The existence of an 'NGO sector' in each country is in itself highly debatable, reflecting as it does official or bureaucratic conceptualizations of what is almost by definition a set of diverse, independent and localized responses to wide-ranging issues. As Tandon (1989) points out: 'It is important that NGOs are viewed . . . as expressions of autonomous, decentralised initiatives, as manifestations of democratic

processes and forms, as non-profit voluntary efforts, as expressions of social commitment for an equitable and just society'.

3 The status of the well-known Grameen Bank as an NGO or as a government agency now part of the formal banking sector is slightly unclear. Some have therefore termed it a 'quasi-NGO'.

4 These categories draw on those of Clark (1991).

Chapter 3

NGO–GOVERNMENT INTERACTION IN BANGLADESH

OVERVIEW

David J Lewis

BACKGROUND

Bangladesh became independent from Pakistan in 1971 after a violent civil war and the new country has faced a range of pressing development problems ever since. Bangladesh has a relatively undiversified economy which has remained dependent upon the production of rice and jute, and is one of the most densely populated countries in the world, with an estimated population of 113 million in 1990. Over 80 per cent of the population live in the rural areas and depend on agriculture for their livelihood, which is still based primarily upon human labour and animal draught power. Bangladesh is highly vulnerable to natural disasters in the form of floods and cyclones, which has ensured that the government and other agencies have been repeatedly forced to steer a difficult course between development and relief.

Government agricultural policy has promoted rapid increases in fertilizer and pesticide use since the early 1980s, while rural impoverishment has led to severe stresses on natural resources, particularly in the fisheries sector. Mechanized irrigation policies have led to the over-exploitation of ground water resources in some areas of the country, while salinity has also resulted from overpumping and from intrusion resulting from reduced flows from the lower Ganges. Bangladesh is also losing its limited forest reserves as a result of increasing poverty, as the Proshika case study presented here illustrates, but state policies are also playing a key role, particularly in the Chittagong Hill Tracts area, where the army is reported to be heavily involved in illegal logging activities.

Despite a moderate increase in the per capita growth rate (about 1.5 per cent per year) during the first two decades of independence, little impact has been made on poverty alleviation, since assets and economic opportunities are highly uneven in their distribution. More than half of the rural population is landless. The country is heavily dependent upon foreign aid (US$1.7 billion per year) and upon remittances from workers abroad. However, the debt servicing ratio remains relatively low due to the comparatively soft terms of foreign financial assistance (Hossain 1990).

Until 1990 Bangladesh was ruled by President Ershad, who came to power in a military coup in 1982. Despite half-hearted attempts to construct a civilian regime, the government remained essentially military and authoritarian, and Ershad gained a reputation for corruption on a massive scale. The popular opposition movement which developed support from 1987 onwards forced his resignation in December 1990, and a democratic government was elected in February 1991. Although the new Bangladesh Nationalist Party (BNP) government does not appear to have policies that are significantly different from those of the Ershad regime, President Khaleda Zia is the subject of widespread expectations in maintaining a commitment to democracy and increasing public accountability. This has been expressed in a formal move from a presidential to a parliamentary system of government. There have also been signs of increasing contact and dialogue between the different actors on the political scene. After the fall of Ershad, the caretaker government convened a series of task forces on economic issues, women's rights, health, education and other sectors in which NGOs, political parties and other community groups were invited to begin a policy dialogue with government.

POVERTY IN BANGLADESH

Bangladesh has become recognized as one of the five poorest countries in the world and the poorest in Asia (World Bank 1991b). The 1991 UNDP Human Development Report puts the average life expectancy at 52 years, adult illiteracy at 68 per cent and infant mortality at 116 per thousand live births. For the large numbers of the rural population without access to land, casual labouring and non-land-based self-employment activities (such as petty trading, rickshaw pulling and raising livestock) have become components of a survival strategy which requires a shifting, seasonal portfolio of income generation activities.

Poverty has important gender dimensions in Bangladesh, where it is experienced differently and disproportionately by women (Kabeer 1989). Women have poor access to educational, health care and legal services. A division of labour based on gender and norms of female seclusion constrains the range of survival strategies open to poor women. Growing numbers of poor women (and particularly those in female-headed households) depend on a limited range of homestead-based income-generation activities such as poultry-rearing, paddy-husking and vegetable gardening (Lewis *et al.* 1993). Although such jobs tend to bring very low rates of return, the role of NGOs in bringing these types of activities (and by extension the material needs of low-income rural women) into the mainstream development discourse in Bangladesh has been significant.

The formal agricultural extension service has tended to ignore the needs of women, assuming that only men are active in productive agricultural activities, and overlooking women's important (but less visible) areas of participation (e.g. seed storage and selection, post-harvest processing, decision-making over crop choice, etc.) as well as women's increasing involvement in non-traditional activities as increasing poverty forces many women into less conventional roles in field crop production (UNDP 1988).

AGRICULTURAL DEVELOPMENT POLICY

Since Independence, Bangladesh's successive governments have tended to ignore the landless in favour of those with resources in a farmer-based development strategy. There has never been a comprehensive national policy for the rural areas which have instead been subjected to a number of competing strategies undertaken by different government agencies, usually under the guidance and patronage of the international donors (Hossain and Jones 1983). The government's continuing Five-Year Plans have prioritized rural development and made self-sufficiency in food production a central objective. In the 1960s public sector corporations were established for input procurement and distribution and a co-operative model was employed for promoting equal distribution of subsidized irrigation equipment, fertilizers, pesticides and high-yielding seed varieties.

The growth of a public sector agricultural research and extension system took place under the British colonial administration (and later the Pakistan government) as the demand increased for exports of raw materials. An elaborate system of organization has emerged for the generation and dissemination of improved technologies for agricultural production and processing (Hassanullah 1991). The government's agricultural extension system is essentially a 'top-down' model based on the departmental system of the USA, where land grant colleges extend information and knowledge about new techniques to farmers. This model remains the basis, with modifications, of the system today.

According to one NGO observer, the system 'assumes that our peasants are backward, and ignorance is the basic cause of their backwardness' (Abed 1991: 2). These assumptions are still evident in the tone of official extension concepts today. A senior director of Agricultural Extension recently wrote: 'The farmers [of Bangladesh] grow crops mainly for themselves with very little surplus. Although hardworking, farmers, by and large, are poor, unlettered and prejudiced'.

The government's approach addressed those in the countryside owning agricultural land – the farmers – and in the 1960s (during the Pakistan period) achieved a measure of success and international recognition with the Comilla co-operative model. The approach was developed primarily by A H Khan at the Bangladesh Academy for Rural Development (BARD), with support from Harvard University Development Advisory Service and Michigan State University in the United States. Its 'two-tier' co-operatives were intended to promote an equitable Green Revolution strategy through farmers' co-operatives (kown as KSSs), assist with agricultural extension and provide a structure for needs-based farmers' organizations. The approach appeared to work under the intensive 'laboratory conditions' of Comilla District, but attempts to 'scale up' the model across the country weakened its innovatory character and left the 'formal co-operative' embedded in the rigid bureaucratic structures of the Bangladesh Rural Development Board (BRDB). As increasing landlessness became the dominant feature of changing agrarian structure, the approach lost its relevance. The co-operative system remains important to government agricultural development programmes in Bangladesh, but has met with only limited success (Faaland and Parkinson 1986). Other critics have been more severe, suggesting that the model 'constitutes a lesson in the futility of "co-operation" in a situation of inequality' (Khan 1979: 397).

The distribution of high-yielding varieties of paddy seed, the expansion of mechanized tube-well irrigation and the increasing use of chemical inputs in agriculture have all been primarily state-sponsored ventures until recently, when the private sector started to play a larger role. The beginning of this shift in public policy, which was encouraged by the World Bank and USAID, took place at the end of the 1970s and was gradually made felt during the 1980s as subsidies were withdrawn. Private companies began to take over the distribution of many agricultural inputs. Against this backdrop of privatization, there has been increasing use made by government of NGOs in agricultural research and extension as donors have tended to favour them as a more efficient alternative to public services (Hassanullah 1991).

Bangladesh is now believed to be approaching self-sufficiency in food production, but it is those with land who have benefited most from this production-led development strategy. The government's emphasis on rice and wheat has led to the relative neglect of other crops along with the fisheries and livestock sectors. It is in these areas of non-land-based production activities that NGOs have tended to work, identifying needs and developing models generally without making use of government research facilities. More recently, NGOs have begun working with the rising numbers of urban poor who face similar neglect from government poverty alleviation programmes.

Under President Ershad, the decentralized Upazila system of local government was established in 1983. This provided the basis for agricultural initiatives through local level officers responsible for agriculture, livestock and fisheries, but a limited resource base and still weak organizational structures have created only moderate improvements in services.[1]

NGO APPROACHES TO THE RURAL POOR

There are probably more NGOs in Bangladesh than in any other country of the same size in the world. NGOs have mainly functioned in order to service the needs of the landless, mainly with external donor funding as a counterpoint to the government's efforts – also with substantial donor funding – to develop the agricultural sector.[2] There has been a long tradition of private organizations in the subcontinent, with voluntary work undertaken by better-off members of the community for organizing schools or mosques. During times of natural disaster, a relief effort was often organized for the victims.

After the War of Independence in 1971 and empowered by the massive relief effort which followed, many embryonic Bangladeshi NGOs recognized that the needs of the landless should be considered as central to an equitable development strategy programme which did not simply rely upon 'trickle-down' benefits. A dramatic growth took place in the numbers and the importance of NGOs. Many activists involved in the liberation struggle saw the need to build the relief effort into sustainable, community development programmes in agriculture, health, education and other sectors (Hashemi 1989).

The 'target group approach' which emerged emphasised the centrality of landlessness to a development strategy, and placed the needs of landless women increasingly to the forefront of its programmes. The activities of larger NGOs such as Bangladesh Rural Advancement Committee (BRAC) and Proshika were based on the twin strategies of income-generation activities combined with 'conscientization' programmes aimed at developing the potential of poor people to challenge existing inequalities through education, organization and mobilization.

A second innovation was the prioritization of non-land-based sources of income-generation for this target group, an area which had been substantially neglected by government. In particular, these income-generation activities are important to the survival strategies of poor women, in both male and female-headed households. This led to a concentration of efforts into small-scale, homestead-based income-generating activities such as cattle and poultry-rearing, food processing, social forestry, apiculture and rural handicrafts combined with provision of formal credit, to which the landless had previously been denied access except from local moneylenders at high cost.

The NGOs' initiatives in establishing income-generating activities proved to be an effective alternative to top-down government rural works programmes, but the extremely low rates of return on such activities have caused many to question their long-term sustainability. In fact, some NGOs in Bangladesh reject the idea of providing credit for income-generation activities in favour of organizing the landless to strengthen control over assets such as land, forests and water-bodies and strengthening their claims on government services. Many of the larger national NGOs continue to combine both approaches, arguing that there are important social benefits to income-generating activities over and above its direct economic value, particularly in the case of rural women (Sanyal 1991).

NGO discourse in Bangladesh is unusual in that it does not include much reference to people's organizations (POs) in comparison, say, with the

Philippines, where NGOs are conceptualized as being service providers to existing grassroots, self-sustaining organizations, such as farmers' groups, or act as catalysts for their formation. In Bangladesh, the term 'NGO' is usually assumed to include the informal groups which NGOs invariably form. Despite reference to this as a long-term objective, few would pretend that these informal groups have yet become sustainable entities outside the framework of the NGO's activities.

NGOs AND AGRICULTURAL TECHNOLOGY

Rural development NGOs have approached the issue of technology development for their target groups in three main ways, each of which has led to a distinctive set of innovative roles:

- NGOs have tried to *develop technologies* themselves to suit the needs of their group members in collaboration with local people. An example of this is the treadle pump, a low-cost bamboo pump developed by the Rangpur Dinajpur Rural Service (RDRS) in the north-west. Similarly, Friends in Village Development (FIVDB) have pioneered the improvement of duck-rearing practices in Bangladesh and adapted technologies for use by its primarily female group members.
- NGOs have worked to *deliver inputs*, by strengthening the existing network of distribution of inputs by the government, often using their closer ties with organized groups of recipients and their more efficient administrative structures. They have tried to achieve this both through directly taking over certain extension roles and often by simultaneously providing training to strengthen existing systems. Examples include BRAC's poultry and Proshika's livestock programmes.
- NGOs have created new forms of *social ownership* of technologies, by reorganizing existing technologies into units of group-based management through which their benefits can be 'captured' for poorer sections of the population. The best known examples of this are BRAC and Proshika's landless irrigation programmes in which landless groups sell DTW and STW water to local farmers on a seasonal basis in return for a proportion of the crop (Wood and Palmer-Jones 1990).

It is perhaps the third role which has proved the most innovative and has led to NGO efforts being directed towards addressing some of the specific technological problems facing rural Bangladesh. In particular, conditions of small-scale farming combined with high levels of landlessness mean that few farmers (except the very rich) have the resources to adopt new technologies on an individual basis and the providers of agricultural services (e.g. irrigation, rice milling, mechanized ploughing, fertilizer distribution) have become key actors in the rural scene (Lewis 1991).

NGO-GOVERNMENT PARTNERSHIP

The government was largely supportive of NGO's efforts in the arena of relief during the 1970s but relations between the government and NGOs gradually deteriorated as NGOs' involvement in development work deepened (Sanyal 1991). Relations between government and NGOs have continued to be polarized and characterized by considerable levels of mutual distrust. This has also been true among opposition political parties, which by and large have remained hostile to NGOs.

As White points out, the opposition between NGOs and government has an almost 'mythic' quality, but nevertheless reflects real struggles for representation as the legitimate voice of the Bangladeshi people and by extension the substantial resources available through overseas aid. At the same time, certain benefits are available to both parties through working together: in the form of a greater role in national development for NGOs, and more effective implementation, the neutralization of potential opposition and a share in 'reflected moral glory' for government (White 1991a).

Clearance by the government has been required for all funds brought into the country for development projects since 1978, and NGOs have been subject to lengthy bureaucratic approval procedures which led to increasing levels of corruption and abandonment of programmes by smaller NGOs (White 1991b: 17). In 1989 the new NGO Bureau was established which provided a 'one stop' service to NGOs within prescribed time limits, but which some observers concluded increased the government's ability to monitor NGO's activities.

The dangers to NGOs of collaboration with an authoritarian government were illustrated in 1989 by an incident at the height of the anti-Ershad popular movement. Dr Z Chowdhury, the leader of the prominent health sector NGO Gonoshosthya Kendra (GK) had attracted international admiration for his organization's emphasis on producing and distributing essential drugs in opposition to the expensive and often irrelevant products of large drug companies and had also alienated powerful interest groups among these (often multinational) companies and among the medical profession. Seeing his chance to work directly with the government health minister to formulate a national health policy, he was then widely criticized for giving credibility to the government by both the NGO community and the anti-Ershad movement at a time when opposition was at its height. NGOs are again currently active in public discussions on health as the government is exploring the possibility of an increased role for the private sector in health.

Formal partnerships between NGOs and government in technology development in Bangladesh have been few. Part of the blame for this may lie with government, which suffers itself from poor co-ordination among its extension agencies. A recent study showed 'a high degree of overlaps in clientele, operational jurisdiction and technologies with little or no resource sharing among agricultural extension organizations' (Hassanullah 1991). Informal rather than formal interactions between government agencies and NGOs have been more common, and have met with varying degrees of success as the case studies presented in this volume illustrate.

In agricultural extension and input delivery, relationships have often developed based on the government's control of scarce inputs (such as livestock vaccines) and its relatively poor contact with the 'grassroots'. Partnership can then arise based upon an NGO's desire to see an improvement in the distribution of inputs to their group members. In addition, NGOs have sought a longer-term strengthening of government services based on direct training and greater 'demand pull' by motivated clients.

A second category of linkage is that of government 'contracting out' service functions to NGOs, which may increasingly play a role in the current privatization agenda being pursued by government and donors. Since the mid-1980s, the government's role has been substantially reduced in the distribution of agricultural inputs, and subsidies are now low or non-existent for many technologies. Unlike the first type of partnership, this does not involve a training component or an assumption that government should in the end perform these tasks, but is the 'substitution' with sub-contracted NGOs of traditional areas of government responsibilities.

Since relatively few NGOs in Bangladesh actually work with landowning farmers, the scope for formal collaborations with government agricultural research institutions (e.g. in field-testing new varieties) has not been as wide as in other countries. However, there are some exceptions. Both RDRS and Mennonite Central Committee (MCC) are prominent NGOs who include small farmers in their target group. The treadle pump has already been mentioned as a distinctive NGO-developed technology. MCC has also been involved with field-testing and varietal research, a relationship which has met with only limited success (pp. 66–73).

One of the most problematic forms of contact between NGO and government has been in the context of so-called 'common property resources' (CPRs) such as forest or government-owned 'khas' land. In fact, as one NGO worker recently pointed out, there is no such thing as a 'common property resource' in Bangladesh: everything is controlled *de facto* by somebody else. When NGO groups have organized to improve their access to CPRs – which in practice are mediated by powerful interest groups in the rural power structure allied with local state representatives – there has been considerable conflict. This is illustrated by Proshika's experiences in its social forestry programme outlined in the next section (pp. 59–65).

The best results of partnership have been in the field of input delivery, but there is a tendency on the part of government to view a relationship in terms of sub-contracting its responsibilities, rather than as a short-term 'role substitution' intended as a training exercise. Such interaction also tends to be local, based on key individuals. A frequent complaint of NGOs is that while government may often be keen to collaborate in this way, they are less inclined to allow NGOs inputs into policy or project design. (For a more detailed discussion of these issues see Lewis 1993.)

OUTSTANDING ISSUES

Government for its part questions the legitimacy of NGOs and their claims to represent the interests of certain sections of the population. How far does the success of NGOs in achieving success in certain activities actually undermine the government's own legitimacy? Do NGOs 'show up' the limitations of government programmes?

Even the largest NGOs in Bangladesh taken together cover only a fraction of the population: some have estimated that they reach only 10–20 per cent of landless households. Primary responsibility for the welfare of the population will always remain with government. Some NGOs point out that their conscientization programmes may have contributed in part to the peaceful democratic movement which ousted the government of President Ershad in December 1990. The objective of 'empowerment' is the ability of people to take economic and political decisions to achieve more control over their lives. Increasing confidence through managing group enterprises and addressing local problems has led some NGO group members to question wider relationships of accountability and participation.

However, the impact of these changes has often been overstated and some NGOs may reproduce only new systems of patronage (Hashemi 1989; White 1991b). There remains a significant gap in many areas between rhetoric and reality. Since patronage remains an essential feature of economic and political relationships in rural Bangladesh (Wood 1981) it is easy for both the state and, by extension, the NGOs to assume a brokerage role in provision of resources such as credit (McGregor 1989).

The government has sought to exclude NGOs from other sensitive issues in Bangladesh. For example, NGOs have been largely excluded from the Chittagong Hill Tracts area in the south of the country, where non-Muslim indigenous peoples are facing a combination of political and cultural repression along with the exploitation and degradation of their local natural resource base by government policies and by forced in-migration from the other areas of the country. Even during relief work (such as after the 1991 cyclone) there were reports that NGOs were prevented from distributing resources to villagers in the Chittagong Hill Tracts by the government authorities.

Some collaboration has been attempted on issues such as land reform, but little has so far been achieved. The government has had laws since Independence which relate to the issue of redistribution of confiscated and deluviated land under government ownership to landless households. But progress has been minimal, with local landlords maintaining effective control of such land and forcing 'tenants' to pay illegal rents. After informal contacts took place during the late 1980s between key individuals in the Land Ministry and Oxfam, NGOs were asked by government to develop proposals for implementing land reform. However, local level government officials proved neither motivated, experienced nor skilled enough to carry out implementation. None of the 484 households interviewed in a sample survey in 1989 had secured full land registration and there were widespread cases of harassment of applicants and misallocation of land (Barrett 1991).

The accountability of NGOs, in terms of their finances and accounts and relationships with donors are also questioned in government circles. If the government is now accountable to people through the newly reinstated electoral process, to whom are the NGOs accountable? It is also difficult for government to quantify the efficiency of NGO programmes in the terms of cost-effectiveness, and compare them with government's own, if their accounts are not easily accessible. However, a more serious barrier to this 'cross-institutional' measuring of effectiveness is the NGO emphasis on a range of activities, many of which (such as 'conscientization') cannot easily be included in a cost-benefit analysis.

Another source of disruption to collaboration has been conflict over organizational structures, styles and culture. This is a point made recently by Fowler (1990), who argues that many post-colonial government structures are concerned with top-down, directive development based on 'control and authority' rather than 'support and equality' (Fowler 1990). When Proshika tried to work with the Bangladesh Krishi Bank (BKB) to provide credit for its minor irrigation schemes, the administrative resources required by the NGO to make the relationship work at ground level led to the establishment of its own system of administering credit. The inefficiencies of the formal banking system proved an insurmountable obstacle to achieving their development objectives (McGregor 1988).[3]

There are also complex official tendering arrangements built into government systems. These rules can slow down distribution of inputs while tendering takes place and can lead to the production of inferior products or inadequate services if tenders are accepted which are too low. A recent example of this was the RDRS's treadle pump project (described on pp. 83–4). Nor are NGOs free of such limitations either, especially as they grow larger, but many government agencies in Bangladesh still tend to begin their initiatives, despite the stereotype, from a position of comparatively entrenched bureaucracy.

There is no doubt that NGOs and government still have differing pictures of each other in Bangladesh. These contradict each other to a large degree. Government tends to see itself as having a monopoly on authority and policy-making; NGOs see themselves as having superior claims to the truth, good intentions and the interests of the poor. And yet some observers have pointed out that the atmosphere of 'antagonistic co-operation' actually serves the interests of all the actors concerned – the NGOs, the government and the donors who coexist and interrelate in the manner of a joint family in which differences are accentuated in order to legitimize their respective, linked roles (Sanyal 1991).

A CONTRADICTORY MODEL OF PARTNERSHIP?

This is certainly the case in technology development, where it appears that two contradictory views of 'complementarity' coexist. In the first (NGO) view, the NGO, through its better needs-based contact with local people and its 'action research' approach, is in a position to develop technologies which are appropriate to people's needs and then 'pass them up' to government for possible replication elsewhere. In the second, (government) view, it is the government with its superior resources, international contacts and high coverage level which is in the better position to develop technologies which NGOs could then help to distribute more effectively, perhaps adapting them for local conditions. It is this contradictory view of the respective roles of NGOs and government in technology development and diffusion which leads to most successful collaboration being restricted to input delivery rather than permanent strengthening of government capabilities. Many NGOs see their relationship with government in terms of lobbying for better services and stimulating effective demand for existing services to be delivered as intended, existing laws to be enforced properly (e.g. rights to CPRs, laws outlawing dowry) and the development of new poverty-focused laws and services.

The FIVDB case study on duck-rearing illustrates the potential benefits to poor people which could be gained from linking NGO action research based on its target group's needs with access to government resources and appropriate mechanisms to allow the communication and discussion of their requirements to take place. NGOs can go some way towards the participatory development of poverty-focused technologies but may falter as higher level resources and better trained personnel are needed.

CONCLUSION: A NEW DEMOCRACY?

During the last year of Ershad's government, the official attitude to NGOs changed with the setting up of the NGO Bureau, designed to speed up the process of NGO registration with the government and ease bureaucratic hurdles to NGO activities. At the same time, the bureau is a means of control: it now exists as an improved means for the government to 'keep tabs' on what the NGOs are doing, and influence NGO programmes which are up for approval. There are reports of the Bureau interfering with NGO activities, and placing pressure on NGOs to strengthen certain aspects of their programmes, such as credit at the expense of other activities.

When the new BNP regime was elected in early 1991 there were indications that government would take a more constructive view of the role that NGOs could play in improving material conditions in the country, strengthening participatory forms of organization and activity and helping with the task of laying foundations for the new democracy. There may be new opportunities for NGOs to influence government policy in education, health and poverty alleviation. As one observer pointed out, the tradition of 'top-down' administration and decision-making have constructed an idea of development which has little meaning for the majority of rural and urban low income people, and the February 1991 elections therefore 'gave an opportunity for Bangladesh to make that first connection between democracy and development' (Samarasinghe 1991: 2142). However, the cyclone disaster of May 1991 immediately drew attention and resources back towards relief and by late 1992 there had been little progress in these new areas over the previous government.

PROSHIKA'S LIVESTOCK AND SOCIAL FORESTRY PROGRAMMES

Mafruza Khan, David J Lewis, Asgar Ali Sabri and Md Shahabuddin

INTRODUCTION

Proshika, which was established in 1976, is now the second largest NGO in Bangladesh, working with both the rural and more recently the urban poor. Proshika began as a project of the Canadian University Service Overseas (CUSO) as part of the national reconstruction effort which took place after the War of Independence in 1971, and Proshika quickly became an independent Bangladeshi organization. Proshika now works with 23,252 informal groups with an average membership of 20 people in 3,415 villages in 26 of Bangladesh's districts; 11,637 of these are women's groups.

Proshika has developed the parallel approaches of addressing the material needs of the poor through income-generating activities and facilitating 'conscientization' (a concept derived from Freire's work) through participatory approaches to education, training and mobilization.

The failure of government programmes to address the needs of the poor under conditions of growing landlessness led Proshika to develop an alternative development strategy aimed at increasing the social and economic self-sufficiency of landless and marginal farmers and workers based on three methods:

- Group formation to build unity and organization, either based on existing relationships (kinship, neighbourhood, etc.) or new groups based on common interests. The groups are single sex and village based.
- Provision of knowledge and skills to group members through human development and skill training.
- Assisting groups with loans and technical support to develop employment and income-generation activities.

Proshika therefore aims to assist the poor, first, to build the institutional structure needed for development, second, to analyse their structural position in society and identify the means to improve it, and third, to utilize both local and external resources efficiently to eliminate their poverty.

LIVESTOCK IN BANGLADESH

Since over half of the rural population of Bangladesh are effectively landless, the rural poor depend on non-land-based forms of production, of which livestock represents a key activity. In the division of labour in the villages, it is women who normally manage the animals. Although the government offers livestock support services through the local government Upazila centre, the government's agricultural research and delivery effort has traditionally neglected the non-land-based farming activities that are vital to the survival of many poor households.

Cattle-raising is an activity well suited to improving the income of landless women. A reasonable profit can be made by rearing from calf to adult animal, while the interim benefits of milk and draught power can greatly benefit rural households, since they are in extremely short supply in many rural areas. Risks, by comparison with other income-generating activities, are relatively low.

Nevertheless, there are many problems faced by livestock farmers in rural Bangladesh, particularly in preventive health-care and dealing with disease. Government efforts had made little or no impact on the need to address important problems arising from small-scale livestock management. For example, the medical care of animals has traditionally been the responsibility of *fakir* (local healers) who, while offering certain advantages to the community, offered little systematic protection against disease and did little to reduce mortality rates.

Proshika's livestock programme

Proshika therefore established a livestock programme, which has grown throughout forty-four area development centres (known as ADCs, Proshika's local operational unit) across sixty-five Upazilas where organized groups have made livestock cultivation one of their main income-generating activities. It is estimated that the programme in 1991 covers about 12,000 households. Most of the group members directly concerned with cattle-raising are poor, rural women.

Each livestock project is undertaken collectively by the groups but managed individually; 50 per cent of each individual's profit is contributed as group savings. Loans are given from Proshika's central Revolving Loan Fund (RLF).

The programme has three major components, all of which have been gradually modified as Proshika's continuous 'action research' programme has led to awareness of new needs and possibilities.

- *Training* Initially, the methods of rearing cattle and controlling disease formed the central components of Proshika's training. Cattle management issues have now been given added emphasis. Feedback from participants indicated that the groups were unable to run the project efficiently due to inadequate knowledge of rearing management.

- *Rearing* In the earlier programmes, there was no specialization of livestock rearing. Calf, dairy cow or bullock cattle were all reared in the same groups. But it soon became clear that different types of cattle required different types of management, which was difficult to provide in mixed cattle groups. Emphasis has therefore been given towards the specialization of cattle-rearing.
- *Vaccination* This has been a major step towards reducing mortality, but is a preventive measure. Alongside preventive measures, curative interventions are also necessary. For example, in the course of the programme, it was found that vaccination alone cannot ensure the productivity of a dairy cow unless it is also regularly de-wormed. The programme has therefore evolved accordingly.

In order for the programme to develop a measure of economic sustainability, the group members have to bear the cost of the vaccination fee, the RLF service charge (at commercial interest rates) and a cattle insurance premium. These costs have sometimes led to disincentives to some group members expanding their participation in the programme, but it is Proshika's policy to encourage this sustainability in order to create self-reliance.

Initially Proshika aimed to provide these services independently to the groups, but a service delivery and training relationship with government gradually developed. A constraint which became apparent was that the distribution of vaccination services from the Upazila Livestock Centres was limited. Vaccinators visited the villages only occasionally and it was difficult for rearers to achieve adequate levels of protection through regular vaccination. Since the supply of vaccines is controlled by the government, Proshika's scope for developing a more needs-based system was limited without entering into an arrangement with government. In some localities, discussions with local Upazila livestock officers yielded promising results.

A partnership with the Department of Livestock (DoL) developed initially through informal ties between local level government personnel at Upazila and district levels and NGO staff. It was later formalized through a contract drawn up in Dhaka with the DoL, after the government invited NGOs working in livestock to attend discussions concerned with improving extension services. In response to problems reported by Proshika groups in securing regular and adequate supplies of vaccines from the Upazila, the government agreed to supply vaccines directly to Proshika for its group members, at subsidized cost. After this, vaccines became available either from the Upazila or the Proshika ADC.

These vaccines were then used directly by Proshika's trained group member vaccinators, who for a small charge of Tk1–2 (to cover the paravet's salaries) carry out regular vaccinations on cattle belonging to both Proshika group members and those in the wider community engaged in cattle rearing.

Proshika's livestock programme has therefore provided the government with a well-networked distribution system for its scarce inputs. This collaboration has served the interests of both government and NGO, allowing the government to distribute vaccines and services more widely, and strengthening Proshika's group-based activities by giving members proper access to inputs and support facilities through which they can generate income from livestock rearing.

From 1977 to 1987 the proportion of the RLF utilized for livestock fluctuated between 3.6 per cent and 24.9 per cent. In 1988–9 it increased from 12 per cent to 32 per cent and is now the highest among all Proshika's Employment and Income Generation (EIG) programmes. This significant change took place in response to the increasing needs for income-generation activities based on cattle as expressed by the groups. Proshika figures indicate that cattle mortality has been reduced to 2 per cent through regular vaccination during 1989–90 in the project working areas (Huq and Sabri 1991). Although no data exist on previous mortality levels, it is estimated that a sizeable impact has been made.

Problems have centred upon contradictory institutional approaches to the provision of vaccination services. There are still government Upazila-based vaccination services providing a free vaccination service in many areas, and although this is by no means adequate for most farmers' needs, it sometimes undermines local commitment to Proshika's service, which levies a small charge.

Box 3.1 The provision of animal production services by rural women under Proshika programmes

Basana Rani Bhaumik of Uttar Jamaya village in Manikgonj District earns Tk2,000 per month by vaccinating chicken and cattle. She charges Tk0.5 per chicken vaccination and Tk1–2 for cattle. Basana gets the vaccines from the Proshika ADC. In addition, she has a contract with five surrounding villages who pay her 18 maunds of paddy at harvest time in lieu of her services each season. Creation of such demand via the extension service is a unique innovation by Proshika. Similarly, heifer-rearing has made Rawsan Ara of Kaunipara of Dhamrai Upazila in Manikganj District an impressive example of a landless woman improving her economic and social position. She has managed to earn Tk50,000 through selling the milk and the calf after three years.

The intensity of livestock programmes especially the vaccination programme within the project areas, has encouraged more participation by village people in rearing livestock. Vaccination, as the core component of the programme, has made a major contribution to livestock resource management.

There are currently around 700 Proshika-trained vaccinators and paravets. Around 200 of these can earn on average Tk1,500 per month and this income has brought about remarkable change in the lives of these poor rural women (see Box 3.1).

Results

This programme has brought about positive changes within Proshika's overall concept of group development. Proshika has found that without the group taking part in economic activities, non-economic activities (e.g. consciousness-raising, functional literacy, lobbying, etc.) will not on their own be effective. Access to income-generating opportunities is needed to create a strong, cohesive group with regular meetings, good management practices and constructive discussion. As an effective EIG strategy, the cattle programme has therefore contributed to wider group development.

The concentration of practical skills within the groups has contributed to empowerment. For example, vaccination is no longer a government-controlled activity. The groups have taken up this practice and de-mystified it, making it more accessible to the poor. The venture has also established a successful link with the Livestock Department, which now uses NGO workers for extension and overcomes the constraints of the limited availability and skills of existing DoL extension personnel.

SOCIAL FORESTRY PROGRAMME

While Proshika's livestock programme represents a successful partnership between NGO and GO over the delivery of scarce inputs, the same NGO has found itself encountering severe problems in developing a poverty-focused relationship with the Bangladesh Forest Department in its social forestry programme. The rich have tended to benefit from government reforestation initiatives which have provided seedlings only to those owning land. Proshika has therefore tried to involve the rural poor in forestry activities on government land.

The social forestry programme grew out of Proshika's long involvement in environmental projects: a tree planting campaign (TPC), for example, was started in 1976. Action research undertaken by Proshika showed that there were three main options for the poor in relation to the forests:

- homestead gardening
- agroforestry
- roadside plantation.

The social forestry programme includes a number of different types of project in different parts of the country, depending on local needs and conditions. Relations with the public sector have developed along different lines in each experience.

Relevant experiences

In Sirajgonj district, seventeen Proshika groups were able to lease fourteen miles of roadside from the Upazila authorities for a period of five years, and with the support of the NGO pulses and trees were intercropped by the groups, raising Tk10,000 for the groups with the result that a supportive relationship between the groups, the local authorities and the NGO gradually emerged.

The success of the Sirajgonj venture strengthened Proshika's growing interest in developing a social forestry programme. The aim of the programme is for organized group members to gain access to available local public and state-owned resources, such as roadsides, ponds and state-reserved forests. However, any conception of common property resources in Bangladesh is complicated by the fact that almost any resource anywhere in the country is controlled by someone, with or without the formal authority of the state.

The struggle to gain secure access to the 'sal' (*Shorea robusta*) forests in Kaliakoir Upazila in Gazipur District – which are managed by the Forest Department – has been a case of negotiation and confrontation by Proshika group members. Villagers are dependent on the forest resource for their daily needs for fuelwood, animal fodder and building materials. An attempt by local poor women to develop a forest protection programme in the area led to the realization that access by the poor to these government-owned resources was actually governed by intermediaries representing the interests of the local elites. The women were not able to lease rights in the forest resources directly with the Forest Department. Confrontation with local elites followed the women's initiative and led to considerable discussion between group members and Proshika staff in order to develop an appropriate strategy.

The focus of the new approach was to concentrate upon the protection of the existing degraded 'sal' forest, and in particular preventing it from being reduced further by illegal felling, which was being undertaken with the knowledge and tacit support of local Forest Department officials.

Conclusion: continuing conflict

The participatory forest management concept posed problems for the government Forest Department officials, who found that they had a conflict of interests in relation to their links with local elites. The government therefore failed to co-operate effectively with Proshika's groups. For the NGO, the reality of local patronage relationships, class differentiation and gender relationships has been brought into sharp focus by the experience.

The government's reforestation projects in the area (in collaboration with international donors such as the Asian Development Bank) were found by the poor women to be singularly inappropriate to their needs, based on fast-growing single tree plantation for cash cropping. This contrasted with the existing 'sal' forest resources, which constituted a multipurpose resource essential to the subsistence requirements of local people.

On the one hand, a clash of interests has developed between the group members and the local level Forest Department officials *vis-à-vis* their allies in the rural power structure. On the other, the group has entered into a new form of negotiation over the newly grown forest resources with members of the community in securing a more equitable distribution of intermediate benefits.

The conflict of interests between the group members, the rural elites and the Forest Department can be resolved only on the part of the state. Similar experiences have been encountered in the roadside forestry projects. As long as the social forestry programme involves simply the extension of services to the rural poor (e.g. tree planting and training) the partnership runs relatively smoothly. However, Proshika's experience of working at the grassroots indicates that appropriate tenurial arrangements are essential prerequisites to successful social forestry initiatives.

We shall argue therefore that while mutually beneficial links are possible between GOs and NGOs over input delivery, a restructuring of existing access relationships for poor people to common property resources requires more challenging initiatives by government agencies.

THE MENNONITE CENTRAL COMMITTEE (MCC)

Collaborative trials, soybean research and securing the adoption of appropriate varieties

Jerry Buckland and Peter Graham

CO-OPERATIVE TRIALS WITH GOVERNMENT AGRICULTURAL RESEARCH CENTRES IN BANGLADESH

Mennonite Central Committee (MCC) is the international service agency of the Mennonite and Brethren in Christ churches of North America. Established in 1922 to provide relief aid to Russian Mennonites after the Russian Revolution, the mission of MCC worldwide has expanded from providing relief and rehabilitation in times of disaster to working in sustainable development, self-reliance and social justice. Since beginning work in Bangladesh in 1970, MCC's technically qualified volunteers and staff have conducted farmer participatory research and collaborated with government researchers and institutions for agricultural development. From 1982, MCC in Bangladesh has pursued a target group approach aimed at the small and marginal subsistence farmer.

The Bangladesh national system of research was relatively underdeveloped in the 1970s and MCC was able to place expatriate agronomists and research scientists at the field level where adaptive testing of new crops was necessary. MCC research was applied research, sometimes undertaken independently and on a small scale, other times in co-operation with national and international organizations on a larger scale.

The Government of Bangladesh (GoB) at that time focused in its research and extension programme on the promotion of rice only, which MCC regarded as being insufficient to address the nutritional needs of the country and building towards food self-sufficiency for the bulk of the population. MCC therefore intended to expand the production and introduction of new crops which would contribute towards an increase in the available quantity of protein plus the vitamins and other materials which are deficient in the Bengali diet.

Early MCC reports show that a wide variety of co-operative trials were carried out with International Research Institutes: the Asian Vegetable Research and Development Centre (AVRDC) on tomatoes; International

Maize and Wheat Improvement Centre (CIMMYT) on wheat and maize; International Soybean Programme (INTSOY) on soybeans; International Crops Research Institute for the Semi-Arid Tropics (ICRISAT) and Purdue University, USA, on sorghum.

Major national agencies and NGOs co-operated also: grains, oilseeds and forage crops were tested in the early 1970s by the Livestock Ministry, BCSIR, (BARI) Bangladesh Agricultural Research Institute, RDRS, BRAC and other NGOs. MCC imported and tested a wide variety of crops itself: screening and adaptive testing of cabbage, cauliflower, broccoli, Chinese cabbage, vegetable legumes (snap bean, bush-type beans, cowpea, English pea and lima bean), carrot and local and imported green leafy vegetables was completed.

In MCC's initial stages, its agriculture programme was seen as a bridge between the government's extension and existing research in an attempt to create a permanent linkage. The cornerstone of the MCC agriculture programme was dry season winter crops with a geographical focus on Noakhali District in the south of the country, where there was predominantly single-cropping and a high population density.

However, the different programmes within agriculture were separated out by 1979 into specialized programmes such as vegetables, soya and low-cost irrigation technologies. Some of these required linkages with government, while others did not and the importance of GoB contacts began to vary. MCC's research programme gradually adopted the role of liaising with the different GoB research organizations. In MCC's 1986 Five-Year Plan one of the main general objectives for its agriculture programme is 'to help transfer the technologies developed by the national research institutions'.

The research programme has since the late 1980s strengthened its ties with the BARI, Bangladesh Rice Research Institute (BRRI), and the Bangladesh Agricultural University (BAU). For example in 1987–8, 3 per cent of overall effort was devoted to co-operative work with GoB; in 1988–9 this increased to 24 per cent and 25 per cent in 1989–90.

One recent collaborative success was with a cucurbit fruit-fly bait system which had been developed over a period of eight years by a senior entomologist at BARI in 1988 but which was still not recognized by government as a viable low-cost alternative for farmers. MCC was able to support his case with further farmer-based experimentation. MCC undertook station research first and then did on-farm trials that showed good results. The bait trap was statistically significant in its ability to reduce fruit fly damage, was low-cost, increased yields by up to 78 per cent and was well received by farmers. MCC reported these results back to BARI for consideration at their annual internal review. The results were noticed by a senior official in the Department of Extension and he has now requested the bait trap to be demonstrated and extended throughout the country. MCC's role in this venture was one of farmer trials and extensive publicity and lobbying.

A survey of present MCC researchers revealed that in general they felt that communications with GoB researchers are important and indicated that the relationship can benefit MCC's target group. Generally they can cite examples of newly released varieties or new technologies that they learned about from a GoB source. Reasons given for the relationship range from co-ordinating joint research trials (e.g. advanced lines rice trials), obtaining and/or checking out new ideas and maintaining collegial relationship. The number of GoB contacts per researcher range from just one or two contacts for some researchers to over ten contacts per year for others who have been here longer and tend to emphasize this relationship.

While MCC researchers' attitudes are quite positive about the eventual outcome of some GoB contacts, it is difficult to see results that actually benefit the poor. In general, the co-operation is one-sided, with MCC pursuing visits and correspondence much more than the GoB researchers and MCC continually trying to draw GoB researchers' attentions to the case of the subsistence farmer, who MCC feels should be the main target for research efforts.

A number of GoB researchers were interviewed recently by MCC and all indicated support for continuing co-operation with MCC, which they felt to be positive. The sentiment was generally that MCC provided them with high-quality field data for their work which required little effort on their part. It appeared that in some cases MCC was even preferred over their particular institute's own on-farm division.

However, from the interviews the general sense arose that although MCC is a partner in the research, it is being orchestrated by the GoB researchers. Thus the idea that communication is flowing from the farmer via MCC to the institution does not appear to be a significant factor in the GoB's researcher working with MCC.

MCC's primary objective in co-operating with GoB institutes is to learn of new techniques for MCC's target group which have been developed by the institutes. This does happen, as evidenced by the fruit-fly bait trap experience, as well as when new varieties are released. It is more doubtful that MCC is acting as a conduit of farmer concerns to GoB institutions.

SOYBEAN RESEARCH AND EXTENSION

Strategic decisions within MCC's programmes tend to develop over time in an iterative process. A particular idea for a programme change may initially come from any number of sources, but becomes a strategic decision after time has allowed it to be checked against the experience of the farmer, extensionist, researcher, and the organization. MCC began to work with soya in the mid-1970s and continued farmer enthusiasm and success that has passed on to extensionist and researcher led to the strategic decision to specialize.

Soybean is the only MCC crop promoted on a commodity basis, mainly because of its vast potential for contributing to Bangladesh's food and nutritional self-sufficiency targets. The soybean has relatively low production costs and high nutritional value. However, until recently MCC has been plagued by the lack of a variety with good seed quality characteristics. Soybean has not been indigenously cultivated in Bangladesh. Its introduction therefore requires the creation of a market demand, which involves changing food preferences and overcoming nutritional ignorance. As with any new crop, soya must then fit into the farming system. Needless to say, these apparently simple problems cause immense complications.

MCC has worked with soybean as long as with vegetables, but it has tended to be more controversial within MCC. The farmer who grows soybeans is often a small or medium farmer, not a subsistence farmer. Soybeans are sold mainly as middle-class products which poor people can seldom afford. There are therefore conflicts with MCC's target group approach.

However, the strategic objective to establish soybean in Bangladesh now appears to be paying off. Soybean fits within MCC's objective of increasing Bangladesh's nutritional self-sufficiency by being an excellent source of protein. However, the particular characteristics of soybeans make it a more indirect route for the development of the poorest farmers (e.g. problems producing good seed, which then must be purchased). This puts soybeans out of the mainstream where researchers and extensionists focus on subsistence farmers, women or the landless. (Other activities have been dropped if they proved unsuitable, such as sorghum, sunflower, credit extension and shallow tube-wells.)

MCC has therefore tried to establish soybeans as a crop and then remove itself from the effort. It is intended that the GoB will eventually take over efforts of seed multiplication, market development, and agronomic research, leaving MCC to use soybeans as one crop of an assortment to be used to benefit subsistence farmers.

Involvement with GoB institutions stems from 1975, when the government set up the Bangladesh Co-ordinated Soybean Research Project (BCSRP), made up of seven institutions, a food corporation, and MCC. This was co-ordinated by the Bangladesh Agriculture Research Council (BARC). GoB involvement in soybean research was spearheaded by BAU, where varietal trials were conducted as well as the development of inoculum for soybean that could be domestically produced. BADC was also involved in seed multiplication, and BCSIR in soyfood development.

BCSRP was terminated as a project in 1981, with two main factors given by the programme: 'difficulty in producing good (soybean) seed, and two, the need for solvent extraction facilities'. These production problems aside, INTSOY's closure of production promotion must also be considered a factor. After the termination of the project, only BAU and MCC continued soybean research to any significant extent.

The termination of BCSRP in 1981 put most of the responsibility for the establishment of soybeans on to MCC in conjunction with BAU. With clear evidence that the establishment process would require an integrated approach including both supply and demand-side interventions, it obliged MCC to attempt a large-scale effort.

Presently MCC's soybean effort is quite large, involving not only agronomic varietal research, seed multiplication and extension, but also market promotion, and soy-food product development: an integrated approach to crop promotion. This approach has evolved over time from primary emphasis on the supply-side factors, combined with homestead level demand promotion through cooking demonstrations. At least two breakthroughs can be seen as critical to bringing the soybean effort where it is today. This includes the introduction of an Indian soybean variety, Pb-1, in 1985 which has vastly improved seed quality and good seed storability over existing Bangladesh varieties. This variety has led to improved plant stand and allowed farmers to store their own seed.

A second area of recent success has been the development of a demand for soybean by private snack-food companies. MCC increased its effort on the demand-side from 1988 by developing an overall marketing strategy which included approaching local businesses to encourage them to use soybean in their process. This has led to a situation since 1990 where supply cannot match the growing demand and farmers are more confident to grow soybeans. In conjunction with rising prices for other pulses, these factors have allowed soybean acreage to expand from 275 acres in 1987 to 1,200 acres in 1989.

Since 1988 MCC has been primarily interested (in terms of GoB co-operation) in having the National Seed Board release Pb-1 as an official variety in Bangladesh. This would be a major step towards establishing soybeans in Bangladesh. However, even with very good agronomic results for Pb-1, MCC could not attract enough interest from the GoB for them to send a team from the National Seed Board to see the crop and to talk to farmers. Such a visit is crucial to gain acceptance of a new variety by the Seed Board.

Finally in 1989, soybean was included in the Crop Diversification Programme (CDP), a CIDA–GoB jointly sponsored programme. A five-year soybean action plan was drawn up, and soybeans finally seem to have become a part of the Ministry of Agriculture's overall crop promotion strategy.

Another milestone occurred in 1988 when a large NGO in a different area (Tangail) introduced soybeans into their programme and joined MCC in the pursuit of establishing soybean as a crop alternative for all Bangladesh. But, as this report is being written, the same organization has withdrawn from extension because a new institutional relationship they are making with the GoB has thrown their whole programme into disarray and they do not know whether they can continue with soybean extension.

Even though MCC would like to privatize soybean production, the variable successes of the co-operative efforts suggest that MCC will continue its work for many years in soybeans. Despite the mixed results drawn from the links established with government in soybean research and extension, it remains one of the main examples of this type of collaboration. In a recent paper, the then director of BARI pointed out that while there had been few cases of such collaboration, the work on soybeans was a successful case of an NGO influencing the government's research agenda (Abedin 1991).

SECURING THE ADOPTION OF APPROPRIATE CROP VARIETIES

Part of MCC's relationship with the Government of Bangladesh has been concerned with lobbying for the release of new varieties of crops which it has researched and tested and has found to be potentially useful to its target group: the poor and marginal farmers. The release of new varieties is primarily determined by the recommendations of BARI, BRRI and BAU.

Vegetables have played a key role in MCC's extension programme since MCC started in Bangladesh. Arguably, in terms of a transfer-of-technology intervention for subsistence farmers, through several years of testing vegetables have proven the most successful of all MCC extended crops. Because of the particular characteristics of vegetable cultivation (high labour and skill requirement, low land requirement with high returns, responsiveness to inputs at low levels) they are a very effective means for increasing subsistence farmers' income.

Subsistence farmers can undertake vegetable cultivation at the risk and investment level they wish. While MCC promotes correct fertilizer and management practices, most farmers cannot afford the investment for purchased inputs. MCC does not therefore promote vegetable production to maximize farm income, but to increase income at a farmer-determined level of acceptable risk. On average, from 1982 to 1986, farm income of MCC subsistence farmers was increased by almost 10 per cent just by the introduction of year-round vegetable cultivation.

MCC's extension programme sets internal performance criteria in relation to the amount of increased income farmers gain as a result of MCC practices: in 1990 the target was set that 50 per cent of second-year farmers should gain the income equivalent of their family food needs for one month.

Popular vegetables in the extension programme include not only cauliflower, cabbage and tomato, but also eggplant (aubergine), radish, kohlrabi, carrot, etc. Presently MCC's extension programme extends to ten types of winter vegetables and thirteen types of summer/rainy season vegetables.

GOB co-operation on vegetables has primarily involved BARI, and to a lesser extent BAU. Station research began on winter vegetables in 1981. From 1976 until 1985 much of MCC's research work for winter vegetables focused on varietal trials using new varieties from overseas and international research organizations (e.g. AVRDC, University of Hawaii), Japanese hybrids, and locally available commercial lines. Since the mid-1980s, as winter vegetable production technologies have reached their adoption potential, farmers have shown more interest in summer vegetable production. Interestingly, research on summer vegetables involves mainly comparison of local varieties.

Until 1985 MCC pursued a strategy to extensively test new varieties from international institutes. It was only after attempts were made to get released what were considered superior varieties that this strategy proved problematic. It was finally argued in 1985, after many failed attempts to have new tomato lines (from AVRDC) released, that 'BARI has also had many years of positive research results from AVRDC lines and yet no lines have been released'. This was primarily because even when new superior lines were found, they could not be extended as obtaining release from the National Seed Board was extremely difficult. As early as 1980 questions were being asked in MCC about the wisdom of working with unapproved varieties, since 'if lines cannot be released and multiplied, we are wasting our efforts conducting variety trials'.

Thus in 1985 a revised strategy was designed to strengthen MCC's relationship with BARI: 'This [relationship] is most essential not only for facilitating the release of varieties, but also for us to be more aware of what is being done [in vegetable research] and sharing what we are doing' (Schwartzenruber 1985: 31).

MCC felt by 1985 that progress in the development of new vegetable technologies was unlikely through importing new line varieties. Emphasis was placed on working primarily with the Vegetable Sector of BARI. This change in emphasis can be seen in terms of the number of co-operative trials and correspondence which have significantly increased in the past four years.

CONCLUSIONS

MCC therefore tries to share the results of its on-station, on-farm and multi-locational FSR extension based crop trials with government agencies and institutions through publications and through the participation of government officials in its annual review of policy. Results have been mixed among MCC's main crops:

- *Potatoes* replication has taken place by government with good results.
- *Tomatoes* poor replication by government and long delays has led to MCC ceasing efforts in this respect.
- *Rice* collaboration is ongoing with mixed results.
- *Soybeans* MCC has had a major stake in R&D since the 1970s and there has been fruitful research collaboration with BAU and BARI.

MCC is now in the process of questioning a primarily technical approach to its agricultural programmes and moving towards a stronger emphasis upon empowerment, an approach which sees people as the partners rather than the objects of development.

FRIENDS IN VILLAGE DEVELOPMENT, BANGLADESH (FIVDB)

Improved duck-rearing practices

Fahmeena Nahas

INTRODUCTION

Friends in Village Development (FIVDB) is a Bangladeshi NGO formed in 1980 to take over and manage an earlier village development project established by International Voluntary Services (IVS). Its work consists of organization building and income-generation among the landless people of rural areas. FIVDB's group-based, integrated approach to development includes functional education, credit and savings, enterprise development, livestock and agriculture extension, duck-rearing, women's development, health and children's education.

This case study describes the duck-rearing component of FIVDB's programme, which has achieved good results for group members – mainly poor women – and has been highly influential among other NGO organizations in Bangladesh.

In the absence of very much government interest or support in this sector, FIVDB has been promoting duck-rearing largely among its female group members in the Sylhet area of north-eastern Bangladesh. It has also been at the forefront of the development and adaptation of appropriate duck-rearing technologies in Bangladesh.

Duck-raising is a profitable and ecologically suitable enterprise in most areas of this mainly flat, wet and tropical country outside the Chittagong Hill Tract area in the south and the Barind areas of the North.

FIVDB AND DUCKS

FIVDB has pioneered duck extension work in Bangladesh by importing foreign breeds and has incorporated duck-rearing into its income-generating strategy for poor rural women. In rural Sylhet there are now thousands of duck farmers raising improved varieties of ducks such as Khaki Campbell, Cherry Valley 2000 and the recently introduced Ding Zeng, in flocks ranging from 10

Box 3.2 Landless women and duck production under FIVDB's programmes

Mayarun Nessa is a landless woman living in Ballutikor village in Khadimnagar, Sylhet. She has been a member of an FIVDB group since 1982. She participated initially in functional literacy classes and now reads a community newspaper. She has also learnt to write letters and keep accounts. Although she learnt duck-rearing skills from her mother, the frequency of disease has increased mortality rates and her traditional remedies no longer seem effective. In addition, she has realized the potential of duck-rearing to the household income. Aside from helping her with the household vegetable garden, her husband can find only seasonal labouring work.

On a twelve-day FIVDB training course in 1987, Mayarun increased her knowledge about hatching ducks, their food and housing, hygiene and prevention and treatment of duck disease. At the end of the course, she received fifteen improved variety ducklings, a syringe for vaccination and a simply written manual on rearing techniques. She now has a flock of thirty-four ducks and requires no loans for duck-rearing, since she generates enough income from them directly. She receives free vaccines from FIVDB or from the Upazila Livestock Office, and has invested in a simple storage container for vaccines (designed by FIVDB) to ensure that the correct temperature is maintained.

She also gains additional income from her vaccination skills, charging Tk0.5 per bird around her village. She gives a free service to relatives and neighbours. She sells duck eggs to visiting traders at Tk4–6 for four. The trader then resells in the market for Tk10–12. Sometimes her husband can sell directly in the market and cut out the middleman.

Since Mayarun's costs are relatively low, she makes a reasonable profit from her ducks without having to leave the homestead. There have been other, less tangible but highly valuable benefits. Mayarun's husband has learnt to respect her for her business endeavours and she handles her income herself. She has also gained mobility within the community as she carries out her 'paravet' work. With her husband, she has been active in campaigning to set up a primary school in the village.

to 500 birds (see Box 3.2). A recent FIVDB survey indicated that about 350,000 improved varieties are now being raised.

FIVDB is a small NGO without an established research capability and is therefore interested in NGO–GO research collaboration. These have so far been on a small scale, such as a research link with Bangladesh Agricultural University (BAU). This is partly due to the fact that no formal mechanisms yet exist for the purpose of establishing this kind of programme.

FIVDB has adapted and established its own hatchery, experimented with new breeds and imparted training on duck-rearing to other organizations and government agencies. FIVDB now considers that it has reached the limits of its own research and development base in several keys areas of duck-rearing, such as disease control, appropriate food research and breeding techniques. FIVDB is interested in establishing a channel through which NGOs can communicate their research needs to government institutions which have access to more specialized R&D capabilities and this paper makes some suggestions for this.

FIVDB has worked to develop and introduce associated technologies into its project area, through establishing contacts with agencies working with ducks in other parts of the world. For example, FIVDB has developed an appropriate hatchery system on a Chinese model.

For research based on limited resources, there have been areas of collaboration between FIVDB and public sector research institutions, e.g. research into duck disease at BAU. FIVDB has also given training to groups of duck-raisers in other areas of the country, and improved rearing methods are now found in belts in other parts of the country.

CONCLUSION: THE NEED FOR CONTINUING PUBLIC SECTOR RESEARCH COLLABORATION

FIVDB has reached the limits of its resource base, but important research is still needed, such as work on local sources of duck feeds and for developing a stable bloodline. FIVDB would therefore like to play a stronger supportive role in the future for the propagation of duck-raising in the following ways:

- *Support for setting up hatcheries.* FIVDB plans to support organized partner groups (target clientele) to help set up mini-hatcheries following the Chinese rice husk incubation system as a collective economic enterprise. This support will include training, technical assistance and credit. FIVDB is also planning to support other NGOs and development agencies to establish hatcheries and duck extension programmes in suitable locations. One such scheme is underway with Swanirvar Bangladesh, a quasi-government agency in Bagerhat, Khulna District.
- *Expansion of the capacity of FIVDB's central hatchery.* This is located in Khadimnagar, Sylhet, and will be developed to increase annual production of high-laying ducks from 30,000 to 100,000 to meet the growing local needs and the demands of other organizations. Other NGOs such as BRAC and RDRS have expressed an interest in buying large numbers of ducks on a continuous basis. The capacity will be enhanced both by expanding the rice husk incubation system and by setting up electrical incubators.
- *Introduction of earthworm culture.* FIVDB is planning to introduce homestead earthworm culture in compost piles. Earthworms contain 60 per cent crude protein and are therefore ideal feed for ducks. FIVDB is currently contacting different groups in order to gain access to the technology of vermiculture. It is also exploring the possibilities of experimenting with the introduction of controlled snail production at homestead level, another important source of protein and calcium for ducks.
- *Upgrading the training programme.* The duck programme is currently being restructured. An improved curriculum will be offered to duck-raisers

and extension workers. Ways of improving the content and the methodology of the training to further meet the needs of FIVDB's clients are being explored.

This case highlights the role of the small, local NGO as an effective source of extension and technology development. It also illustrates the need for a mechanism through which NGO-identified needs can be communicated to larger-scale government research facilities.

BANGLADESH RURAL ADVANCEMENT COMMITTEE (BRAC)

Backyard poultry and landless irrigators programmes

Shams Mustafa, Sanzidur Rahman and Ghulam Sattar

INTRODUCTION

BRAC, the largest NGO in Bangladesh, is a national development agency with multifaceted programmes and activities spread across the country. It has been active since 1972, when relief efforts were begun in the aftermath of the War of Independence. The relief approach was soon replaced by a long-term development approach and currently the objectives of BRAC are poverty alleviation and the empowerment of the very poor, particularly women. BRAC has a target group approach, and those who are functionally landless and/or have to sell their labour comprise the target population.

BRAC takes a holistic approach to development, with people as the subject and the different sectors as the object. After the first few years certain areas of intervention were identified, that is human and organizational development, employment and income-earning opportunities, and health-care, particularly for mother and child survival. These areas of actions have remained constant with changes taking place in emphasis, strategy and implementation designs. Interventions are programmed for specific fields to attain the objectives of poverty alleviation and empowerment of the poor.

Different existing technologies (both hardware and knowledge) are being used as tools in economic activity. BRAC does not see a role for itself in advance research into technology, which is better left to specialized agencies. Instead, BRAC defines – in conjunction with its target groups – simple, low-cost technologies which can be used by the poor for their own benefit and to the advantage of society at large. The aim is for the poor to own, operate and manage such technologies in order to generate employment and income, increase the productive efficiency of different sectors and to empower themselves.

BACKYARD POULTRY PROGRAMME

In the late 1970s BRAC identified poultry-rearing as a source of income for the landless, particularly destitute women. A high mortality rate for poultry in Bangladesh, combined with its relevance as an income-generating activity for poor women, led BRAC to carry out participatory 'action research' aimed at increasing productivity.

Initially, efforts were made to increase the productivity of local varieties of cockerel, but an exchange system with improved cockerels for cross-breeding failed since the improved birds tended to be sold and mortality remained high. The focus changed from 1981 to supply of improved chicks, common disease prevention and improved scavenge-rearing training.

Key rearers (KRs) from BRAC groups were trained (jointly, by the Department of Livestock (DoL) and BRAC staff) in each programme village to demonstrate good rearing practice with chicks supplied from a BRAC farm. Research indicated that regular vaccination was the key to improving production. Another set of female group members were trained as poultry workers (PWs) to vaccinate for a fee, using vaccines provided free of cost from the government.

The poultry programme is a collaboration with the government. The goal of collaboration is to orient and activate local level functionaries toward the specific needs of the poorest by involving them in meaningful implementation of different programmes.

Since the government controls the supply and distribution of vaccines, and since BRAC aims at improving the services provided by government to the poor, relationships have developed between BRAC and DoL and the Department of Relief and Rehabilitation (DoRR). This relationship has the short-term aim of securing adequate supplies of inputs for BRAC group members and the longer term aim of orienting government workers towards the specific needs of the poorest. Relationships have so far been generally positive.

Structures have therefore been set up to provide technical options to poultry keepers. Access to day-old chicks from government farms and village breeders provides better supply, and an opportunity for demonstration of improved techniques. The chicks are reared by specially trained women group members on their homesteads before being sold to KRs. There is now also access to two-month-old chicks for poultry keepers less confident in their ability to raise day-old chicks. BRAC has ensured the supply of vaccines from the respective Upazila Livestock Centres to the participants.

The case illustrates how contacts with government at local level can be mutually advantageous for the distribution of scarce, government-supplied inputs to the poor. The BRAC approach has been taken up and adapted by other NGOs and has been replicated by BRAC in a number of government programmes in which it participates.

Replication of the BRAC model by government at first took place informally. Between 1983 and 1985 the government District Livestock Office at Manikgonj agreed to supply vaccines and technical advice, despite scepticism from some government personnel of the ability of poor women to carry out the task of vaccination. Today the system is part of BRAC rural development programme (RDP) in thirty-two Upazilas around the country.

At the same time, a parallel partnership took place in the vulnerable group development programme, organized by the World Food Programme (WFP) in conjunction with the DoRR. Destitute women receiving food/relief also receive training on poultry-rearing and vaccination. Over 33,000 KRs are now operating through the programme on a commercial basis and almost 5,500 PWs have been trained as part of the Income Generation for Vulnerable Group Development (IGVGD), a joint project of BRAC, the DoRR and the World Food Programme. In the programme, the income-generating potential of destitute women receiving food/relief is strengthened through provision of poultry training and inputs (including credit) using the BRAC model.

These partnerships with government have generally been successful. However, although the government appears convinced of the viability of the BRAC approach to poultry production, a number of administrative procedures need to be agreed before BRAC can safely move on. One area is the distribution of vaccines, which poultry workers currently collect from local livestock sub-centres. The DoL wishes to reduce its costs by making the vaccines inputs available only at Upazila Livestock Centres, but BRAC considers that such a change would have a powerful disincentive effect on procedures by imposing additional travel costs.

This case study therefore underlines the vulnerability of the structures required to reach the rural poor, and the sensitivity required by government if such initiatives are to be sustained.

LANDLESS OWNED AND OPERATED IRRIGATION

Rural unemployment is a critical problem in a country in which over half of the rural population is landless and therefore effectively barred from productive agricultural work, except as wage labour. There is no real sign that such vast quantities of 'surplus' rural labour will be absorbed by new forms of industrial employment in the towns. However, as new technologies are starting to transform agricultural production in rural Bangladesh and in particular the expansion of mechanized irrigation equipment, it is possible to see such technological change as offering potential opportunities to the organized landless, if appropriate institutional interventions are made.

The social and economic activities required to make use of a technology relate to ownership and control, to development of appropriate management and organization. Be it powered pumping tilling machines or fish production,

BRAC works for the ownership to be in the hands of the poor who will also operate and manage the techniques.

BRAC and Proshika both pioneered this approach in the irrigation sector in Bangladesh (see Box 2.3 for an outline of Proshika's work).

The irrigation case study presented here illustrates how the NGOs in the early 1980s identified the possibility of reintegrating the landless into agricultural production through their ownership of non-land means of production and their provision of a technology-based agricultural service to farmers. The landless had traditionally been neglected by the state-sponsored institutions and programmes such as the 'Comilla Model' farmer co-operative system development in the 1960s to widespread acclaim (see pp. 49–50).

The most significant aspect of the BRAC irrigation programme is the creation of a process effecting changes in the existing resource distribution pattern and power structure through ownership of means of production by the poor and by their gained access to resource and power.

The landless irrigation strategy is based on the argument that several forms of rural property rights, other than land, do exist in the rural areas. The irrigation groups were formed by BRAC to manage and operate deep tube-well irrigation facilities and to secure an income by selling water to local farmers. Water is marketed on a seasonal basis to farmers whose land lies within the 'command area' of the well in return for a share of the crop.

During the 1970s and most of the 1980s the deployment of irrigation equipment and other technologies was controlled solely by the government parastatal known as the Bangladesh Agricultural Development Corporation (BADC). After developing its landless irrigation programme BRAC entered into a formal contract with BADC in 1979 for the installation and servicing of its groups' deep tube-wells. At the same time, credit arrangements for the groups to purchase the equipment were made with the public sector agricultural bank known as the Bangladesh Krishi Bank (BKB).

However, attempts to work through the public sector in this way proved difficult in view of the difference in working methods, approaches and structures between the NGO and the other agencies. BRAC found itself unable to construct an efficient and realistic framework within the severe limitations of these rigid public sector bureaucracies. For example, when agreement on a banking plan was achieved at the senior level, it proved impossible to implement at the local level.

The government agencies also had little credibility in the eyes of the group members who were unused to dealing with this section of the rural population. Therefore the NGO decided to develop its own capacity to supply certain inputs more efficiently, without having to rely on government. Where a working relationship with government could be established quickly and without serious problems, it was attempted. Today, BRAC is able to provide its own credit to group members, while working arrangements for the efficient installation of irrigation equipment have been worked out with the BADC.

BRAC provides information, skill and training to the irrigation groups. The farmers or water buyers require more technical assistance in improving farming methods. There is scope for collaboration in this area.

BRAC's involvement with irrigation groups (along with that of Proshika) has helped to restructure the irrigation agenda in Bangladesh and raises important possibilities for the social ownership of large-scale 'lumpy' technologies by the poor. This involvement has coincided with the government's present privatization policy in which an increasing number of deep tube-wells are being taken over by landless groups. BRAC's programme has development impact and an empowerment potential in the long term.

RANGPUR DINAJPUR RURAL SERVICE - THE TREADLE PUMP

Manual irrigation for small farmers in Bangladesh[4]

The treadle pump is a cheap, human-powered twin cylinder pump-head with a bamboo or PVC tube-well, which has been developed by Rangpur Dinajpur Rural Service (RDRS), a German Lutheran NGO working in Bangladesh with small farmers. It has been hailed as a highly successful innovation but, as Box 3.3 argues, government procedures have not been up to the task of promoting its replication.

The need for small farmers to gain access to cheap, small-scale irrigation equipment became apparent during the late 1970s and shallow and deep tube-wells were the main forms of irrigation technology being promoted by government. The cost of these pumps placed them beyond the reach of most farmers, who were then made dependent on buying irrigation water from the larger farmers who became well-owners.

The treadle pump can lift about 1.5–3.0 litres of water per second, which means that an adult operator could hope to pump about 20,000–40,000 litres per day. This makes it the most efficient system available for manual irrigation for lifts of up to 3.5 metres (see Figure 3.1). In 1988, a treadle pump and a bamboo tube-well could be installed for US$20. A sales network of private dealers has been created with the help of International Development Enterprises (IDE), an NGO founded to assist with the development of small business enterprises.

The pump has increased cropping intensity, employment and income to small farms. The index of cropping intensity on land irrigated by the treadle pump was found in a recent study to be 248 per cent compared to 211 per cent on unirrigated land, mostly due to the farmer's increased ability to grow winter crops such as tobacco, vegetables and 'boro' rice. The pump also increased the employment of family labour during the slack winter season, and the work was shared by men, women and children. Farm incomes have also increased. It is estimated that over the pump's lifetime, the farmer will earn a net return of Tk3.4 for every Tk1 invested.

The treadle pump was introduced in 1979 and 185,000 have now been sold in Bangladesh. Sales are now running at around 65,000 a year. However, the scale of mechanized irrigation has also expanded rapidly and it is estimated that only 0.01 per cent of the total irrigated area is covered by manual irrigation technologies. The treadle pump therefore remains a supplementary technology which widens the choices available to small farmers. The pump is therefore viewed by RDRS as an important 'complementary' technology for use alongside mechanized irrigation. For example, small farmers can use the pump to irrigate if they cannot secure the quantities they need at the right time from a local mechanized well-owner.

Box 3.3 Conflicts between NGO and government working procedures: the case of the treadle pump

This case study illustrates the bureaucratic problems which can undermine a joint, flexible response to a localized crisis by government and NGOs. During July–August 1989, in what should have been the rainy season, a drought occurred in areas of Rangpur and Dinajpur Districts. The 'amon' (rainfed) paddy was severely damaged by the failure of the monsoon, causing great hardship to small and marginal farmers.

The Ministry of Agriculture attempted to deal with the crisis with a two-year Post-Drought Agricultural Rehabilitation Programme, based on a decision by the Council of Ministers in 1989. The total outlay of the project was Tk10,000 million and there were two major components: (1) free distribution of seeds and fertilizers for winter crops and (2) free installation of treadle pumps among 100,000 small, marginal farmers and sharecroppers. The programme was administered through a District Agriculture Rehabilitation Committee headed by the then Minister of Social Welfare, whose constituency was in that region. RDRS had been promoting the treadle pump for two decades, as a technology well suited to the needs of small farmers. Marketing was being carried out in conjunction with International Development Enterprise (IDE), another specialized NGO. An order was placed by the government with RDRS/IDE for 100,000 pumps through a Memorandum of Understanding (MoU) for the supply and installation of the pumps during the winter of 1990: 1,595 pumps were installed within a short time. The price charged by RDRS was Tk280 for the equipment and a similar charge for sinking each pump.

However, the Ministry of Finance would not endorse the contract since there were open tender rules within GoB for orders of this scale. According to these rules, the MoA therefore opened the contract to outside tenders and the lowest rates offered were Tk178 per pump and Tk260 for sinking the well. The three lowest bidders were given the order, and all were below the rate charged by RDRS/IDE. The contractors supplied 13,690 pumps and sunk 590 of these, but the Tender Committee on inspection found that those supplied by the contractors fell short of the required specification. The entire supply was rejected and the project was delayed further. The MoA, by diverting Tk10 million from an integrated development project in the south, revised the programme and secured special permission from the president to enter into a new contract with RDRS/IDE, with a second MoU.

Shortly afterwards, there was a change of government after the removal of President Ershad in December 1990. The funds were diverted from the programme in the north, since it was considered that the farmers had by this time recovered from the effects of the drought. The MoA has been able to pay for the work already carried out by RDRS/IDE to the tune of Tk7.2 million and a revised budget is currently being prepared under the new government. It remains to be seen whether or not the order is completed.

Source: Hassanullah (1991)

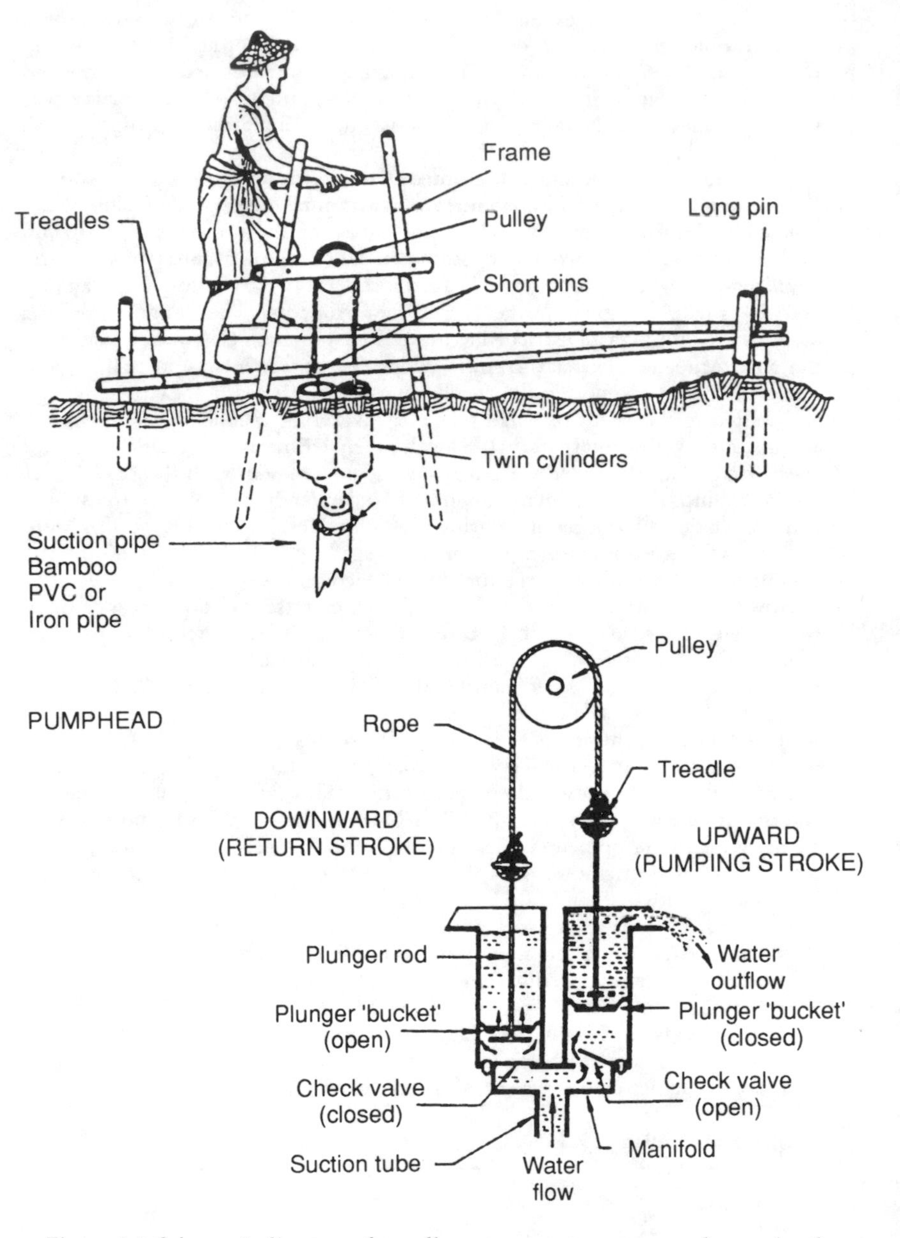

Figure 3.1 Schematic diagram of treadle pump superstructure and pumphead
Source: Orr *et. al.* (1991)

CARE's LIFT PROJECT

Sustainable techniques for vegetable production[5]

CARE's LIFT (Local Initiatives for Farmers' Training) project is a production technology transfer programme for a technology which does not require significant capital investments and can be undertaken mainly without credit. The project has its roots in the agricultural development programme (ADP), which aimed to produce 'exotic' varieties of vegetable for sale in the high-value Dhaka market. However, the high cost of inputs required led to a shift in focus towards local varieties for sale in local markets. This also created an interest in regenerative agriculture in the form of low-cost, local resource-dependent gardening techniques. The target group shifted to landless and marginal farmers. There was a new emphasis on providing a mix of inputs to growers, rather than simply credit and marketing as had been provided before. The need for training and the transfer of sustainable, appropriate technologies became apparent.

This new emphasis became part of the new LIFT project in Gaibandha. The project now works with 2,150 participants in 55 villages in Gaibandha Upazila. In 1987 staff from CARE attended a course in bio-intensive gardening and regenerative agriculture techniques at the International Institute of Rural Reconstruction (IIRR) in the Philippines (see Chapter 6). The technologies of raised beds, compost and multiple cropping were selected as being relevant to the needs of landless farm families growing indigenous vegetables in CARE's project area in Gaibandha District. These techniques also were familiar to most farmers and easily adoptable.

The project was developed with the local Upazila Parishad as its counterpart, under the understanding that the Upazila would manage the project after the NGO's withdrawal. In addition, there was a further regular link with the National Institute of Local Government (NILG) to conduct twice-yearly training sessions for Upazila staff. This training was composed of management skills for administrative staff managing development funds, and technical assistance in bio-intensive gardening for training for the Upazila's block supervisors and block officers. These officers are in regular contact with CARE-trained LIFT progressive farmers' (PFs).

In Gaibandha, the average LIFT participant has recorded a five-fold increase in vegetable and fruit production with a ten-fold increase in its productive value on an average of just 0.25 ha. This was found to be the result of bringing more land into horticultural production, changing cropping patterns, year round production instead of single season and more labour and time invested in homestead gardening using many of the LIFT techniques.

NOTES

1 Upazilas were recently rechristened 'thanas' by the BNP government.

2 Although figures are not available showing the overall picture of NGO funding sources, there is evidence that it is increasing sharply. White (1991b) has estimates that NGOs absorb US$85 million to 100 million each year, or about 6 per cent of total oda received by the Government of Bangladesh in 1988. This contrasts with an estimated 1 per cent share of foreign aid in 1972 (White 1991b: 11). The NGOs have essentially remained dependent on foreign funding despite attempts to reduce this. While BRAC recently announced that one-third of its income was self-generated (through among other sources, a print business, handicraft centres, a garment factory and potato cold storage), other NGOs' attempts to provide transport services and marketing of produce have apparently proved less successful.

3 The Grameen Bank, which is now a part of the public banking system and not strictly speaking an NGO, has become internationally known for its successful provision of credit to poor rural women and has been an important model for the credit programmes of many NGOs. However, attempts to replicate the Bank in other countries has met with varied results, illustrating the need for 'institutional breeding' rather than direct institution building in other settings (Hulme 1990).

4 This section has been drawn from Orr *et al.* (1991).

5 Drawn from Bear and Rahman (1989) and Dean (1990).

Chapter 4

NGO–GOVERNMENT INTERACTION IN INDIA

OVERVIEW

Mark Robinson, John Farrington and S Satish

ECONOMIC CONDITIONS AND GOVERNMENT ANTI-POVERTY PROGRAMMES

Although still a mainstay of the economy, agriculture in India is of declining relative importance. Between 1965 and 1989 the share of GDP attributable to agriculture declined from 44 per cent to 32 per cent, with corresponding increases for industry (22 per cent to 30 per cent) and services (34 per cent to 38 per cent). Agriculture currently accounts for 39 per cent of exports, and provides livelihoods for over 70 per cent of the population. Population growth in India averaged 1.93 per cent per year in 1985–90, some reduction from the 2.3 per cent average in 1965–80. Total population is expected to exceed 1 billion by 2000. India is the ninth most industrialized country in the world, yet, with 322 million from a total of 816 million estimated to be living in poverty in 1988, it contains more poor people in absolute terms than any other country. Although the Government of India (GoI) predicted a trend decline in poverty incidence in the 1980s and early 1990s, the data suggest that there has been no significant trend away from 40 per cent since the 1960s. This is reflected in India's poor performance in terms of social development. For example, the adult literacy rate in 1990 was 62 per cent for men and 34 per cent for women, and life expectancy at birth 59 years. Although the under-5 mortality rate has halved since 1960, it remains at 130 per 1,000 live births.

While rural–urban migration is swelling the numbers of the urban poor, poverty in India remains largely a rural phenomenon. It is strongly linked to access to land: while the number of operational holdings rose from 70.5 million in 1970–1 to 89.4 million in 1980–1, the number below 1 hectare grew from 36.2 million to 50.5 million over the same period. From 1973 to 1983, the proportion of casual labour (i.e. those generally without any access to land) rose from 22 per cent to 29 per cent of the total work-force, at a time when 55 per cent of casual worker households were below the poverty line.

There is a significant spatial and social dimension to poverty in India. Central (Madhya Pradesh, Gujarat, Karnataka) and eastern (Bihar, eastern Uttar Pradesh, Orissa, West Bengal) states are characterized by higher concentrations of poverty than the north-east (Punjab, Haryana, Western Uttar Pradesh). In 1986–7, for instance, 53 per cent of the population were below the poverty line in Bihar, against 15 per cent in the Punjab. These differences are attributable not only to the ineffectiveness of land reform in the central and eastern states, but also to previous government expenditure on irrigation, agrochemical subsidies and research into Green Revolution varieties of rice and wheat, all of which are more relevant to the agro-ecological conditions of the north-west.

A large number of poverty alleviation programmes have been undertaken in India since the early 1950s. It is impossible in the present context to do more than review the main outcomes.[1] Early efforts (the community development programme, the intensive agricultural district programme) did not distinguish clearly enough between interventions accessible to all, and those restricted to identifiable groups of rural poor. As a consequence, many of the benefits they brought were appropriated by the rich.

The integrated rural development programme, launched in 1978, is one of the largest efforts, aiming to reach 45 per cent of households living in poverty during the Eighth Plan period. Despite its ambitious objectives the programme has had some difficulty in reaching the specified target populations (especially women) and in identifying viable income-generating technologies lying within the management capabilities of the poor. Its impact has therefore been rather limited, even if it has proved popular politically.

Large-scale employment generation programmes include those promoting the acquisition of artisanal skills (e.g. training of youth for self-employment), through which almost 1 million individuals passed in 1980–5, and various schemes for the employment of unskilled labour in (largely) the construction of community assets. Both have had some impact on poverty, but the latter have been plagued by low involvement of women, low wages and delayed payments, and failure to generate a sense of 'ownership' among the communities of the assets created, and the former by the lack of demand in many of the agriculturally underdeveloped areas for the types of skills in which training is being given.

On the whole, while demonstrating some government commitment to poverty alleviation, the above programmes have been characterized by weak targeting, poor follow-up, low repayment levels and problems of asset retention. They have also been subject to local political pressures and petty corruption which has in turn undermined their potential appeal for the poor. Access to credit from banks on a recurrent basis in the form of crop loans also continues to be problematic.

NGOs IN INDIA: APPROACHES AND FUNDING

Estimates of the number of NGOs active in rural development in India range from fewer than 10,000 to several hundred thousand depending on the type of classification used. Some 15,000–20,000 are actively engaged in rural development.

Wide variations in the densities of NGOs exist among states. To some extent these reflect not only differing patterns of poverty, but also historical factors (such as areas of Christian missionary influence or Gandhian activity) and the priorities of foreign donors. Within these states certain districts possess dense networks of NGOs, which overlap and compete for clients, while in other areas there are hardly any NGOs active on the ground.

The most common type of NGO in India is the small agency working in a cluster of villages in a particular locality with a handful of staff; as one moves up the scale there are comparatively few organizations which possess the staff or financial resources to work intensively at the state and national levels, although it is these organizations which are the most well known in government and donor circles.

Annual NGO revenue from abroad is in the region of Rs9 billion (US$520 million), up from Rs5 billion in the mid-1980s and equivalent to approximately 25 per cent of official aid flows. A further Rs500 million–700 million is provided annually by government. When individual and corporate donations are added in, annual NGO income amounts to around Rs10 billion. This figure is equivalent to 10 per cent of the government's annual poverty alleviation expenditure provision.

NGOs and government share many of the same poverty alleviation objectives. Major points of difference lie in the smaller scale but more focused nature of NGO interventions, a greater commitment to social uplift, and an explicit concern with participation. Some NGOs have chosen to tackle the symptoms of poverty – low educational standards, ill-health, poor sanitation and inferior housing – by means of targeted programmes of assistance. Others have concentrated on enhancing the asset position and income-earning potential of the poor through land improvement schemes, credit and skills training. Both of these centre on a consensual approach to development in which the existing social and economic structure, although inherently important, is not directly challenged. An alternative 'social action' approach for some NGOs, particularly since the late 1970s, has been to politicize poor people, thereby challenging directly many of the social and economic structures established by the state.[2]

NGO approaches have evolved from early relief efforts sponsored largely by Christian organizations, through the 'village uplift' of the Gandhian movement, to a professional development approach stressing sound management planning and co-ordination. A particular feature of the 1980s has been the emergence of NGOs providing support services to other NGOs in the form of training, evaluation and documentation.[3] These agencies are usually financed by core grants from foreign donors and from payments for staff training from individual NGOs. Intermediary agencies responsible for channelling funds from foreign donors to small national NGOs in some cases provide qualified staff for undertaking evaluations.

Since the early 1980s there has been an element of convergence between these various approaches, in which NGOs seek to combine project-specific development work with active organization of the poor. In a reflection of this greater uniformity of approach, community organization is now treated, for the most part, as an essential prerequisite of participatory problem diagnosis and of institutional sustainability. An important corollary now widely recognized among NGOs is that organizational work cannot be sustained unless programme activities generate material improvements and this, importantly in the present context, requires that such programmes be supported by technical skills, appropriate technologies and, in many cases institutions and channels geared to input supply and marketing.

Most NGOs work through groups, although there is considerable variation in both their purpose and in the approach of individual NGOs to group formation. Some, notably the Gandhian agencies, chose to work through existing village institutions but these are invariably dominated by the rural elites. Most NGOs therefore prefer to form new groups which can be organized along class, caste and occupational lines.

Despite problems of competition for clients or for scarce government resources, some networks have been established at the state level, which seek to present a common front in negotiations with the government over legislation and policy formulation.[4] Other networks have been formed by foreign donors, although these can also have the effect of creating or widening divisions between groups of NGOs. More recently, there has been an initiative to establish an NGO network at the national level. A number of NGOs, especially the larger and more established organizations, formed the Voluntary Action Network India (VANI) in April 1988 as a common platform for NGOs. VANI provides resource materials, and organizes conferences on strategic issues of broader concern to NGOs. Despite this initiative, co-ordination between NGOs remains relatively weak, which results in a duplication of effort and limits their potential impact on a larger scale.

The Seventh Five-Year Plan (1985–90) provided for an active involvement on the part of NGOs in the planning process together with a massive increase in the volume of government funds (i.e. to Rs1,500 million per year, or US$170 million) assigned for use by them in rural development programmes. The work of voluntary agencies was considered supplementary to that of government in offering the rural poor a range of choices and alternatives, at low cost and with greater participation. Most of this allocation was for NGOs to work in government programmes of the type outlined above, in the areas of social forestry, ecological development, primary health-care, the provision of safe drinking water, education, rural housing, land ceilings implementation, enforcement of minimum wages legislation and bonded labour rehabilitation.

Two further types of relationships between NGOs and government deserve mention:

- There are numerous examples of government 'scaling up' of the ideas generated by NGOs, mainly in the health and education sectors.
- Many attempts have been made by government departments to involve NGOs in project or programme implementation in the expectation of reaching beneficiaries more cost-effectively.

However, relations with government are not always cordial, and it has introduced legislation designed to control and monitor the activities of NGOs. All voluntary agencies with seven or more members are compelled by law to register under the Societies Registration Act 1860 or as a trust under the Indian Trusts Act 1982 in the case of religious organizations. This is usually a formality which takes up to six months to complete and rarely poses a problem. Registration is essential for NGOs wishing to apply for grants from government or receive foreign funding.

The Foreign Contributions Regulation Act 1976 has been a major instrument in regulating the receipt of funds from abroad not only by NGOs but also by commercial companies. Established in response to allegations that foreign funds were financing subversive activities, it compels all organizations wishing to receive funds from abroad to register with the Home Ministry, to submit audited accounts on a half-yearly basis and to provide details of each individual contribution. The provisions of the FCRA were further tightened in 1984 and a number of NGOs had their registrations suspended. While the amendments to the Act have increased the delays and reporting requirements faced by NGOs, and have made it easier for government to conduct investigations or impose sanctions, they have done little to stem the flow of illicit foreign funds.

The Council for Advancement of People's Action and Rural Technology (CAPART), established by the GoI in 1986, has the twin aims of promoting NGO involvement in rural development and promoting technological innovations through NGOs. In practice, it serves as a channel for distributing some of the GoI funding for NGOs, including part of that which comes through official bilateral assistance.

Approximately 90 per cent of NGO funding in India comes from foreign sources, mainly from international NGOs, via four principal mechanisms:

- An intermediary organization or umbrella grouping in India identifies projects on behalf of the external partners, and undertakes monitoring and evaluation.
- Several churches adopt a consortium model, in which an intermediary in India pools incoming resources (usually in the form of block grants) and distributes them to selected projects.
- Some international NGOs (e.g. Oxfam, Novib) have in-country offices

staffed mainly by Indian nationals and with a high degree of autonomy, which function largely as intermediaries.

- Some NGOs (e.g. Christian Aid) work directly with local NGOs through periodic tours, i.e. without resort to an intermediary.

Many organizations seek funding from the government in preference to foreign donors for a number of reasons. Government funding does not require NGOs to apply for registration under the Foreign Contributions Regulation Act which can be cumbersome and time consuming. There has always existed a strong nationalist tradition among NGOs which abjures foreign funding on the grounds that it undermines their independence and limits their freedom to determine programme priorities. Another argument is that the legitimacy derived from government funding can provide NGOs with a degree of protection from harassment from local vested interests.

Against this, dependence on official funds leaves NGOs susceptible to changes in government policy, and can result in programme modification to accommodate official funding priorities. A more profound concern is that government funding can lead to co-option, whereby voluntary agencies tone down their social and political objectives in order to secure financial support. In practice, however, there is not such a sharp dichotomy between organizations in receipt of government as opposed to foreign funds, and most organizations seek a blend of funding from both sources.

PUBLIC SECTOR AGRICULTURAL RESEARCH

The Imperial Council for Agricultural Research – renamed the Indian Council for Agricultural Research (ICAR) after Independence – was set up in 1929. Major points in the evolution of its structure and management since the early 1950s include:

- the incorporation of all Central Agricultural Research Institutions under ICAR in 1965
- the positioning of ICAR within the Union Department of Agricultural Research and Education in 1973
- the creation of a unified Agricultural Scientists Recruitment Board in 1973.

Further changes are currently being considered in response to the report of the GVK Rao commission (ICAR 1988) in relation to rationalization of research planning and policy formulation, strengthening of monitoring and evaluation; streamlining and rationalization of research projects and institutions to enhance efficiency and reduce overlap, and reorganization of responsibilities at headquarters level.

As a registered scientific society, ICAR is legally autonomous, but follows

GoI rules and regulations. ICAR's mandate is to undertake, aid, promote and co-ordinate agricultural, animal husbandry and fisheries education and research, and the application of research, and to undertake clearing-house, library and support functions in relation to the above.

The Minister of Agriculture is the president of ICAR. The director-general of ICAR is also the secretary to the Government of India for the Department of Agricultural Research and Education (DARE). ICAR draws on a number of specialist committees, including scientific panels which consider research schemes in different disciplines; eight regional committees which review the research and training needs of the eight major agro-ecological regions; and a number of joint panels formed between the ICAR and related research organizations such as the Indian Social Science Research Council.

India's public system of agricultural research consists of 44 central research institutes (including four national bureaux), 25 national research centres (15 more are planned), 71 All-India Co-ordinated Research Projects (AICRPs), 26 State Agricultural Universities (SAUs) and 127 zonal research stations. In addition, a number of small research institutes under state departments of agriculture focus largely on location-specific problems. ICAR also supports 530 *ad hoc* research schemes submitted by universities and research institutes over and above their programme funding. Over 20,000 scientists work in the system. The ICAR fully funds expenditure at its own institutes and the costs of any AICRPs they run. For AICRPs implemented by SAUs the ICAR normally provides 75 per cent of the costs and the state government provides the balance. The ICAR also provides substantial grants-in-aid to the SAUs for their general development, thus fulfilling for the SAUs the role performed by the University Grants Commission for the general universities, but without its statutory powers.

India thus has a public sector agricultural research system of considerable pedigree and size – second only to China in overall size and to Brazil in aggregate expenditure. India currently spends almost 0.4 per cent of the value of its agricultural production on research (i.e. approximately US$200 million annually). The central government provides some 60 per cent of the total, state governments some 20 per cent, private companies 12 per cent and foreign donors (through the ICAR) the remainder. ICAR institutes perform about 40 per cent of the research, SAUs about 30 per cent, the private sector 15 per cent and international centres 8 per cent.

AGRICULTURAL RESEARCH OBJECTIVES

Some of the objectives of agricultural research seek primarily to enhance the overall level, value and efficiency of production while others are geared to varying combinations of production and equity.

Production

By the end of the 1990s India's annual foodgrain production will have to rise by 60–70 million tonnes over its present level of some 170 million tonnes if the increasing demand for food deriving from population growth and rising incomes is to be met. Unlike in the past, much of this increase will have to be met from intensified production in rainfed areas as the marginal costs of opening up new irrigated land increase. In this production context a recent World Bank review lists the 'main underlying resource questions' that need to be addressed by research as:

- the fate of an expanding livestock population in the face of diminishing forage resources
- the long-run problems of managing soil and water resources and irrigation systems in a sustainable manner
- the opportunities for higher product quality, product specialization and agro-processing in the wake of rising domestic demand and the need to increase exports
- the opportunities to ameliorate growing environmental constraints, and increase crop and animal production (possibly at a lower real cost) by vigorous development of the 'new biology'.

Production and equity

Central government concern that research should increasingly pursue joint objectives of research and equity grew in the 1980s, with the realization that rural per capita income levels were some 50 per cent higher in the north-western states than elsewhere. Accordingly, the Seventh Five-Year Plan (1985–90), in addition to listing the customary production-related objectives also gave priority to e.g. 'resistance against . . . saline and alkaline soils, droughts and floods . . . upland rice production . . . dryland technology . . . in the predominantly rainfed states . . . human resource development with special emphasis on the weaker elements of society'. Many of these concerns are carried forward into the Eighth Plan (1991–95).

A number of weaknesses in Indian public sector research and extension have been recognized (Clay 1988; ICAR 1988; Mukhopadhyaya 1988; Biggs 1989c) which have inhibited its response to both production and equity objectives. These are discussed below.

GENERAL CONSTRAINTS

Size, diversity and co-ordination problems

The large number of institutes to which ICAR provides financial support evolved gradually over a long period. At no point has it been possible to stand back from the system and to allocate responsibilities in a coherent and complementary fashion. As a consequence, institutional mandates now seriously overlap. Duplication generated by the institutionalization of certain AICRPs and by the spawning of regional sub-stations (166 by 1988) has been particularly severe. Senior staff in ICAR remain largely uninformed of the degree of overlap among institutes and programmes, and of performance in relation to the inputs provided. They are also overwhelmed with detailed management responsibilities to the detriment of strategic planning and professional leadership. Weak mechanisms for monitoring and evaluation mean, in effect, that most research institutes and programmes receive financial allocations, often for multi-year periods, the results of which are rarely subjected to close scrutiny. ICAR is discussing with major donors the possibility of instituting a computerized management information system to address some of these shortcomings, and to strengthen the monitoring and evaluation aspects of the research project cycle.

Personnel and facilities

A high proportion of research budgets is allocated to staff emoluments which leaves little flexibility for materials and equipment to meet new needs as they arise. For every member of scientific staff there are 4.5 support staff, indicating substantial over-staffing in the lower grades.

Buildings are often inappropriate to the research being undertaken, much equipment is obsolete or malfunctioning, and repair and maintenance facilities tend to be inadequate, prompting donor agency suggestions that capital investment needs over the next few years will be at least US$100 million.

Quality and relevance of research

Much of the Indian agricultural research establishment sees itself in the role of 'advancing the frontiers of science'. The divorce between research (*de facto*, largely a central government responsibility) and extension (a responsibility of the individual states) is not merely attributable to the divide between the institutions involved, but also arises from a deep-seated view among many scientists that they do not need to co-ordinate with extension services. It is up to extension services to make the most of whatever research results are published. ICAR's own mandate to transfer technology directly to farmers has added to confusion. It has allowed an elaborate and much-criticized set of programmes to be set up, including Lab-to-Land programmes and Krishi Vigyan Kendras (farm science centres). Training and Visit is the predominant extension system but, despite its organizational strengths it has provided only weak feedback to researchers (Clay 1988; Mukhopadhyay 1988).

Weak structures to ensure that research projects are client-oriented and do not overlap mean that research is often conducted without prior reviews of literature or of work in progress elsewhere. For many of the same reasons, research is often designed to produce publishable output, which is virtually the sole criterion of performance among scientists, the generation of technologies adoptable by farmers becoming *de facto* a secondary objective. This, together with the high proportion of total emoluments made up by benefits (sickness, housing, retirement) which are not performance related, makes it extremely difficult to make the overall system more productive, and its output more relevant to farmers' needs.

Constraints specific to the achievement of equity objectives

While the research service has had considerable success in developing Green Revolution technologies for the north-western states, it has faced numerous limitations in developing new agricultural technologies for those parts of (predominantly) eastern and southern states which can be grouped under the following headings:

1 *Resource requirements and availabilities* CDR areas are characterized by diverse farming systems in which inputs to and outputs from crops, trees and animals are intrinsically linked. Such systems contrast starkly with the monoculture (e.g. irrigated rice) widely evident in more homogeneous areas. To raise agricultural productivity through research in CDR areas is a more complex and expensive task than to achieve a similar increase in homogeneous areas, simply because of the range of crops, trees and animals involved. Further difficulties are posed by the complexities of interactions among them, by the importance of off-farm resources (trees; grazing land) and by the socio-institutional complexities governing, e.g. access to common land and reciprocal labour obligations.

To achieve a certain percentage increase in agricultural productivity in CDR areas therefore requires more resources per unit of land area than would a similar increase in more homogeneous areas. However, in reality, far fewer resources are allocated to CDR areas.

2 *Research focus* Much of the reputation of public sector research institutes in India has been built on research into high-yielding varieties of cereals and oilseeds with their concomitant high levels of agrochemical inputs. This more easily and more reliably generates publishable work than the sustainable low-input technologies more relevant to CDR areas.

3 *Research methods* Conditions in homogeneous areas are easy to replicate on research stations. On-station experiments under controlled conditions therefore have good prospects of producing technologies adoptable by farmers operating in these areas. By contrast, in CDR areas much more time and effort needs to be invested in understanding farmers' objectives, constraints and practices in the context of the agro-ecological and socio-economic complexities in which they operate. Researchers' relations with

farmers need to be participatory and collegiate. These approaches and techniques, however, are still in their infancy in many public sector research institutes, including those in India.

4 *Research-extension linkages* Major empirical studies drawing on experience from numerous countries have demonstrated that research-extension linkages are weak (e.g. Kaimowitz 1990). Feedback from extension to research on the performance of technologies disseminated has been particularly weak. India is no exception to these general findings: while the introduction of the training and visit extension system has improved management procedures, it does not have sufficient flexibility to tailor the type or timing of messages to areas characterized by unreliable rainfall, for example. Nor does it necessarily improve feedback (Howell 1988). A powerful lesson from reviews is that research-extension linkages require consistent effort from researchers to identify not only ways in which their technologies can be better disseminated, but also how far the technologies have met farmers' requirements and in what ways they can be improved. It appears that only a few individuals within Indian public sector research institutes have begun to take steps in this direction for CDR areas.

5 *Linkages across disciplines and institutions* The methods and approaches outlined in (3) are necessarily interdisciplinary, often requiring strong input from social as well as natural sciences. Several reviews (Coulter and Farrington 1988; World Bank 1989b) have drawn attention to the small numbers of social scientists employed in ICAR institutes. An alternative approach, relying on inter-institutional collaboration, has so far been little explored in the Indian context. Two ramifications of limited inter-institutional collaboration go far beyond the restrictions they impose on interdisciplinarity: first, contacts between research and implementing institutions (such as special agricultural projects, integrated rural development projects and the plethora of agencies charged with watershed management) have been limited, thereby restricting researchers' knowledge of the implementation constraints faced in the use of existing or new technology. Second, with the exception of the case studies reported at a recent workshop,[5] links between public sector research institutes and those (non-profit) institutes operating outside the public sector have been very limited. Given the potential complementarities between the different types of institute – i.e. the availability of specialist skills and 'lumpy' facilities such as libraries and laboratories in the public sector which NGOs generally do not have, and NGOs' skills in identifying farmers' needs, opportunities and constraints, in participatory approaches and in community organization – some loss in overall efficiency seems to have been incurred through failure to exploit this potential more widely.

PROFESSIONAL ASSISTANCE FOR DEVELOPMENT ACTION (PRADAN)

An NGO de-mystifies and scales down technology

M Vasimalai

Professional Assistance for Development Action (PRADAN) is a non-governmental development organization registered in 1983. Its policy guidelines are formulated by a general body, comprising representatives from other voluntary agencies, academics, public sector administrators and the private commercial sector. It is financed by contributions and fees for the provision of its services to voluntary agencies and decentralized government agencies (e.g. district rural development agencies). It also receives project-specific grants from government agencies and donors, and is in the process of setting up a Corpus fund in support of its training, apprenticeship and experimentation activities.

PRADAN's initial mode of action was to place its professionals into local NGOs to strengthen their capacity to implement poverty alleviation programmes by addressing constraints in technology and social organization, and in the interface with government and project management agencies. This mode has now largely been replaced by development support teams organized on an area basis and servicing small NGOs and informal village groups in the technical and management aspects of programme design and implementation, and institutional development. The teams also provide support to Panchayat Raj institutions in implementing poverty alleviation programmes, and have regular contact with national and state-level line department agencies at block and higher levels (see Figure 4.1).

One of PRADAN's central concerns is to draw socially concerned young professionals to work in villages by providing 'learning space' for them through its 'field university' concept. This provides guided field experience for fresh graduates and, through links with their universities, stimulates curricular change.

Three examples are given of PRADAN's mode of operation: the production of 'tasar' (raw silk) and of mushrooms, and village-level processing of hides and skins.

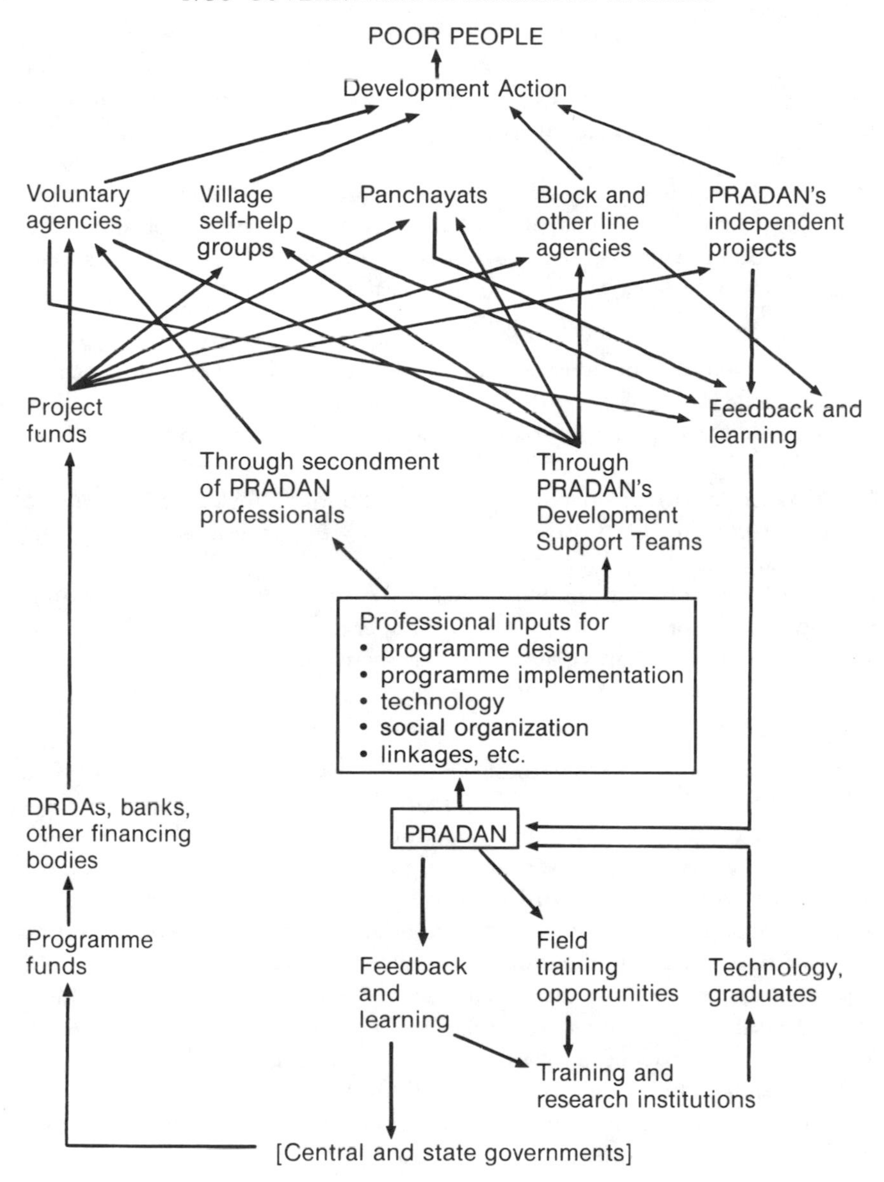

Figure 4.1 PRADAN's perception of its role as an intermediary between government and the rural poor

TASAR

The Central Tasar Research and Training Institute (CTR&TI) based at Ranchi, Bihar, under the central Silk Board (Ministry of Textiles) is responsible for technology design and promotion through its branches in the main tasar-producing areas. The model of tasar production which it has developed, and which is being promoted through state-level sericulture departments, relies on high-density plantings of arjuna (*Terminalia* sp) which, with fencing, cost some Rs17,500 per hectare and so represent a high-cost option for the rural poor.

PRADAN's principal objective was to make the advantages of improved technology (managed arjuna plantations; disease-free rearing of silkworms and post-cocoon processing) more accessible to the rural poor, with special emphasis on scheduled tribes, scheduled castes, backward classes and poor women. The main components of its approach were:

- establishment of a demonstration plot of arjuna at lower cost than official recommendations (i.e. with a plant density which is lower than recommended, but is appropriate for the wastelands to which the poor have access)
- low-cost and participatory management techniques, including live-fencing and the creation of off-season employment opportunities (pit digging; transplanting)
- establishment of arjuna on what had hitherto been open-access wastelands
- establishment of village-level grainages operated by village youth
- establishment of participatory management of physical and financial transactions, and of a community-managed marketing service catering for both improved and traditional producers.

A programme incorporating these elements was launched in twenty-five villages in Bihar in 1988 with funding from a Netherlands-based NGO ICCO (Inter Church Co-ordination Committee for Development Projects). Despite reservations about the use of arjuna to rehabilitate wastelands, the National Wastelands Development Board provided financial support for the nursery component. Some initial difficulty was experienced in securing scheduled tribe and caste participation in the project, owing to their fear of losing land once it is planted as part of a productive enterprise. However, by 1991, over 1,000 acres of wasteland in almost 100 villages, 70 per cent of which belonged to scheduled tribes, had been planted to arjuna. PRADAN's support to the Mahila Mandlas (women's saving and credit groups) ensured strong representation of women in managing the plantation programme and 40 per cent of employment opportunities in nursery and plantation work were taken by women.

Improved technology in the form of disease-free layings (DFL) of silkworm eggs were also adapted to facilitate their sustainable management at village level. This commenced through a barter system in which rearers received one DFL from the local government-managed grainage (hatchery) and subsequently paid back eight cocoons in kind, three of which covered the relevant overhea l expenses. Subsequently, unemployed village youths were trained in DFL production techniques (microscopic examination of eggs and chemical fumigator), provided with a stock of tools and chemicals, and involved in the establishment of three village-level grainages. Government inspection of these certified that they adhered to recognized norms, and resulted in agreement that they would be eligible for long-term technical support from the sericulture department.

Existing marketing arrangements for cocoons relied largely on private traders whose monopsony power had led to low producer prices, a situation exacerbated where production loans commanding high interest rates had been advanced by the traders. A Raw Materials Bank had been established by government near Ranchi to address these difficulties but, in practice, remained inaccessible to the majority of producers. In response, PRADAN supported informal groups of rearers (Tasar Vikas Samity – TVS) by providing loans to allow them to acquire cocoons at a 'floor' price (usually below the price given by traders), but sell to the terminal market at a much higher price (adding value by stifling the cocoons), paying out much of the difference in the form of a dividend, but using the remainder to cover the TVSs' own costs. The Raw Material Bank, presented with evidence of this experience, is now considering the provision of working capital loans to the TVSs.

PRADAN has also helped to create value-added by making spinning technology more accessible to those women wishing to process pierced cocoons (i.e. those from which the larva has emerged and which command a lower market price) into yarn. While the spinning wheels (*charkas*) cost some Rs1,000, official credit is available only in a minimum amount of Rs6,200 which envisages a 'package' including the construction of a work-shed. This arrangement has proven unpopular with village women, being both high-risk and inconvenient since it creates a division between the work-place and home. PRADAN has therefore begun to provide on credit a more modest kit of spinning wheel and cocoons, and will press for a change in government credit policy once these arrangements have proved themselves.

MUSHROOMS

Technology 'packages' for mushroom production generated by the universities and research institutes rely on techniques which are high cost, require high levels of management skill, and are potentially risky. PRADAN has succeeded in 'unpackaging', simplifying and 'de-mystifying' these and so has made them more accessible to the rural poor.

The modifications that PRADAN has made to standard technology include the following:

- The use of closed instead of open bags for mushroom production. This has the basic advantage of conserving moisture, and PRADAN has demonstrated that comparable levels of mushroom fruiting can be achieved by cutting holes in closed bags.

- The use of closed instead of open bags allows ambient humidity to remain at levels which can be achieved simply by regularly pouring water on to the floor of a mud hut instead of having to invest in electrically powered (and therefore potentially unreliable) humidifiers.
- PRADAN has devised management practices appropriate to village-level individuals and institutions for two broad stages of spawn production:
  (a) spawn culture, requiring strict hygiene and advanced skills, is purchased from universities and research institutions
  (b) the processes of multiplying culture into mother spawn, and deriving production spawn from this, are simplified and training courses held for village youths having no more than high school education.

  Dividing the spawn process in this way has made it accessible to village-level skills, and has reduced investment from Rs150,000 (for the entire package) to Rs10,000 (for the processes now managed at village level).
- Management practices have been simplified: standard recommendations are that the straw substrate used in mushroom production should be sterilized either by passing steam through it, or by brief immersion in boiling water. Both of these practices are cumbersome and prohibitively costly at village level. An effective and widely accepted alternative introduced by PRADAN is to soak the straw in water for two to three hours, drain off excess water, stack the straw and cover it with a plastic sheet or moistened gunny sack. Fermentation then produces temperatures of some 60°C inside the heaps, ensuring that undesirable organisms are eliminated within forty-eight hours. This process is also amenable to small batch spawning of a few mushroom bags at frequent intervals, to ensure a regular distribution of workloads and income.
- Involvement of government staff to facilitate administrative processes. Initially, PRADAN assisted beneficiaries in completing bank-loan forms under the integrated Rural Development Project for the purchase of materials and for hut construction, and followed up the progress of these applications.

HIDES AND SKINS

Poor telecommunications, inadequate transport and the limited capacity of artisanal flayers mean that the hides of the majority of bovines that die in India go to waste. Flayers belonging to the Chamar caste predominantly occupy the village of Haddiganj in Uttar Pradesh (UP) and, as throughout India, rely on artisanal methods of processing skins (i.e. salting and drying), which results in a low-value product. This group registered as a co-operative in 1983 with the UP Khadi and Village Industries Board. The co-operative, now comprising 101 members, is licensed to collect skins from the entire block, which is divided into clusters of villages manageable by groups of four to six flayers relying on bicycles for transport. When cattle die, the relevant group is informed, carries the carcass outside the village, removes the hide to the co-op's central store, and returns next day for the bones, by which time scavengers will have removed the flesh.

Government support for the flaying and tanning industries has taken several forms, practically all of which have been restricted to urban areas.

- The central Leather Research Institute has developed technologies - some of which are intended for small-scale industries - for chrome tanning, the utilization of by-products and effluent treatment, and provides on-site training courses.
- The UP Government Model Training cum Production Centre provides training courses of four to twelve months on carcass recovery, flaying, tanning and footwear manufacture.
- The Khadi and Village Industries Board provides for the registration of co-operatives, the provision of grants and loans at concessional interest, and the sponsorship of co-operative members to attend training courses.
- The UP State Leather Development and Marketing Corporation is engaged in both commercial and development activities; this institution provides marketing and processing assistance to small-scale tanning units.

The effectiveness of these government initiatives is, however, limited by the wide dispersal of local-level tanners, and the weakness of institutional outreach capacity. A primary objective of PRADAN's intervention was to act as a bridge between village-based artisans and the existing institutional support structure.

Jointly with members of the Haddiganj co-operative, PRADAN initiated a review of the features of commercial leather tanning technologies, and of the various possibilities for processing by-products. Matching these against the skills and aspirations of the co-operative led to the conclusion that attention should focus on the introduction of wet-blue chrome tanning technologies. The principal reasons for this decision were that the technology would add considerable value to the hides, which already constituted 60 per cent of carcass recovery values; that the technology was already commercially available, but would need scaling down; that existing government institutes could assist through their mandates in training and that investment, at Rs100,000, was manageable for the co-operative, whereas the Rs300,000 required for a finish-processing plant would not have been.

To proceed from identification of the technology path most acceptable to the co-operative to the start of activities on this path took over six months (i.e. from April to November 1987). This time was required for the co-operative to satisfy itself of PRADAN's credentials, and to become familiar with the technical, financial and skill implications of the technology; and for PRADAN to secure initial financial assistance (in this case, from OXFAM (India) Trust). PRADAN then provided the specifications for construction and equipment while the co-operative took on the construction tasks (completed in October

1989) and underwent training on a four-month programme designed by the Model Training cum Production Centre with advice from PRADAN.

Even so, wet-blue tanning is complex, requiring five chemical and two mechanical processes from raw hide to wet-blue product. The skill levels that these require are particularly high, given that tanning is a biochemical process in which the chemical concentration and tanning time depend on ambient temperature and on hide characteristics. For these reasons, PRADAN provided a leather technician to assist in setting up the tanning process, and further technical assistance was obtained from the Training cum Production centre, and from the Leather Development and Marketing Corporation. Feedback from the private sector on hide quality was also obtained.

With PRADAN's assistance, the co-operative began to structure and organize itself to maintain quality control and keep costs down: the literate members noted down details of the treatment of individual batches which could then be checked against the quality achieved; the most skilled among the tannery workers was made technical supervisor, and market enquiries were made of private tanneries some 100 km distant before the sale of the first major batch was agreed. Withdrawal of the PRADAN technician in April 1990 led to no loss in production quality – indeed, in response to market feedback, the co-operative subsequently managed to reduce operating costs by 10 per cent without loss of quality.

The co-operative is now independent of external assistance, apart from calling occasionally on PRADAN to assist with marketing. The investment of PRADAN's time and effort in assisting the establishment of this co-operative has been substantial, not least because of the complexity of technical processes involved. Nevertheless, its value goes beyond the level of empowerment it has generated among the members of the co-operative: it has already served to demonstrate to the Khadi and Village Industries Board the feasibility of the approach, and KVIB has begun a programme to set up forty tanneries on similar lines throughout Uttar Pradesh. PRADAN is itself identifying two further groups of flayers in other states interested in developing similar processing plants.

LESSONS

PRADAN has summarized the main lessons from its experience as follows:

- Technologies are not class-neutral: many of the technologies developed by universities and government institutes are too costly, complex and risk-prone to be accessible to the rural poor. PRADAN has identified a need to adapt these to the capabilities and aspirations of the rural poor by demystifying and simplifying them. For instance, in the case of mushrooms, commercial-scale production can follow laboratory practice by keeping whole rooms in which spawn production takes place free from contamination. This key principle – the avoidance of contamination – can be adapted to the circumstances of the rural poor by applying it only to the immediate environment (i.e. the straw substrate) in which spawn production takes place. The identification of how key variables affect the technology process is a critical component of simplification.

- The change agent has to spend three to six months working with local groups to elicit their interest in the technology, identify their actual and potential levels of capability and to allow them to 'own' the idea.
- The implementation of the technology by local groups is lengthy, requiring training and support not only in technical skills, but also in management and marketing. Government organizations generally have a single function focus, and are ill-equipped to take a broad-fronted and flexible approach. Once skills have been developed in a local organization, it becomes a powerful source of learning which visits from others can exploit.
- PRADAN sees for itself two broad objectives in this process: first, the control by local organizations of technical processes previously believed too complex for them provides 'hands-on' evidence of models and processes which government agencies can learn from and replicate. Second, skill enhancement, the use of local resources and the creation of self-confidence among local organizations through the command of new technologies are empowering, not least in the sense that they allow them to draw on the 'single function' government services relevant to individual parts of the technology processes.

ACTION FOR WORLD SOLIDARITY AND INTEGRATED PEST MANAGEMENT OF THE RED-HEADED HAIRY CATERPILLAR

S Satish and T J P S Vardhan

Castor in India – the world's largest producer – is predominantly a rainfed smallholder crop, over 300,000 ha being grown in the dry parts of Andhra Pradesh alone. Red-headed hairy caterpillar (RHC) (*Amsacta albistriga*) is a severe pest of castor grown in Andhra Pradesh where yields, at 200 kg/ha, are less than half the national average. Economic losses are caused by direct damage to young seedlings, by the need to re-plant and by the delayed planting that occurs in an attempt to avoid the main period of damage. Extensive observations in Andhra Pradesh suggest on average a 50 per cent crop loss, valued at Rs2,400 per hectare, attributable to this pest (see Figure 4.2).

The pest's thick coat of hairs impedes the effectiveness of conventional sprayed insecticides, which most farmers would, in any case, be unable to afford. A wide range of indigenous control practices has therefore developed, including hand-picking the larvae and the planting of 'trap' crops such as cucumber, but these have been of limited effectiveness and/or have required large amounts of labour.

A number of NGOs and individuals had been screening these practices and brought them together at a People's Science Conference in Bangalore in the early 1980s. This provided a forum for contact among staff from NGOs, public sector institutions and universities. Subsequent work proceeded at several institutions: investigations at the Andhra Pradesh Agricultural University led to better understanding of the pest's life-cycle; work at the Transfer of Technology Unit of the ICAR's Zonal Co-ordination Office for Andhra Pradesh and Maharashtra involved field-testing of a number of indigenous technologies, including summer ploughing to unearth and kill pupae; manual collection of egg clusters; planting cucumber as a trap crop and arranging a network of bonfires to attract and kill moths shortly after emergence from the pupal stage. These were tried with farmers on 100 ha in Nalgona District in 1988, and revealed that, while some of these practices (especially bonfires, on which subsequent collaboration was to focus) appeared to achieve certain impact, their further testing and imple-

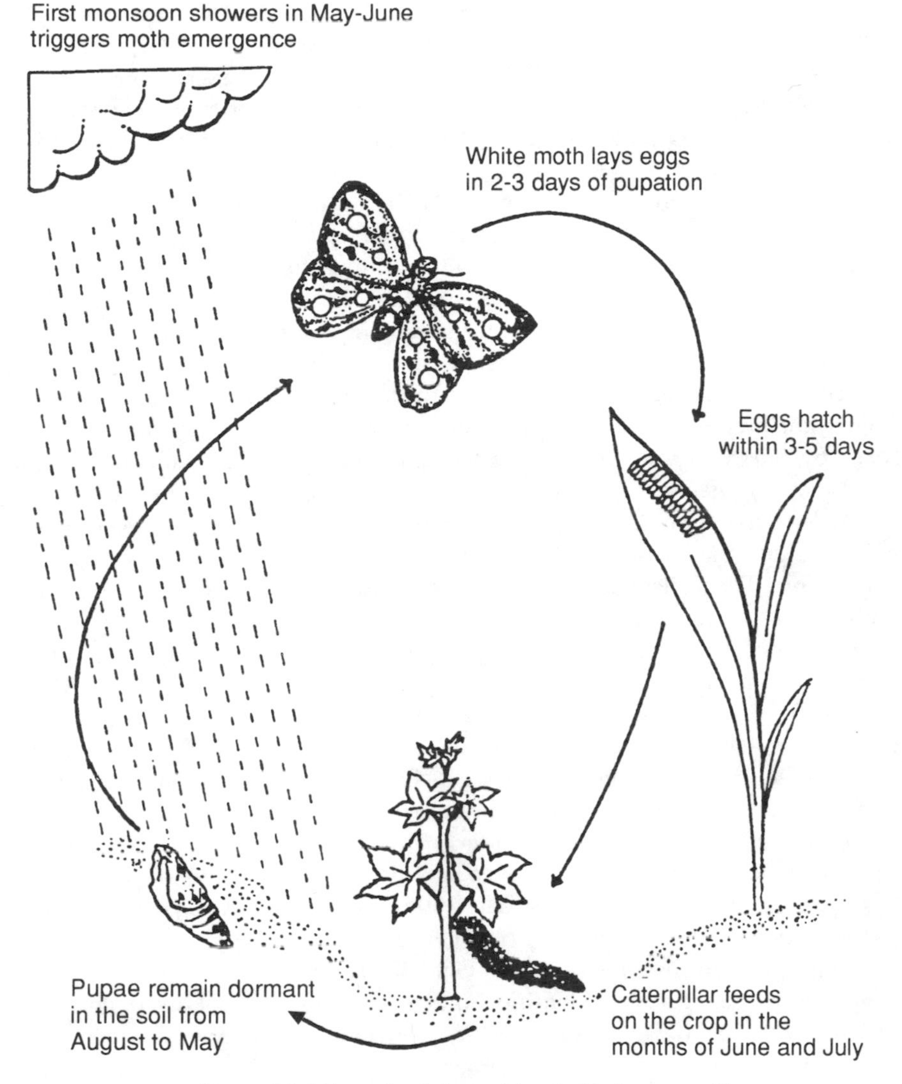

Figure 4.2 Life cycle of the red-headed hairy caterpillar

mentation required more resources than NGOs could provide, and, to be effective, required levels of awareness creation, commitment and organization among farmers that GOs had found difficult to achieve.

Given NGOs' perception of the opportunity for drawing on government resources in refining and scaling up traditional RHC control methods, and government scientists' perception of NGOs' potentially powerful role in organizing group action, the stage was set for NGO–GO collaboration. This began informally, through discussion between two key individuals: Dr Vittal

Rajan, a social scientist associated with the Deccan Development Society (DDS), an NGO based at Hyderabad, and Dr N K Sanghi, ICAR Zonal Co-ordinator. Dr Vittal Rajan was also instrumental in opening discussions with Action for World Solidarity (AWS) concerning the possibilities of its acting as a link between grassroots organizations and public sector research and extension agencies.

AWS's initial intention had been to play a background role in supporting local NGOs, but it took on a more significant co-ordinating role once the magnitude of necessary efforts in training, input supply and institutional finance became clear. These efforts included finance by the Agriculture Department of Rs250 per hectare for arranging bonfires and trap crops;[6]; training for NGO staff at the State Institute of Plant Protection and Pest Surveillance (arranged by the DoA); further in-field training of farmers by NGOs; the preparation of visual aids and literature; and the provision of crop loans by the National Bank for Agriculture and Rural Development, with special arrangements to include previous defaulters in these provisions.

AWS set up a series of consultative meetings with seventeen local NGOs in early 1989, of whom five subsequently dropped out, finding the work too time-consuming. The collaborative programme was to have covered 2,500 ha (subsequently reduced to 2,000 ha) and involved the co-ordinated provision of rain-gauges and light-traps (to permit monitoring of moth emergence from the pupal stage) and bonfire material (waste rubber provided by the DoA). On the 2,000 ha covered during the first (1989) season, average yields were estimated to have doubled, the overall incremental value at 1990 prices being Rs4.8 million against programme costs of Rs1.5 million. While these figures must be treated as indicative, they indicate strongly positive rates of return. To reduce the administrative effort of co-ordinating a large number of NGOs, and in response to the views of a number of local NGOs which had found difficulty in implementing activities in 1989, the 1990 programme envisaged only four participating NGOs each covering a larger area.

Spin-off from collaborative NGO–GO efforts in 1989 has led, first, to proposals to test recently identified indigenous pest control technologies, including the spraying of neem to control RHC, and to test various attractants (including rice pellets mixed with tumeric) for birds in the expectation that they will eat RHC larvae once attracted to the crop, and second, programmes of castor seed production: in response to the high price and poor quality of seed available from GOs, a local NGO (Weaker Sections Development Service Society) in 1990 organized farmers in one village to allocate land distant from other castor (castor being an open-pollinated crop) to the multiplication of seed of an improved variety. Despite a number of difficulties, the NGO succeeded in eliciting collaboration from seed inspectors of the Directorate of Oilseeds and from the Andhra Pradesh State Seeds Development Corporation to inspect and purchase the seed produced. A number of difficulties (chiefly pertaining to the pricing of seed) remain, but the seeds corporation has asked

AWS and other NGOs to organize farmers to produce seed on a further 1,000 ha in 1991–2.

CONCLUSIONS

This programme is an unique inter-institutional experiment in which research institutes, extension agencies, financial institutions, non-governmental organizations and rural people have participated in developing and managing technology for the successful control of RHC.

Important ingredients from the government side were as follows:

- Openness on the part of the Zonal Co-ordinator to the notion of working with NGOs as equal partners in development efforts.
- Reorganization of ICAR leading to the creation of the extension directorate and the zonal co-ordination units to provide not only financial assistance but also technical leadership to the state agencies in efficient technology transfer and also to link with NGOs.
- The GoI's oilseeds technology development mission which aims to explore technological options for enhancing the productivity of oilseeds. The monthly reviews of the performance of the oilseeds technology mission conducted by the Prime Minister's Cabinet secretariat brought pressure on the Ministry of Agriculture which, in turn, induced ICAR to seek means of enhancing castor yields.

Important ingredients from the NGO side were as follows:

- Perception by key individuals of the scope for complementarity between NGOs and GOs, and of the need for an NGO to act as co-ordinator on behalf of others.
- The willingness of AWS to take on this co-ordination role.
- AWS also provided information on the pest control strategies traditionally undertaken by farmers, and worked with government researchers in screening them for suitability in the present context.

Informal links had evolved among key individuals over several years: these formed an essential basis for the subsequent more formal links in which GOs and NGOs worked together in overcoming the very considerable obstacles imposed by orthodox administrative procedures to collaborative efforts of this kind.

Issues requiring further consideration include the following:

- Should this arrangement, involving NGOs in technology management, be institutionalized for the longer term? If the answer is yes, the modalities will have to be such as not to erode the flexibility and autonomy of the participating grassroot-level NGOs.
- The nature, roles and responsibilities of an intermediary non-governmental

organization, such as AWS, and ways in which these roles can be strengthened.

- The role of NGOs in bringing to light and testing farmers' traditional technologies, and the scope for their wider dissemination.
- The limited extent to which government research agencies have taken up research into the field problems encountered, and the scope for NGOs to influence their agenda.
- The possibilities of stimulating more widespread openness among other branches of ICSR and among central and state-level GOs to the concept of working with NGOs.

Some moves are already being made in this direction within ICAR: in the Eighth Five-Year Plan, it is proposed to integrate all 'Transfer of Technology' (ToT) projects (NDP, ORP, LLP and KVK) into Integrated Krishi Vignana Kendras (KVK). These KVKs will have triple functions – farm advisory service, vocational training and on-farm research. This integration is expected to enable farming systems approach for development and dissemination of technologies particularly for resource poor areas. By design, one-third of these KVKs will be with NGOs.

RAMAKRISHNA MISSION

Research, extension and training in a farming systems context

S Chakraborty, B Mandal, C Das and S Satish

INTRODUCTION

Ramakrishna Mission (RKM) was established in 1897 by the Swami Vivekananda with its headquarters in Howrah, near Calcutta. Although initially a religious organization, RKM has developed a number of welfare and developmental activities since the early 1950s. It became, for instance, India's first adult education centre in 1952, and now has programmes in education, health, relief and rehabilitation, publication and integrated development. RKM now has 167 branches in various countries, with 93 in India alone. RKM, Narendrapur, is notable for both its size and the wide range of activities it undertakes.

RKM, NARENDRAPUR AND LOKA SIKSHA PARISHAD

Loka Siksha Parishad (LSP) is a specialized institute of social welfare and integrated rural development under Ramakrishna Mission which provides training in the full range of RKM's programmes, with about 6,000 participants per year, but is also responsible for agriculture-related research, development and dissemination.

The Ministry of Human Resources Development (MoHRD) in 1979 formally recognized RKM/LSP as a national training centre, opening the way for substantial links with government. Since 1982 RKM/LSP has been implementing the Lab-to-Land programme of the Indian Council for Agricultural Research (ICAR) aimed at transferring adoptable technologies to resource poor farmers and landless agricultural labourers. Since 1987 a Farming Systems Research unit in RKM/LSP is focusing particularly on low-income farmers through farmer/family-oriented participatory technology development and management. Besides these, RKM/LSP is working in close collaboration with the Government of West Bengal (GoWB), on several programmes such as training the agricultural technical assistants (ATAs) in the Department of Agriculture; training village-level workers (VLWs) and farmers and their sons, the promotion of entrepreneurship development and training of Block youth officers.

Spread over 1,500 villages in 12 districts of West Bengal, RKM/LSP's programmes are staffed by about 250 professionals seeking to help the villagers to help themselves through collective/community action. In all this, its emphasis has been on utilizing the locally available resources to the optimum.

RKM/LSP's programme organization

The search for an appropriate development model led to the 'youth club' approach. West Bengal is traditionally known for forming informal and temporary groups or associations to achieve a specific purpose in sport, cultural or religious activities, for example. Following this model, RKM/LSP identified youth clubs (YCs) to play a catalytic linking role between villagers and external agencies (including RKM/LSP). There are at present 483 registered and affiliated YCs in 1,500 villages, spread over twelve districts of West Bengal. The YCs of a region (about twenty-five, spread over a maximum of sixty villages) join together to form a 'cluster club' (CC). The CC, which comprises the secretaries of YCs, is the link between RKM/LSP and YCs. This has recently been supplemented by the establishment of a District Level Youth Council (DLYC) with the aim of strengthening the linkage between YCs and the district administration.

Every YC has an executive committee, sub-committee(s) specific to a programme, and *para* (hamlet) committee. The *para* committee implements needs-based projects and arranges local financial contribution to these. The secretaries of the YCs throughout the state assemble for a five-day workshop at Narendrapur each January or February. Workshop activities include presentations by RKM/LSP of new ideas across a wide area, the sharing of experiences among participants and the discussion of relations between themselves and RKM/LSP.

Programme planning and implementation

RKM/LSP aims to select operational areas having a high degree of poverty and limited attention from government. It also prefers to select areas on a 'cluster' basis. Once the selection of areas is complete, several orientation programmes are conducted for YC members to sharpen their capability to identify and address socio-economic problems. Help is also provided, wherever necessary, in completing the legal formalities for the registration of clubs.

The members are encouraged to organize open discussion meetings in the village to build rapport and to begin to identify villagers' needs. Subsequently RKM/LSP, along with the YC, undertakes a detailed survey of the village in order to

- ascertain the needs of the individual families
- identify local human and natural resources
- ascertain the extent of resources offered by other agencies
- ascertain the extent to which RKM/LSP's efforts can lead to self-reliance and development.

A project outline is derived from this survey, and suitable funding arrangements are made on either a grant or loan basis. If the requirement is modest, the fund is provided from RKM/LSP's own revolving capital. Otherwise, external agencies such as state government, Khadi and Village Industries Commission, commercial banks, or, in some cases, central government or international agencies are approached.

Projects financed by RKM/LSP have beneficiaries' committees responsible for planning, implementation and evaluation, with an advisory committee largely comprising representatives from government and technical specialists.

In the context of this organizational structure we present case studies of RKM/LSP's experiences in agriculture-related research, development, extension and training, with focus on first, training agricultural technical assistants, second, vegetable production, and third, farming systems research.

AGRICULTURAL TRAINING PROGRAMME

In the 1960s the Directorate of Agriculture, GoWB, sought to establish a training centre in this coastal zone to complement those already operational in the five other agro-ecological zones of the state. Rather than set up its own school, the directorate in 1967 sponsored a training centre alongside the existing RKM vocational agriculture school at Narendrapur.

While training was provided to a broad spectrum of village extension workers from 1967–80, since then it has focused specifically on agricultural technical assistants (ATA). Two types of technology transfer training programmes are being conducted: first, pre-service or induction course for six months, and second, upgraded/refreshers' programme for six months, conducted after two years of service. Between 1980 and 1990 1,069 ATAs in eighteen batches were trained .

The training involves a blend of theory and practice; it covers pre-kharif vegetable crops, such as brinjal, bhindi, chillies; kharif crops of paddy and jute, and rabi crops of wheat, potato, vegetables, oilseeds and pulses. The RKM/LSP's trainees have consistently fared better than those in the other centres (run by the GoWB) in the examinations conducted at the end of the training programme. This is attributable to the holistic approach taken by RKM, and to the stimulus that staff and students receive as a result of the fortnightly meeting held at RKM, involving agricultural development officers, village-level workers and subject matter specialists in the planning and monitoring of crop production programmes.

VEGETABLE PRODUCTION PROGRAMME

The villages around Narendrapur are traditionally known for growing vegetables, but face the constraints of saline and unreliable water supplies. Discussions among farmers at the agricultural fairs held annually at RKM since 1957 have focused on their need for commercial vegetable varieties capable of coping with these constraints. RKM/LSP's interest in vegetable research was thus farmer-driven and focused initially on chillies in view of their buoyant market and the good prospects of identifying suitable varieties.

Research on chillies sponsored by Government of West Bengal

RKM/LSP and the Departments of Genetics and Plant Breeding and Horticulture of Calcutta University (CU) had been collaborating on vegetable research from 1970, when, in 1975 the GoWB was successfully approached to fund a research project entitled 'Evolution of commercial varieties of chillies' as part of the RKM–CU collaboration. CU's role in the research was mostly limited to imparting technical guidance; collection and screening of cultivars; design of field experiments, and analysis and interpretation. Biochemical analysis (pungency and vitamin C content) were conducted at CU as RKM/LSP lacked facilities. The research project was designed for five years (1975–80). In the first instance, 105 different strains of chillies were collected from sources including: Horticultural Research Station, GoWB; Vegetable Research Station, Himachal Pradesh; Indian Institute of Horticultural Research; farmers' fields and seed traders from different parts of the country.

As a result of the screening trials, the number of varieties under experimentation was reduced from 105 to 8 for the second year. Agronomic, biochemical and genetic trials ultimately narrowed these down to three varieties:

- *Sundare* (for dry chillies)
- *Kalolanka* (for green chillies)
- *Surya mukhi* (dry and green; grown round the year).

GoWB accepted the recommendations of the RKM/LSP–CU research, incorporating the three varieties into its official handbook on agricultural practices. The RKM/LSP promoted dissemination and adoption of the varieties and relevant cultivation practices through its youth clubs.

In an effort to screen other vegetable crops in a similar way, RKM approached the Ford Foundation, New Delhi, for financial sponsorship. In 1980 the Ford Foundation agreed to contribute financing to this collaborative work with the objective of promoting vegetable farming as a component of cropping systems in coastal South Bengal. Both the principal and co-investigators were from CU and were mainly responsible for technical guidance and providing laboratory facilities for biochemical analysis. The research was conducted on the farms of RKM/LSP under the overall supervision of its director.

The project comprised a study of the genetic, agronomic, biochemical, and plant protection aspects of three major vegetables – egg plant (aubergine), lady's finger (okra) and cauliflower. The aim was principally to upgrade to the genetic base through selection, hybridization and mutation. The X-ray facilities for inducing mutation were provided by the Jute Institute of ICAR, and the Bose Institute. The agronomic trials were aimed at developing technology packages for both seed and table purpose. The research efforts resulted in an increase in the yield by some 20 per cent.

Besides conducting research, fifty-four programmes of one to six days' duration were conducted: about 3,000 women were trained in improved methods of growing vegetables in home gardens. By contrast with the earlier work on chillies, the results of this research were not published as part of GoWB's official recommendations. This was partly a consequence of the more limited involvement of GoWB in this latter project.

The vegetable programme was high on the agenda of RKM/LSP for fifteen years (1970–84) but had been essentially a commodity-based programme. In an effort to understand more fully the role of vegetables in farming systems and to identify possible areas requiring technological change, RKM decided to pursue a farming systems programme.

FARMING SYSTEMS RESEARCH

The farming systems research (FSR) project was started in January 1987, with financial assistance from the Ford Foundation. The main objective was to develop a sustainable and improved livelihood system through innovative location-specific technologies for resource poor farmers who depend upon lowland rainfed rice-based subsistence farming. However, the specific objectives originally proposed were found too broad-based for focused implementation. This, coupled with frequent external interventions, has resulted in the considerable reformulation of the sub-components in a short span of four years. Additionally, RKM/LSP's strategy of setting up an altogether new FSR Unit divorced this activity from those of other aspects of RKM's work. These issues are examined in detail below.

History of farming systems research

RKM's interest in FSR began in 1980 when an agronomist, returning from training in the Philippines, suggested setting up integrated rice-based crop and animal production farms for research and demonstration purposes on farmers' fields. A proposal to this effect was received with interest by the Ford Foundation, which at the time was seeking a centre to co-ordinate networking for the Eastern India Universities participating in its FSR programme. To allay doubts about its research capacity, RKM agreed to seek technical support from ICAR institutes, appointed an internal adviser on FSR, and established a Scientific Advisory Committee.

Farming systems research programme

The FSR programme has been underway in nine villages (grouped into three clusters) since January 1987. A diagnostic survey of 2,526 farm families in the nine identified villages was conducted in collaboration with the youth clubs. Some 98 per cent of the families owned less than 2 ha of land and more than one-third (37 per cent) were landless agricultural labourers, dependent upon wage-employment and home-garden production. The survey reflected a predominantly rice-based (mono crop) economy with below-average yields and low income levels. The survey identified climatic, biotic, edaphic, socio-economic, and institutional constraints, as well as opportunities for diversification. A number of action plans were prepared jointly with the youth clubs to be tried in 250 households (10 per cent sample) selected on a random basis.

The four major components of the FSR programme were as follows:

- *On-farm technology verification trials* The diagnostic survey and interaction with farmers suggested that high priority should be given to increasing rice productivity. Yields in coastal Bengal, at 20–22 qtl per hectare are below the state average of 27 qtl/ha. However, farmers' decisions to cultivate traditional varieties are rational since these are tall and of long duration, and therefore can tolerate submergence in areas where 10 per cent of the land had up to 30 cm water depth, 40 per cent 30–50 cm and the remainder 50–100 cm.

  Initial trials were extended to investigate fertilizer application. Split doses were recommended to avoid losses due to leaching in higher water regimes. The varietal research has resulted in an increased yield of between 10 and 30 per cent, and the fertilizer experiments showed increases of up to 40 per cent. Discussions during field visits confirmed that these increases could realistically be obtained by farmers. At Jelerhat, starting from little more than a hectare, improved seeds have now spread over 40 ha in a three-year period mainly through farmer-to-farmer dissemination. GoWB and ICAR subsequently included these varieties in their recommendations.
- *On-station experiments* These farmer participatory on-farm trials are supported by adaptive trials conducted on RKM/LSP's farms. Participating farmers visit the farms once a year to assess the suitability of the varieties for their specific locations. These are then incorporated into the on-farm trials programme. The Central Rice Research Institute (CRRI), the State Rice Research Station, and Tamil Nadu Agriculture University (TNAU) have been the major source of germplasm for on-station trials, and

assistance from Vishwa Bharati University has been obtained for weed control.

On-farm trials with vegetables have recently been incorporated into the FSR programme in order to diversify away from crops having a high water requirement.

- *Land-shaping and irrigation* Small tanks are traditionally dug near homesteads in many villages in coastal Bengal to collect rainwater for domestic use. Vegetables are also commonly grown on the bunds. FSR staff are seeking to expand this practice into systematic land-shaping to improve both irrigation and drainage and to increase the scope for aquaculture and horticulture. The fields are generally small (0.1–0.3 ha) and trials are being conducted with tanks covering about 20 per cent of surface area. The spoil is used for raising the level of the adjacent area so that short or medium duration HYVs can be cultivated during the monsoon season with vegetables, pulses, and oilseeds in the dry season. As this technology requires some initial investment, the Council for Advancement of People Action and the Rural Technology (CAPART) have come forward to support 200 marginal and small farmers to dig small tanks on a pilot basis.
- *Women in development* On the recommendation of the Ford Foundation, a homestead vegetable production component catering (initially) for 250 women has been integrated into the FSR programme. Some success in supplementing family nutrition and cash incomes has been achieved. Efforts are also afoot to stimulate interest among women in the rearing of goats, ducks and poultry. Since, mainly for socio-cultural reasons, the youth clubs attract few women, separate efforts have to be made to mobilize women into groups to facilitate decision-making in areas within their purview. While some modest successes have been achieved in the women's programme, no mechanism has been worked out to ensure that incomes from their activities will remain under their control.

Farming system research network

Nine organizations pursuing FSR in eastern India are linked by an FSR network with support from the Ford Foundation, including five universities and four other research institutes. RKM/LSP acts as the co-ordinating centre of the Eastern India FSR network whose responsibilities include

- organizing training programmes and conducting workshops and seminars
- arranging inter-institutional visits
- organizing visits to other countries where FSR is carried out
- collecting and disseminating FSR literature
- publishing FSR newsletters and information bulletin and educational aids
- preparing guidelines for the implementation of FSR.

However, networking has not led to the establishment of complementary operational inter-institutional links similar to those which RKM/LSP previously enjoyed with Calcutta University (see pp. 118–19).

RKM/LSP, FSR and linkages/interventions

During its four-year life, the FSR project at RKM/LSP has been the subject of a large number of official consultancy visits and reviews. These are over and above the regular meetings of network participants and interaction with the local ICAR institutions and universities, and have led to the incorporation of components – cropping systems, women in development, computers and management information systems, research methodology – not spelt out in the initial terms of reference. The introduction of such a large number of initiatives is, despite the best intentions, bound to disrupt regular activities. The mode of their introduction (i.e. through the recommendation of (often) expatriate consultants) raises wider questions. It is clear, for instance, that several of the early technologies introduced by RKM/LSP were identified on the basis of a sound understanding of local farming systems, nutritional needs and market opportunities. However, this early work was not labelled 'farming systems research', and it should be asked whether setting up a new FSR unit oriented more towards contacts with external institutions distracted attention from the perspectives already being pursued within RKM/LSP.

CONCLUSIONS

Almost two decades (1957–75) of contact with the rural poor led to the identification of technological gaps which resulted in RKM/LSP's initiatives to undertake research on vegetable crops, RKM taking a practical development approach which was complemented by technical help from various local government research institutions. The village network of youth clubs ensured quick and wider dissemination of the RKM/LSP–CU research results. The GoWB's sponsorship also helped in official recognition. The desire on the part of RKM/LSP to extend its research operations from vegetables to rice-based farming systems brought it closer to the Ford Foundation, and the eastern India FSR Co-ordinating Centre was set up at Narendrapur.

THE AGA KHAN RURAL SUPPORT PROJECT (AKRSP)

Influencing wasteland development policy

Armin Sethna and Anil Shah

INFLUENCING POLICY

This section describes how AKRSP, an NGO supporting income generation by the rural poor through improved land and water management in Gujarat, attempted over three years, in loose coalition with two other NGOs, to change government procedures which were impeding local participation in wasteland development.

The process of influencing government described in this section included extensive correspondence; meetings with central, state and local officials; the development of workable afforestation programmes; the setting up of village-level peoples' organizations for forest protection, and the process of maintaining effective links with departmental officers and staff at the local level. It was characterized by periods of official silence or indifference; periods of unofficial empathy and encouragement; and finally, a level of understanding and agreement.

AT STAKE: DWINDLING RESOURCES

Official efforts to protect India's forest lands have gone on for several decades. Under the Forest (Conservation) Act 1980, authority for permitting any individual or organization outside government to develop or manage forest lands rests with the Ministry of Environment and Forests in New Delhi.

Despite this attempt at official vigilance, about half the country's designated 75 million sq. km of 'forest lands' remains devoid of trees and wildlife. The half that is under tree cover is under heavy pressure from expanding human and livestock populations and from extraction of fuel and timber. Since 1960 Gujarat State, for instance, has lost 3,760 sq. km of forest area to dam-building, encroachment and illegal felling.

The restrictions imposed by the 1980 Act have meant that numerous programmes to afforest wastelands have met with delay, restriction or rejection. Meanwhile, local forest communities and business interests have tended to ignore the regulations.

AKRSP: A BRIEF BACKGROUND

The Aga Khan Rural Support Programme in India was set up with the broad mandate of promoting income-generating activities related to land and water through people's institutions. In all three AKRSP programme districts in Gujarat examples abound of earlier failed efforts by government agencies in land and water resource development. AKRSP's philosophy has been to take up projects, at first on a pilot basis, in line with the people's expressed needs, drawing on government funding where available. The fact that AKRSP's development initiatives attempt to improve on earlier performance by government leads to relations which have to be treated sensitively, especially given AKRSP's mandate to make available its experience to GOs and other NGOs working in the same field.

Initial efforts by AKRSP to encourage more flexible government approaches to wasteland development began in the Bharuch District, where strict enforcement of government regulations had prevented villages from acquiring fuel and fodder, and had led to violent confrontations. AKRSP's efforts in Bharuch began with a wasteland development project on a 17-hectare plot of Revenue Department land in Soliya Village. Much time had been spent earlier in organizing the villagers into a Gram Vikas Mandal (GVM) (Village Development Association) and eliciting a spirit of co-operation and common interest. Employment was generated by the clearing of land, digging of pits, and planting of saplings. Then, village co-operation was sought in protecting the re-planted area from cattle grazing and illegal felling.

As the programme grew, however, it ran into several stumbling blocks. First, Revenue wasteland was not always available in and around villages. Second, though there was much Forest Department land that was denuded and dry, access to such areas by villagers or NGOs was restricted by the provisions of the Forest (Conservation) Act 1980.

PROPOSALS AND PROCEDURES

In 1986 a local forest officer helped AKRSP to identify further areas of degraded forest that could be taken up for development through planting and protection. This led to two detailed AKRSP project proposals requesting the handing over of degraded forest lands for development by village communities. The proposals took a long time to reach the MoEF, being scrutinized at various stages in the departmental hierarchy by officials who were generally reluctant to hand over part of the area under the department's jurisdiction on lease to village communities.

After three years of unsuccessful appeals, a meeting was arranged by AKRSP to discuss its proposals with a senior officer in the Ministry of Environment and Forests in New Delhi. The meeting eventually led to an invitation to AKRSP to draft a scheme that would allow for increased access by villagers to forest products.

THE IMPORTANCE OF 'MODELS'

In early March 1988 a team of government forest officials, AKRSP staff members, and villagers held meetings in several villages in Bharuch where AKRSP's forestry programme was being implemented. Detailed discussions with the villagers allowed the group to gain a comprehensive idea of the programme's approach and functioning. It then produced several recommendations which were forwarded to Delhi as well as to officials in the Gujarat State Government, including:

- granting usufruct and not lease rights to a village organization (either a co-operative or registered society) that agrees to undertake wasteland development
- keeping membership of the village organization open to all families
- ensuring that officers not below the rank of range forest officer have the right to attend all such organization meetings and have unrestricted access to the areas
- limiting land use to forest produce (including grass, trees, and fruit trees) and in no case allowing the cultivation of agricultural crops.

INTERPRETING OFFICIAL SILENCE

AKRSP(I) went ahead with plantation and protection work in the tribal villages of Bharuch District, despite some reluctance from local forest officers, who were wary of endangering their careers by allowing activities not expressly permitted by the government, since AKRSP's proposals had still not been approved.

Meanwhile, encouraging news was received from the other side of the country: a local forest officer in the West Bengal area of Arabari had been developing a method of protecting forests with the involvement of villagers. Encouraged by his success, the West Bengal government had taken several supportive measures. These initiatives helped to support AKRSP's case.

Political factors also came into play at this time: promises made during an election campaign had led to widespread commercial logging in south Gujarat, prompting the Forest Department to try to protect the remaining resource through popular participation in forest management. Villagers and Forest Department officials from south Gujarat visited AKRSP's Bharuch projects in order to see at first hand the results of participative development and disciplined protection of forest lands. This led to practical support from the Conservator of Forests in integrated development planning, for example, but further progress was slowed by senior staff changes in Delhi: the secretary of the Ministry of Environment and Forests was replaced, as was the chairman of the National Wastelands Development Board. Both of the new officials took initially unsympathetic attitudes towards AKRSP's requests.

It was not until December 1988 that progress was made in the form of the National Forest Policy Resolution. This document committed the government to 'meeting the requirements for fuelwood, fodder, minor forest produce, and small timber of the rural and tribal populations'.

More important, the document emphasized that 'The holders of customary rights and concessions in forest areas should be motivated to identify themselves with the protection and development of forests from which they derive benefits'. However, in apparent contradiction, an amendment to the Forest (Conservation) Act (FCA) was passed which virtually prohibited any local discretion over the use of forest, including tree planting on barren forest land by a non-governmental organization.

Many have interpreted the emergence of such contradictory documents as an indication of a split between environmental and developmental camps within the GoI, each seeking to gain the support of the Cabinet and Prime Minister. In early 1989 AKRSP, in a loose partnership with two other NGOs, PRADAN and the Society for Promotion of Wastelands Development, met senior officials to seek guidance on how this apparent contradiction might be resolved.

Anil Shah of AKRSP was asked to prepare a paper suggesting how the 1988 Forest Policy resolution could be implemented through NGO involvement. This provided the agenda for a meeting of central and state representatives in March 1989, but resulted in deadlock over the exact nature of forestry or wasteland-related activities allowed by the 1988 amendments. Nor was it possible to make much progress at state level in Gujarat; newly appointed senior staff were taken on field visits to AKRSP's projects and were favourably impressed, but proposals for participatory approaches to forest management were blocked by the state Legal Department.

During the general elections in 1990 the forests once again became the targets of populist pronouncements made by local politicians in Bharuch district, who offered the tribals and, through them, commercial companies, the freedom to extract timber. As a result, incidents of illicit wood felling increased sharply. In one case, a group of woodcutters were caught and one was killed while attempting to escape. Much damage was done to forests during the ensuing protests, but certain areas of forest remained unaffected, including those protected by the village institutions that AKRSP had been supporting. Impressed by the discipline of AKRSP-supported groups, the State Conservator of Forests set about the task of organizing villagers on a large scale for protection of 'their own' village forests. Lessons were incorporated into this process from AKRSP's experience, and the Forestry Department made resources available for village labour to be recruited in land-clearing and protection works. An order was issued by the conservator formally endorsing the tripartite working arrangement through which villagers where motivated and organized into village institutions (Gram Vikas Mandals) by AKRSP and were given the task of taking up necessary works for the sustained regeneration and development of degraded forest lands near their village.

THE POWER OF THE PRESS

Disturbances caused during the 1990 elections had drawn media attention to the area, which AKRSP was quick to enlist in its own cause. Journalists from *The Times of India* were taken to see various AKRSP forestry projects, and one (Mrs Usha Rai), combining that information with her own research and interviews, published two in-depth articles on the issue. AKRSP Chief Executive, Anil Shah, contributed occasional pieces to the *Indian Express and Economic Times*. And the environmentalist, Mr Anil Agarwal, published a case story about the forest-tribal link in *The Times*.

This publicity enhanced the interest by senior central government officials in AKRSP's work. In early 1990, the new Minister of Environment and Forests heard the case put by Anil Shah and in April 1990 called a meeting of NGO representatives to discuss participatory forest management. The NGOs worked out their position in a prior meeting, and their argument that, without the involvement of communities, forest protection had little hope of success, won acceptance.

The minister then constituted an *ad hoc* committee to draft a policy document along those lines. Accepted by the minister in May 1990, the policy document, along with initiatives taken by the GoWB, influenced the preparation of guidelines issued by the MoEF in June 1990 (see Box 4.1) which emphasized the importance of joint forest management. State governments have followed these guidelines to varying degrees: as a result of further pressure from AKRSP and others, the government of Gujarat formally approved them in March 1991.

LESSONS LEARNED

As this abbreviated case study shows, a number of conflicting forces come into play when trying to reconcile micro-level development with micro-level policy and aims. To recapitulate, they include the following:

- differing interpretations of the same policy at various levels of the government hierarchy
- sensitivity in like departments to pressure from politicians and from powerful lobby groups, notably the pro-industrial and pro-environment factions
- a tendency toward the 'government knows best approach'

- an NGO's tendency to view problems in narrower terms – either issue specific, or geographically specific

5 the difficulty of overcoming government defensiveness caused by previous exploitation of resources by villagers and others having entrenched interests.

AKRSP's strategy had three strands:

- *Building goodwill and confidence* In order to do this, it helps to have a clear understanding and appreciation of the problems facing the government. Only then can informed and empathetic contact be pursued with officials at all levels. These efforts at communication should include but not be confined to senior policy-makers, given their rapid turnover. By building up a network of communication with all levels in the government hierarchy, AKRSP(I) helped make the wasteland issue a fixture on departmental agendas.
- *Evolving sustainable 'models'* Since 'seeing is believing', the team visits to AKRSP(I)'s wasteland development projects in Bharuch went a long way in persuading officials of the viability of 'people's participation'. Even without overt government support, where possible NGOs should strive to create and refine workable models to prove the point of their development strategy.
- *Staying persistent and calm* Obvious, but easily forgotten in the face of bureaucratic intransigence, procrastination, or rejection. The policy of knocking on as many doors as possible, and more than once if necessary, helps ensure that somewhere within the official machinery the issue is being kept alive.

Box 4.1 The GoI policy instructions: towards a user-friendly system of forest management

1 June 1990 was a watershed in the history of India's forest management. On that day, the Government of India's Ministry of Environment and Forests issued policy instructions to all state forestry departments supporting greater participation of village forest communities and NGOs in the regeneration and management of degraded forest lands. The implications of this document are powerful and wide-reaching in support of the joint forest management strategy.

In accordance with the National Forest Policy of 1988, the recent guidelines for action emphasize the increasing importance of jointly managed forest systems which are centred on the needs of forest communities. In order to help ensure community participation, the document spells out the need for Forestry Departments to work out the operational mechanisms by which forest communities are given *usufruct* rights on regenerating forest lands. Successful mechanisms have emerged in a growing number of Indian states in the form of mutually developed microplans – detailed management agreements which are then authorized by enabling state level orders.

The GoI policy instructions also strongly encourage Forest Departments to enlist the expertise of local NGOs to serve as interfaces between forest agencies and communities. To achieve the goal of full people's participation, with particular sensitivity to women and the most marginalized community members, experienced NGOs are proving highly effective actors as motivators and organizers of village groups. Highlights of the programme guidelines promulgated by the Ministry are summarized as follows:

1 *developing partnerships* between communities and forest departments, facilitated by NGOs when helpful
2 *access and benefits* only to organized communities undertaking regeneration, with equal opportunity based on willing participation
3 *rights to usufruct* all non-wood forest products and percentage share of final tree harvest to communities, subject to successful protection and conditions approved by state
4 *ten-year working scheme* microplans detailing forest management institutional and technical operations should be developed by community management organizations with local foresters
5 *funding* from Forest Department social forestry programmes for nursery-raising, with encouragement to communities to seek additional funds from other agencies
6 *use rules* strict adherence to no grazing, agriculture or cutting trees before maturity, except as outlined in working scheme.

Source: Poffenberger (1990)

THE AGA KHAN RURAL SUPPORT PROJECT (AKRSP)

Participatory approaches to agricultural training

Parmesh Shah and P M Mane

AKRSP is an NGO working in Gujarat, Western India, to promote and catalyse community participation in natural resources management. A number of village institutions supported by AKRSP viewed the strengthening of extension and training in agriculture and animal husbandry as important steps in raising farm incomes. AKRSP's response falls into three broad phases.

PHASE I 1985–7

Given the large number of GO training and extension agencies already existing, for example those under the Gujarat Directorate of Agriculture and the State Agricultural University, AKRSP saw its role as linking these up with relevant village institutions, and as facilitating programme design, pre-programme visits for scientists to become familiar with local conditions (in which scientists, in the event, did not participate), and post-programme evaluations by participants.

Farmers' evaluation of the first course included several criticisms:

- Trainers made too little effort to understand prevailing agro-ecological and socio-economic differences. Courses and the curricula were prepared well in advance and delivered without regard to differences in farming practice and resource availability.
- Some farmers had already experimented along the lines suggested by trainers, but no attempt was made to draw on their experience.
- Lectures were the predominant mode of teaching, little time being allowed for informal two-way exchange. Farmers found that they learned much from each others' experiences outside the scheduled training sessions.
- Inadequate attention was paid to the economics of introducing certain practices.
- Much of the training was based on implementation of recommended 'packages', ignoring the possibility of stage-wise 'try it and see' approaches.

- Inadequate attention was paid to the economics of introducing certain practices.

PHASE II 1987–9

AKRSP began to realize that improvements to the quality of training programmes conducted by GOs required better understanding of farmers' opportunities, aspirations and constraints, and the articulation of these by farmers themselves into learning processes. However, it also realized that it would be in a weak position to persuade GOs of the merits of this strategy if it had not experienced them itself at first hand. AKRSP therefore began to develop a methodology of participatory training and extension, the key components of which were appraisal and technology identification, technology adaptation and testing, and technology diffusion.

Appraisal and technology identification

Farmers and AKRSP jointly make a village resource inventory, identifying the type and extent of diversity in natural and human resources and the options they offer for technology and management practices. The village is requested to nominate two or three extension volunteers, who take the resource inventory further through relevant participatory mapping, transect, diagramming, ranking, scoring and discussion group techniques. These volunteers are also given the key task of representing villagers' viewpoints in more formal training programmes, and, as part of the preparation for these, elicit farmers' priorities regarding information exchange and training needs.

The volunteers engage in 'experience sharing' at two levels: first, within their own villages they present an overview of their findings and invite feedback from villagers; second, the volunteers from several villages meet in each village in turn, presenting findings from their own village and highlighting any innovative or unusual practices that they noted. In this way, volunteers' capacity is developed to analyse, discuss and articulate a viewpoint. They also learn how to stimulate a dialogue in the context of an 'enabling' environment.

Training programmes are then organized to address the priorities identified by farmers against a background of detailed knowledge of resource endowments, opportunities, aspirations and constraints. Each training session relies heavily on audio-visual aids and on practical exercises in observation and analysis, and is followed by participant feedback on individual components of the programme.

Technology adaptation and testing

Further aspects of the training programme include visits to typical fields in each in order to identify with villagers those technologies that are feasible from the range of options recommended by research and extension; discussion in each village of technologies being explored in other villages and the selection from these of options likely to be relevant; the setting up of small experimental plots on farmers' fields having different characteristics in order to test new recommendations from researchers; the division of responsibilities for managing and recording these experiments, and the discussion of group action for those technologies (e.g. integrated pest management) where individual action would be costly or ineffectual.

Technology diffusion

Farmers become aware of the performance of these experiments through casual observation, but end-of-season field days and group discussions provide a more rigorous assessment and stimulate decisions among farmers on how far they will implement specific technologies. Group-based decisions are particularly important in those operations which require community action in order to be effective (e.g. integrated pest management; watershed management) or to be economic (e.g. shared use of large equipment).

AKRSP is working with this three-stage approach in 120 villages of widely differing agro-ecological and socio-economic characteristics. The success of the approach lies in the fact that it is iterative, participatory and process-oriented. A high degree of accountability is also built in: extension volunteers are paid by village institutions, not on a fixed salary, but according to the village's assessment of their performance.

The village institutions are formed around a village natural resources management plan, prepared by the community as a result of a participatory appraisal exercise. The village institution decides to implement the activities identified as a part of the plan and comprises different groups of villagers who decide to come together for implementing the plan. These institutions are watershed development associations, water users' associations, forest protection and development committees, village development associations, women's small farm development associations, lift irrigation co-operatives, dairy co-operatives, savings, credit and marketing associations and federations. These village institutions are involved in an ongoing process of implementation, management, monitoring and evaluation. These institutions have been led by individuals who command widespread respect on account of their expertise in specific areas, and in three-quarters of the villages where AKRSP is working, have been able to prevent domination by political authorities.

These village institutions work through a number of para-professional, village extension volunteers who undertake the roles of facilitators, service providers, managers and networkers for lateral extension and farmer-to-farmer learning. These volunteers are paid by the village institutions based on their performance in delivering services, appraisal, generating financial resources, managing village institutions and increasing income for the members. These institutions evolve the performance-oriented incentives in relation to the work done and the likely economic benefits for the programme. The village institutions generate a capital fund for the village as a part of the savings from increased income and make payments from these reserves. This is accompanied by a process of participatory monitoring and analysis which allows incentives to be linked to income generation on a continuing basis. The extension volunteers have to make regular presentations to the village institutions on their performance.

PHASE III 1989 ONWARDS

Once AKRSP had developed participatory methods for stimulating farmers' awareness of technology options to meet their needs, it began to introduce this methodology to GOs responsible for extension and training. The principles it stressed included pre-training of trainers at village level and their involvement in participatory appraisal; introduction of participatory, dialogue-oriented training methods; use of farmers as extensionists and trainers; the need to enhance farmers' capacity to articulate requirements and provide feedback, and increased client-orientation of the training and extension process.

Realizing that GO trainers did not have the authority to introduce such radical changes, AKRSP began to demonstrate the advantages of these changes to senior administrators in the state-level directorates of agriculture and of extension education, and in the university. In coalition with other NGOs working in Gujarat (including BAIF and Lok Bharti Sanosara), AKRSP then ran an 'exposure workshop' on the training of farmers (May 1989) for senior administrators of all thirty-four training institutes within the state. The workshop, also involving a number of farmers, sought to dissuade trainers from promoting a single technique or practice by lecturing and, instead, to identify options with farmers, stimulating processes of experimentation, dialogue and feedback, and recognizing the value of farmer-to-farmer extension.

At a follow-up meeting in September 1989, a number of practical outcomes from the May workshops were discussed: the state government of Gujarat had formally sent the recommendations of the May workshop to departmental heads, requiring them to follow up; the director of agriculture proposed to increase the Eighth Plan budget for recruiting trainers in order to raise overall standards. The director of extension education at the Gujarat Agricultural University agreed, pointing out also that the better recruits almost invariably wanted to become research scientists, not trainers. Other participants stressed the desirability of a research–extension–training continuum; several experiences were reported in which training institutes had benefited by observing how nearby villages had responded to various recommendations made by training staff. It was resolved that each training centre would identify in a preliminary way the diversity of their client groups on agro-ecological and socio-economic criteria and develop a range of training programmes according to needs.

A second 'exposure workshop' was held over two days in January 1990 at Gujarat Agriculture University. Participants were divided into three groups, each focusing on an aspect of agriculture (seed technology; cotton production; animal breeding) and devising training programmes for different farmer groups within these themes. Particular attention was given to criteria for deciding which parts of a training course should be held at a training centre, and which at field level; selecting farmers to stimulate farmer-to-farmer extension and deciding which farmers should receive training at the centre as against in the field.

The status of this exposure workshop was enhanced by the presence of national-level dignitaries at the opening and closing ceremonies. In addition to consolidating the resolutions of earlier meetings regarding more participatory and sensitive approaches to training, the workshop resolved:

- that the thirty-four training centres in Gujarat should be responsible for training, extension and input supply in a co-ordinated fashion
- that some one to three villages should be adopted by each training centre to demonstrate how productivity can be increased in a practical manner
- that each training process should involve components of marketing and production economics
- that the trends towards more focused and village-based training, and greater use of farmer-to-farmer extension should continue
- that trainers should seek out innovative practices introduced spontaneously by farmers and communicate these to research centres
- that farmers should be used as resource persons in all future training programmes.

At a third exposure workshop in January 1991, a Krishi Vignan Kendra based at Deesa gave a presentation on the participatory approaches to identifying training needs that it had developed. AKRSP gave a presentation on participatory rapid appraisal methodology, and five farmers from various villages gave their reactions to earlier training efforts. Group discussions led to the following recommendations:

- further consolidation of participatory approaches to training
- the establishment of thematic networks among trainers, crossing conventional NGO–GO barriers, which was intended to consolidate newly emerging professionalism in approaches to training, since contacts among trainers within specific institutes had hitherto tended to concentrate on the achievement of targets and administrative issues
- changes in the organizational structure of training institutes
- changes in evaluation mechanisms.

The principal lessons derived from AKRSP in introducing to GOs more participatory approaches to farmer training and extension are as follows:

- First-hand experience of the methodology is essential before attempts are made to introduce it elsewhere.
- Time must be spent explaining the 'why', especially to senior GO administrators, in order to create an enabling environment before the 'how' can be addressed.
- The creation of a semi-permanent consultative committee bridging across NGOs and GOs was an important factor facilitating the design and implementation of 'exposure' workshops.
- The creation of a number of informal possibilities for networking on a thematic basis among professionals from different types of institution was important in stimulating readiness to discuss and experiment with new approaches.
- While process-orientation is important in changing the approaches and ethos of GOs, there comes a point at which these have to be tested on a large scale and, if successful, scaled up and replicated.

BHARATIYA AGRO-INDUSTRIES FOUNDATION (BAIF)

Research programmes in livestock production, health and nutrition

S Satish and John Farrington

INTRODUCTION

Origins

BAIF, a non-profit development research foundation, was founded in 1967, at Urulikanchan, near Pune. It seeks to raise incomes and employment among the rural poor through the application of science and technology. From modest beginnings with a dairy cattle production programme, BAIF later diversified to embrace animal health, nutrition, afforestation, wasteland development and tribal rehabilitation, and currently employs a staff of almost 3,000. BAIF sees its development research as 'based on technological research designed in response to field realities'.

In its cross-bred cattle programme, BAIF pioneered research into aspects of artificial insemination (such as the freezing of semen) but, equally importantly, developed village-based delivery systems, known as cattle development centres, to carry out the inseminations, provide feedback on the progeny of individual bulls and supply related services in animal health care and nutrition. Through excellence in problem-focused research, BAIF has gained recognition by several international and Indian government agencies.

This recognition has been instrumental, first, in the forging of links between BAIF and the wider scientific community, and second (as we discuss below), in enabling it to secure a funding base within and outside India. Particularly important for fund-raising within India is the fact that it is recognized as a research organization by the Indian Council of Agricultural Research (ICAR) for conducting both *ad hoc* and co-ordinated research.

Organizational set-up

BAIF's trusteeship is vested in a board of trustees. The overall operational responsibility including issues of policy and strategy, the approval of new research projects and of papers for publication rests with the President's Council (PC), which derives a strong technical and scientific support from the

Advisory Council of Scientists. The PC operates informally through discussions in small sub-groups at different points of time. Review of progress and corrective measures, if need be, are the responsibility of the Central Management Council (CMC) which consists of all the heads of divisions and associated organizations. BAIF's structure at village and area level is discussed on pp. 138–9. BAIF currently employs over 3,000 persons, of whom one-third are professionally qualified, including 8 at PhD level, 45 with other postgraduate qualifications, 170 graduates and a similar number of diplomats.

Since the early 1970s, the recruitment of professionals to BAIF has undergone several shifts. The initial focus on veterinarians was, first, broadened to include other agriculture graduates. Difficulties in retaining graduates in the face of attractive offers from elsewhere led to the recruitment of technicians and the provision of in-house training.

We shall review BAIF experience and its links with public sector research and extension institutes in three related areas of R&D: cross-breeding for dairy improvement, vaccine production, and animal nutrition.

CROSS-BREEDING FOR DAIRY IMPROVEMENT

Evolution of the cross-breeding programme

BAIF's interest in raising the productivity of dairy cattle began with the work of its president, Manibhai Desai, in the 1950s at what is now the location of BAIF's research farm near Pune. Desai's early work focused on improved nutrition and health care for local, nondescript cattle. He subsequently switched attention to the genetic limitations inherent in local stock. Of the twenty-six recognized breeds of cattle in India, four are noted for their milk production, but 80 per cent of cattle in India are nondescript and produce only around 30 per cent of the yields of these breeds. Encouraged by his early success in cross-breeding, Desai established BAIF in 1967 and began to seek funding from internal and external resources for a wider effort.

Cross-breeding induces faster reproduction, early maturity and high milk production in the nondescript cow. A cross-bred cow at village level yields 1,600–2,200 litres per lactation of about 250 days (Deore 1990). However, the optimal percentage of exotic blood has to be a carefully researched compromise between yield and resilience, and will vary according to local conditions affecting health and nutrition. Cross-breeding on a wide scale requires adequate artificial insemination (AI) facilities. In the late 1960s in India, these used unreliable liquid semen techniques, hence the plans to develop frozen semen alternatives.

The establishment of BAIF in 1967 marked the beginning of a period of intense activity to secure land (eventually donated from local councils), to design and construct office buildings and laboratories (with the assistance of the India Dairy Corporation), to identify sources of staff salaries (from *ad hoc* donations by commercial companies in the early period) and to travel abroad in order to acquire the breeding stock (initially 200 heifers from Denmark) and subsequently the equipment and knowledge for frozen semen production on a commercial scale.

The semen freezing laboratory established by BAIF in 1975 with DANIDA assistance meets international standards for quality control, and some 1.5 million doses are produced annually, the majority of which is used in BAIF's own cross-breeding programmes at village level, but with the sale of some surpluses to GOs. Parallel research has been conducted on such other issues as semen collection and freezing methods, and influence of nutrition on semen quality, and on 'feedback' information on the field performance of AI.

BAIF has a unique advantage over GOs in so far as three functions (research, dissemination and implementation) which are often administratively and institutionally distinct in the public sector, are integrated into a single organization. BAIF implements AI through over 500 local cattle development centres in six states covering 1.5 million families. Their origins lie in a model developed at the Pravranagar Sugar Co-operative Factory in Maharashtra in the early 1970s. Farmers (i.e. co-operative members) were assured such services as artificial insemination, pregnancy testing, clinical diagnosis and training in calf management. In return, a farmer was required to pay one rupee per animal per month. The development centre was designed to have 1,500–2,000 breedable cows within a radius of 10–12 km. Each staff member at the centre sought to carry out ten AIs per day with a 60–70 km round trip. The initial conception rate was estimated at about 40 per cent and has now risen to over 60 per cent. BAIF relied heavily on local sponsorship – from sugar cane factories, dairy societies and milk federations, especially in the 1970s. The pattern of this early sponsorship was such that BAIF's cattle development programme was initially confined to the irrigated areas inhabited by relatively rich farmers.

The areas proposed for cattle development by BAIF in the six states in which it operates are initially identified by the respective state government. BAIF examines them against criteria such as facilities for transport, marketing and education, and the scope for operating programmes in clusters of adjacent villages.

Performance targets for each cattle development centre are fixed under a tripartite agreement among the BAIF, and administrations at state and district levels. The farmer is not required to pay for AI services: funding is through a range of development programmes and its level is linked to the performance of individual centres. BAIF's field programmes operate through a structured hierarchy:

- *Cattle development centres*, managed by a rural development officer, are the smallest operational unit, catering for 1,500–2,000 breedable cows in a 10–

15 km radius and responsible for AI, extension programmes and field level data collection.

- *Area-level programmes* co-ordinate the activities of several centres and are responsible for implementing demonstrations and training.
- *Zone-level programmes* typically cover five areas. They act as resource centres for area programmes and follow up the needs and opportunities identified during monitoring.
- *Regional-level programmes* co-ordinate input procurement for area and zonal programmes and are responsible for social science surveys of field activities and documentation.

Progeny testing within the cross-breeding programme

The Progeny Testing Programme (PTP) is an *ad hoc* project on a co-ordinated basis, sponsored by ICAR. Participating institutions include BAIF, Punjab Agricultural University (PAU) and Kerala Agricultural University (KAU). The programme started in 1985, and aims at assessing the performance of progeny to help in the selection of bulls. It is an important element of AI quality control. The BAIF programme involves recording the milk yields of over 3,000 cows, standardizing the procedures of data collection, and identifying and estimating the contribution of non-genetic factors to performance. Average milk yields recorded among the BAIF, PAU and KAU cows under the PTP were 2,400 kg/year, 2,100 kg/year and 1,300 kg/year, respectively. Under a semen exchange programme recommended by the Indo-US Subcommission on Agriculture (GoI 1988), farmers showed a distinct preference for semen provided by BAIF over that of PAU and KAU.

Research sponsorship and relations with the public sector

BAIF's links with ICAR facilitate its fund-raising from private commercial companies. The GoI provided a 100 per cent rebate of income tax on donations made towards approved R&D (recently raised to 133 per cent) providing that the recipient has been certified by a competent authority (in BAIF's case, the ICAR) as having adequate skills and facilities.

Having obtained this approval, BAIF subsequently received donations from a number of companies (e.g. Mafatlal Industries and Asian Paints) to conduct research on e.g. genotype–environment interaction, the utility of cross-bred bulls and the evaluation of trace element status in dairy cattle and bulls. Much of this research was conducted in a combination of on-farm and on-station research, an approach to livestock research which BAIF has pioneered in India.

BAIF research challenged conventional wisdom in several ways. For instance, the ICAR view in the 1970s had been that imported semen performed better in AI than locally obtained semen. BAIF's data on conception rates and progeny testing demonstrated that this was not the case at the small-farm level. BAIF's data withstood the scrutiny of a high-level ICAR team, and their conclusions were upheld, albeit after almost three years of debate with ICAR.

VACCINE PRODUCTION

In response to the susceptibility of cross-bred animals to foot and mouth disease (FMD) and the high prices of vaccine (Rs14 per dose) produced by a multinational company, BAIF succeeded in obtaining a DANIDA grant for a programme which allowed it to produce 8 million doses by the late 1980s at Rs3.50 per dose.

At present, altogether sixty-five different biological products are manufactured by BAIF's animal health subsidiary, the Bharatiya Research Institute for Animal Health (BRIAH) with an annual turnover for 1989–90 of Rs50 million. The future possibilities of developing vaccines for sheep-pox, goat-pox, infectious bronchitis, infectious laryngotracheitis and infectious bursal disease are currently being researched. A disease surveillance programme involving 50,000 bovines has been initiated to standardize the field reporting system for screening and monitoring diseases. As a part of the programme, 8,175 farmers have been trained in disease surveillance and control through 226 training programmes.

Numerous other technologies have been developed by BRIAH since the late 1970s (see Satish and Farrington 1990) including vaccine potentiating agents, vaccine concentrating, freeze-drying and reconstituting technology, standard inoculants for legumes and delivery systems for biological products. One of these technologies – aluminium hydroxide geladjuvants – saves Rs10 million in foreign exchange each year.

BRIAH products are used by BAIF, and by a number of GOs, and since 1980 have been exported to over ten other developing countries. Their development involved collaboration with several government research institutes which has generated practical outcomes including, for instance, the adaptation of a polymer-based product (*jalashakti*) originally designed by the National Chemical Laboratory for moisture retention in plant production into a thermally insulated packaging for vaccines, resulting in a cost reduction from Rs40 to Rs7 for the packaging of 200 doses of FMD vaccine.

ANIMAL NUTRITION

Cereal straw has conventionally been a major dietary component of Indian cattle. However, it has numerous nutritional shortcomings. Urea treatment helps to overcome some of these difficulties, but is uneconomic at farm level. BAIF is one of the fifteen participating institutions in the ICAR's All-India Coordinated Research Project on forage crops. The project has been instrumental in stimulating the inclusion of edible by-products as a criterion in screening new crop varieties.

This and other BAIF efforts focus on non-traditional fodder crops and multipurpose trees to meet the requirements of year-round production; minimal managerial requirements; high crude protein and dry matter content; easy digestibility and drought resistance.

BAIF was among the first organizations in India to conduct trials on *Leucaena leucocephala* as a multipurpose tree. Planting material was obtained from Hawaii for observation plots in 1976, and following the assembly of technical information (including a visit by senior staff to the Philippines) BAIF progressed to varietal trials and large-scale multiplication on farmers' fields. BAIF's programme included a trial in which bull-calves were fed exclusively on *Leucaena* for over two years in order to identify whether mimosene or dihydroxypyridine (DHP) toxicity was likely to be problematic. No ill-effects were observed. After visiting this trial in 1978–9, scientists from CSIRO, Australia, initiated joint experiments with the University of Hawaii and succeeded in isolating DHP-degrading bacteria from the rumen of goats maintained on diets of *Leucaena*. A subsequent survey suggested that these naturally occurring bacteria are found in Hawaii, Indonesia, Philippines, India and several other countries, but absent from Australia, Papua New Guinea and parts of Africa. In 1982 the bacteria were successfully introduced to Australia and by the mid-1980s a series of papers had been published internationally outlining simple methods for testing for mimosene and DHP toxicity, and for introducing the necessary bacteria into areas where persistent toxicity indicated their absence.

Despite the pioneering efforts of BAIF to test for toxicity through long-term feeding trials in India, and despite the publication of some 300 research papers on mimosene or DHP toxicity in the 1980s, articles are still appearing in India (e.g. in *Wastelands News*, 1989) conveying alarmist views of toxicity problems in *Leucaena*, and suggesting that 'the possibility of presence of micro-organisms [i.e. toxin-degrading rumen bacteria] is highly controversial'.

This episode illustrates numerous problems within the Indian scientific establishment:

- Its reluctance to accept the findings of research conducted by an NGO.
- Its apparent incapacity to test for the presence or absence of relevant bacteria in India through large-scale surveys.
- The resultant absence of clear guidelines published under the auspices of the ICAR on the toxicity 'problem'; its incidence; detection methods and methods of stimulating the natural spread of relevant bacteria.
- The low-level of familiarity among many scientists, whether in government institutes or NGOs, with current international literature and the limited ability to place their own work in this wider analytical context in order to avoid uninformed conclusions.

5 A tendency to make inappropriate suggestions for future research: the 'Wastelands Development' article suggested as a priority the development of low mimosene varieties of *Leucaena*, whereas work elsewhere has shown that no stable lines of *Leucaena* are known with less than 2 per cent mimosene (against 4–5 per cent normally), which is not low enough to avoid DHP toxicity in animals lacking the appropriate bacteria.

Overall, the episode illustrates the capacity of BAIF to apply research to specific issues in India in a fashion well-informed by wider literature and underlines the potential value of BAIF's Information Resource Centre.

This work on multipurpose tree species has been supplemented by investigations into soil micro-organisms (specifically, *rhizobium* culture) capable of promoting nitrogen fixation under conditions of low soil fertility such as are found on wastelands. BAIF's *rhizobium* suitable for *Leucaena* has been officially recognized by the GoI, and over 0.5 million doses have been distributed. Further work in this field is being conducted on *mycorrhiza*, and inoculants have been developed for a range of oilseed crops. The Ministry of Agriculture has provided funding for the large-scale production by BAIF of biofertilizers and inoculum.

CONCLUSIONS

BAIF's cross-bred cattle development programme, with related nutrition and animal health components, covers some 1.5 million households in six states of India and produces some 200,000 cross-breeds through AI each year. In aggregate, it has produced almost 10 per cent of the current national stock of some 9 million cross-bred dairy cows.

Its particular advantages as an NGO include the following:

- The integration of research, dissemination and implementation into a single organizational structure.
- The ability to tailor research to field-level problems in an integrated fashion. The clearest illustration of this is the integration of breeding, nutrition and health-care.
- Strong emphasis on feedback – both positive and negative – from village level. The progeny testing programme is a highly structured and quantified example of feedback and has been managed better than those of many government institutes concerned with AI (GoI 1988). Other, less-formalized, examples of feedback exist which have influenced subsequent research.

BAIF's relations with government have been mixed. On the positive side, many BAIF senior scientists command wide professional respect, and have good contacts with ICAR institutes, not least because some of them worked there previously. These strengths have been instrumental in gaining the necessary official recognition for BAIF to receive private sector donations under tax-relief schemes. Similarly, BAIF's strengths in integrating research with implementation have brought substantial funding from state-level development programmes. At the scientific level, certain BAIF findings (e.g. concerning the acceptable performance of locally produced semen) have been accepted by ICAR, and some of BAIF's proposals (e.g. for disease monitoring) have subsequently been taken up.

There are, however, two negative aspects to the relationships between BAIF and government. First, learning by BAIF from the public sector has been limited, partly because research co-ordination in India continues to be inadequate, partly because of the limited relevance of much work conducted at ICAR institutes and in Agricultural Universities to the practical context of development. This disjointedness becomes particularly costly in the more advanced techniques currently being researched. Thus, BAIF plans to research Multiple Ovulation Embryo Transplant technology, but so do several other institutes, and the benefits of co-ordination in a way more imaginative than that characterizing most All-India Co-ordinated Research Projects seem self-evident.

Second, there remains in some government research institutes considerable mistrust of work done by BAIF. Some see it as primarily a development organization lacking research pedigree. Certainly, the higher priority given by BAIF to achieving implementable results than to producing publishable material is in marked contrast to behaviour patterns in many Indian public sector research institutes. There appear, however, to be few (if any) examples of sub-standard research conducted by BAIF.

In many respects, BAIF is an unusual NGO – exceptional in the charismatic qualities of its founder, and in the research capability that it has been able to develop. Some would argue that these characteristics cannot be emulated by other NGOs. But such an assessment of BAIF is excessively restrictive: perhaps the most important difference between BAIF and other NGOs lies in its confidence in technological interventions as a means of achieving sustainable improvements in livelihood, with minimal attention to the social organizational issues central to the philosophy and practice of many NGOs. Whether such improvements reach the rural poor and can, in fact, be sustained, requires a fuller evaluation than can be provided here, but even the *possibility* that they may have done so suggests that two distinct lessons can be learnt from BAIF.

First, NGO programmes seeking technical improvement in agriculture (and there seems little assurance that social organization and conscientization programmes – though important – are alone sufficient to achieve livelihood improvement) require professional skills and careful, systematic monitoring and evaluation. If they cannot provide them from their own resources, NGOs must carefully consider whether, and, if so, from where, such skills can be drawn in from outside. They must also consider the structural and organizational measures necessary for implementation of systematic procedures.

Second, government research and extension services will achieve less than their potential impact unless they are organized increasingly around problem-oriented research, involving greater collaboration among disciplines and stronger integration within the research-dissemination-implementation chain.

These are likely to be the most enduring lessons of the BAIF experience.

GREENWORK AT AUROVILLE

From survival to inter-institutional collaboration

Edward Giordano, S Satish and John Farrington

INTRODUCTION

Auroville – the 'city of dawn' named after the Indian Philosopher Sri Aurobindo – is an experiment in living dedicated to the promotion of peace and international understanding. Auroville was founded in 1968 on the Eastern Coast of India, near Pondicherry. The 1,000 ha of land occupied by Auroville's 750 adults, initially severely degraded, has now been planted with over 2 million trees. Land management approaches integrate the use of trees, animals and annual crops over individual watersheds. As the trees grew, new micro-climates were formed, and many animal species returned. The first bird count made in Auroville listed 40 species in 1972 against 104 confirmed species in 1990.

Twenty years after the first settlement at Auroville, the Auroville Greenwork Resources Centre was established as a means of testing the principles and practice of sustainable resource management that Auroville has gained against the opportunities and constraints – whether agro-ecological or socio-economic – existing within local communities. AGRC's mandate includes the development of silviculture and agroforestry in relation to seasonal agriculture; the reintroduction of indigenous rice strains in the context of a biological agriculture programme; the development of techniques for land reclamation and integrated soil and water conservation, and their promotion through nursery establishment, seedling supply, farmer training, programmes for schools and local associations. AGRC's wider mandate includes the publication of an *index seminum*, the provision of advisory services to and running of workshops for GOs and other NGOs, and the international search for appropriate seed varieties.

Greenwork challenges

Numerous challenges have been encountered by Aurovillians in resource management over the 1970s and 1980s:

- *Technological challenges* include the identification and testing of technologies for soil and water development and conservation.
- *Social and cultural* soil and water management technologies introduced by AGRC treat each watershed as a single entity. Hence, the social, cultural and economic conditions of the twenty local villages situated within Auroville's watersheds have to be taken into account.
- *Organizational challenges* include developing a network of contacts with research, extension, and development agencies, both GO and NGO, in India and abroad, for first, drawing on ideas, skills and methods, and second, for providing wider dissemination of Auroville technologies.

We shall document Auroville's responses both to the wider challenges and specifically to those posed by afforestation and biological agriculture.

AUROVILLE'S AFFORESTATION PROGRAMME

An overview

In 1968 the Aurovillians initiated programmes of fencing (using mainly sisal plants), borehole sinking (to 24 m depth), nursery establishment and tree planting. Experiments were also conducted on means of catching and controlling monsoon runoff to check soil erosion and facilitate infiltration, including the establishment of bunds, check dams and catchment ponds. Numerous other soil and water conservation efforts are now underway in India. Many of them started in the mid-1980s, that is several years after Auroville's work began. Unfortunately, exchange of information among those efforts has been extremely limited, so that early errors continue to be repeated elsewhere.

Many ornamental tree species tried in the initial stages proved unsuccessful. Eventually, 100 species from the 400 initially tried were selected for multiple characteristics such as drought resistance, fast growth, capacity for soil enrichment, and timber, fodder and green manure. In the early period exotics were favoured because certain species proved to be excellent pioneers. More recently, however, the focus shifted back to indigenous trees that temple scriptures and old Tamil poetry showed to have been common until about 200 years ago. Auroville's international *Index Seminum* exchange programme has helped in obtaining the species that became extinct. However, experiments revealed that not all the old varieties can be readily reintroduced due to environmental changes. Experience has led to concentration on two broad tree management strategies: agroforestry and silviculture.

Auroville and government programmes

An eminent scientific adviser to the GoI, Professor Madhav Rao Gadgil, visited Auroville in 1981 as part of his mandate to monitor forestry activities. Gadgil was impressed by Auroville's mixed species afforestation programme which contrasted with official programmes of planting exotics – particularly eucalypts – and his visit led to collaboration between Auroville and the Department of Environment (DoE), through a Rs1 million (then US$100,000) five-year project (1982–7), to plant trees and monitor performance so that the most appropriate techniques and species for different situations could be identified.

This was the beginning of a new orientation for greenworkers in Auroville, since it demonstrated that they had something to offer outside their own boundaries. Under the project some 300,000 seedlings belonging to twenty different species were planted (with a survival rate of 60–70 per cent) and their performance monitored. The project provided for collaboration with the Indian Institute of Sciences in Bangalore. However, this was limited by logistical problems, and most of the work had to be undertaken by Aurovillians.

Other collaboration has also been implemented; for instance, the Ministry of Environment recently approached the AGRC to take up a project on environmental awareness, but approved only Rs30,000 against the requested Rs150,000. Auroville's experience reveals that both priorities and assessment criteria change frequently in government circles. However, as with BAIF (see pp. 136–44) projects and eligibility for tax exemption from the GoI are sought after by AGRC as these imply government recognition of Auroville as a research and development organization, and this helps in mobilizing funds from other sources.

While some success has been achieved in obtaining foreign funds for Greenwork, severe delays have also occurred (e.g. a decision is still awaited on a Dutch 'Global Forest Foundation' proposal submitted in 1988), and others have failed to get off the ground through lack of agreement with local agencies, including, for instance, a social forestry project submitted to the Swedish International Development Agency (SIDA) by the Tamil Nadu Forest Department, with dissemination and feedback to take place through local NGOs such as Auroville. However, numerous attempts by SIDA and local NGOs to define modalities were rejected by the Forest Department, partly because of its reluctance to cede control of resources, partly because foresters perceived NGOs' work as threatening to their own status and aspirations.

By contrast, AGRC's informal relations with GOs tend to be cordial. The Forest Departments of Karnataka, Kerala, Bihar, Tamil Nadu, and Andhra Pradesh, for instance, obtain seed supplies regularly from Auroville. The botanical garden (Department of Agriculture) in Pondicherry and AGRC have a long history of seed exchange. Fencing plants and bamboo seeds from AGRC are exchanged for the botanical garden's mango and African mahogany. Following a visit by the project director, District Rural Development Agency (DRDA), Pondicherry, AGRC was requested to conduct programmes to train social forestry officials of the DRDA. In addition, Auroville is also represented in some of the advisory bodies and working groups of Pondicherry government.

Auroville, NGOs and R&D institutions

AGRC's desire to develop technologies relevant to low-income farmers in the surrounding villages led to a project comprising training programmes for fruit tree production, environmental education, soil and water conservation and health and sanitation with assistance from the Threshold Foundation (USA) and the Commonwealth Human Ecology Concern (CHEC) funded by the Overseas Development Administration (ODA, UK).

In preparation for the project, AGRC staff received training from private horticulturists in Bangalore, and a one-day course was provided by the Department of Horticulture in Bangalore, with a further three-day course at TNAU.

AGRC has sought not only to use these methods and technologies on its own and on neighbouring village land, but also to stimulate their introduction and adaptation by others. For instance projects sponsored by Swiss Aid brought two groups of twenty-five tribal and field-workers representing different NGOs from six states – Andhra Pradesh, Madhya Pradesh, Maharashtra, Orissa, Rajasthan and Gujarat – to Auroville for intensive training programmes in 1987 and 1988. The trainees from Rajasthan representing People's Education Development Organisation (PEDO), on their return planted over 2 million trees between 1987 and 1989. In 1989 AGRC was requested by PEDO to train fifty-four tribal people in Rajasthan.

Research and educational initiatives

The research and educational initiatives taking place in Auroville include AGRC's association with 'Agriculture, Man and Ecology' (AME), a Dutch-based NGO in Tamil Nadu seeking to train NGOs, forest department officials, academics, and administrators from Asian countries in biological farming and afforestation. AME draws on the staff of AGRC and uses the infrastructural facilities in Auroville for its training programmes. For effective technology transfer, eco-clubs (at present totalling ten) have been organized in the local villages and a series of illustrated textbooks and posters for environmental teaching in primary schools have been produced.

Auroville's technological success has created a demand for the seeds of such trees as teak, rosewood and acacia species, to the extent that seeds sold by Auroville are often resold at a premium. The increased demand for seeds prompted the publishing of *Common Trees of Auroville*, which lists 120 commonly grown trees in Auroville with a short description and details of propagation. The *Index Seminum* lists over 300 tree species, the seeds of which are offered for exchange. This index has 450 correspondents (botanical gardens and universities) participating internationally in a seed exchange programme.

Pondicherry Central University has requested AGRC to be advisers in horticulture and ecology. This association has had wider spin-off in the form of collaboration between AGRC and the University's School of Ecology in undertaking a research study 'Ecological, Economic and Energy Auditing of some Fast Growing Trees' in Auroville. The study's objectives include monitoring of growth rates, and energy input and output assessment.

AUROVILLE'S BIOLOGICAL AGRICULTURE PROGRAMME

Objectives and approaches

Agriculture in Auroville was initially limited to areas where enough top soil still existed. Integrated soil and water conservation measures were a first step towards rehabilitating agricultural lands, and comprised the planting of leguminous hedges, the use of farmyard manure, and systematic rotations. Attempts to grow leucaena, despite initial problems, have now succeeded with the use of rhizobium culture products by BAIF (see pp. 136–44).

Experimentation with indigenous rice strains

High-yielding varieties of rice proved impossible to grow at Auroville because of inadequate water supplies. A literature review revealed that 4,300 indigenous strains of rice had been identified in India. Efforts were therefore made to collect indigenous strains to test for capacity to withstand moisture stress; disease and pest resistance; palatability; nutrition, including protein content; digestibility, and medicinal quality; cooking quality and cooking time.

Systematic experiments are now being conducted with seventy-five varieties under a range of management practices, including zero tillage and land regeneration practices through mulching; soil salinity correction through biological methods; variations in seed rate and plant spacing; variations in irrigation scheduling and water supply during critical growth stages. Many ideas have been drawn from local farmers, including the spreading of casuarina needles over the soil to reduce salinity. The experiments on upland rice are being conducted with technical assistance from the Dry Land Research Institute of the TNAU and some of the results have caught the attention of breeders over a wider area.

AGRC and local farmers: participatory action research programme

The need for systematic efforts to develop technology packages for use on the fields of the surrounding villages, has led to a participatory action programme.

Twelve 'nodal' farmers were brought to AGRC in 1990 to be trained for twenty-one days spread over three to four months.

The training was highly practice-oriented and broad-based, covering basic principles of organic farming, soils, seeds, interplanting tree crops, mulching, cover crops, farm planning and budgeting, energy usage, and allied activities. The second phase of the programme envisaged observation and identification of which technologies farmers adopted and why. A cluster of farmers was chosen around each nodal farmer for discussion of technology options. The technologies/approaches emerging out of the discussions are tried out on the nodal farmer's land. It is postulated that the proven technologies will be disseminated easily and quickly by farmer-to-farmer contact within and beyond the cluster. In this programme, funded by German Agro Action, the cluster approach will be replicated widely if it performs adequately.

Food co-operatives

A further scheme, the 'Organic Food Coop' has been devised to test and promote organic farming technologies on farmers' fields. Farmers are mobilized into a co-operative and each member gives an undertaking that no agrochemicals will be used for a minimum of five years. In turn, AGRC provides technical services, compost and manure, helps to bund the fields and guarantees to buy the product at remunerative prices. During the five years, if any crop failure occurs for two consecutive years, some payment will be made to ensure that basic needs can still be met. The scheme is based on the confidence in organic farming technologies developed by AGRC and on a growing demand for organic products locally in Auroville and Pondicherry and also in Europe and the USA. Organic cashew production has become popular under this scheme, but a number of problems persist, including the difficulty of ensuring that agrochemicals are not used, and of certifying the 'organic' character of the produce in ways acceptable to international markets.

CONSTRAINTS, PERSPECTIVES AND CONCLUSIONS

Auroville's Greenwork has slowly evolved into a compendium of appropriate methods and technologies which in recent years it has increasingly shared with GOs and other NGOs.

To some extent Auroville is a victim of its own success: its lands are no longer eligible for 'wasteland' grants, and efforts are now in hand to cover the Rs300,000 per year expenditure on maintenance of AGRC infrastructure and experiments by charging for its services on a 'full-cost' basis.

The principal remaining question is how Auroville's experience can be 'scaled up'. This can be broken down into several components:

- What mechanisms are needed to ensure that Auroville's knowledge on species and management practices can feed into government programmes for wasteland rehabilitation at national, state or district levels?

- What mechanisms need to be set up to feed this information into government research programmes, whether at ICAR institutes, SAUs or elsewhere? What hypotheses for research does it raise? How relevant is 'official' research to some of the issues faced by Auroville and other NGOs? What skills or facilities could Auroville usefully draw upon from specialist institutes? Can NGOs and GOs work together in collaborative programmes of technology testing and feedback?
- What mechanisms need to be established to ensure stronger flow of information among NGOs? Auroville's experience is 'on offer' to other NGOs. Do they have experience that would be useful to Auroville and others? Is there a need (and possibility) for some specialization of tasks/activities among NGOs – or perhaps for some informal umbrella group to provide interaction among them?

Answers to these questions will not be found overnight, but they need to be at the forefront of discussions and inter-institutional negotiations if valuable experience of the type described here is to be used to maximum effect.

MYSORE RELIEF AND DEVELOPMENT AGENCY (MYRADA)

Participatory rural appraisal and participatory learning methods

Aloysius P Fernandez and James Mascarenhas

ABOUT MYRADA

MYRADA is an NGO which has been involved in rural development since 1968 but now works in approximately 2,000 villages in South India, in the states of Karnataka, Andhra Pradesh and Tamil Nadu. It initially started as an organization which resettled refugees from Tibet. Its mission statement of 1985 is:

> To foster a process of ongoing change in favour of the rural poor in a way in which this process can be sustained by them through:
> - Supporting the rural poor in their efforts to build, appropriate and innovative local level institutions rooted in traditional values of justice, equity and mutual support.
> - Working towards recreating a self-sustaining habitat based on a balanced perspective of the relationship between natural resources and legitimate needs of the people.
> - Influencing public policies in favour of the poor.

Today MYRADA has six major programme thrusts.

- Participative resource development and management projects (particularly in semi-arid areas). These include wastelands and watershed development programmes.
- Resettlement and rehabilitation of released bonded labour and landless families.
- Development of women and children in rural areas.
- Development of rural credit systems.
- Development of appropriate institutions and management systems in the rural areas.
- Training – evolving training methods which are appropriate to the Indian context – particularly the rural areas.

In order to foster a process of change and influence public policies in favour of the poor, MYRADA evolved a strategy which sought:

- *To target the system* for structural change in the long run, and for the creation and implementation of *new* policies and legislation in favour of the poor in the medium term.
- *to support relevant components in the system* and responsive officials in implementing government programmes and policies to eradicate poverty.
- *to support the emergence of alternative innovative and appropriate institutions* of the people in their efforts towards self-reliance.

All these thrusts are related to government: the first in a constructive manner, where MYRADA is careful not to polarize the relationship by avoiding publicity and a position of moral superiority; at the same time MYRADA tries to adopt a professional approach based on detailed analysis of the problems and careful selection of possible solutions jointly with people and government; the second thrust relates directly to the government and draws on its strategies; the third focuses on supporting appropriate and innovative people's institutions (e.g. credit management groups) which prove that the poor can mobilize and manage part of their credit requirements; once public institutions like NABARD and banks studied this experiment and realized its potential they were willing to support these groups.

The provision of funds plays a major role in the relationship between NGOs and government. The government is aware that it has to back up its support of NGOs with resources, and is willing to give funds directly to them. Some NGOs however, perceive this as the beginning of government control and recall the experience of the co-operative societies which started by accepting funds and ended up under the control of the government bureaucracy and politicians. Other NGOs feel that dependence on foreign funds tarnishes the images of self-reliance; besides, these funds can be cut off at any time; government funds they feel are more reliable. Others take the position that the NGO should have several sources of funds, both indigenous and foreign, to enable it to have a degree of independence.

There are yet others who are reluctant to receive government funds directly but are actively involved in mobilizing and managing funds given directly to beneficiaries through government anti-poverty programmes. They organize local groups, assist them in developing skills and attitudes to manage and use funds and ensure that all the designated funds reach the beneficiaries in time. These funds do not pass through the NGO accounts; these NGOs see their role as providing the added services required to make these programmes achieve their objective. A large number of government programmes have been mobilized by MYRADA in this manner.

To what extent can NGOs collaborate with government without losing or diminishing their voluntary features? Can NGOs perform their role effectively if they are too closely integrated with the government? The debate continues.

In a way it has helped to keep options open and to create opportunities for government and NGOs to meet, work together and build up mutual confidence. In a limited way, the debate helps to keep the official system flexible and resilient enough to absorb the consequences of involving NGOs especially in organizing the poor to participate effectively in their own development.

The importance of people's participation as the key to the success of anti-poverty or minimum needs programmes has been accepted officially, but its implications have still to be worked out; meanwhile the official system has to be kept flexible and resilient to absorb the consequences of this acceptance. Once again each experience will differ from the other. MYRADA believes that it has to develop a strategy not only to build effective participation of people but also to make the official system especially at the interface more open, flexible and responsive. MYRADA does not adopt a rigid position that all the government does is against the development of the poor; to do so would be to close all doors to constructive dialogue. It is, finally, the poor who would suffer as a result of such ideological pride.

A FEW CAVEATS

While MYRADA does collaborate with government in implementing these programmes it has been careful to avoid the image of being a 'turn-key' operator or a contractor. If this image of a contractor is allowed to grow, MYRADA would lose its flexibility and the ability to press for change where it is necessary. Often government officials, pressured into attaining targets, find MYRADA non-co-operative since achieving physical targets in disbursement is not a guiding norm or the major indicator of success in MYRADA. Other pressures to install infrastructure and 'models', which MYRADA knows from experience to be unsustainable by the people, were also avoided. While in the short run, therefore, MYRADA may be considered non-co-operative, its long-term strategy has given it sufficient space to involve people in the planning and implementation of programmes which ensure that all infrastructure is maintained by the people and not left unused after an initial burst of enthusiasm and publicity.

There is an emerging trend which is causing concern. Several international donors who provide soft loans or grants to the governments are now insisting that NGOs should be involved. This requirement is often based not on a real appreciation of the role of people but are gestures made to appease pressure groups abroad who have been sharply critical of programmes formulated by experts. Translated in practice, this demand for NGO participation often turns out to be as follows: a team of experts – both from abroad and from India – formulate a proposal. Sometimes an NGO representative (seldom from a genuinely operational NGO) is included on this team. The government is then expected to implement this proposal in partnership with an NGO who is brought in after the agreement is signed between the donor and the government. This pattern of operation is becoming common and once again reduces the role of the people and the NGO to one of implementers. MYRADA has been drawn in to one such project but is making its involvement conditional to a new formulation of the proposal based on interaction with local groups which MYRADA has organized in the area.

We shall briefly describe two activities in which MYRADA has a formal relationship with government departments.

COLLABORATION WITH GOVERNMENT IN PIDOW (PARTICIPATIVE INTEGRATED DEVELOPMENT OF WATERSHEDS) GULBARGA

There are certain programmes which by their nature require formal participation with government at high levels; the management of micro-watersheds is one example. But first the numerous government departments relating with a single micro-watershed must learn to work together among themselves. A further problem which MYRADA had to overcome was how to make peoples' participation effective in this collaboration, so that it is their institutions who gradually take over control of the watershed and its resources, in order to manage, use and generate its resources in a sustainable manner.

In a watershed, lands and resources come under various authorities which exercise control with varying degrees of effectiveness; for example, while the Revenue Department owns the tank foreshore, the Forestry Department or Public Works Department own the trees planted in the foreshore; the supervisor is either the Junior Engineer PWD or the Forest guard, and the beneficiaries change depending on who funded the programme and the policy behind it. The Gomal land is under the Animal Husbandry Department or the Revenue Department, which claims rights to the trees, and is now being transferred to the Mandals. Parts of the lands which are Revenue lands were handed to the Forest Department in Karnataka, but the latter have not surveyed them in most areas prior to formally taking them over, hence their status is not clear. The Mandal Panchayats have now been given charge of maintaining the community forests. Though the Mandals have been given a greater role in management of resources at local level the rules and regulations governing each resource and area are still not clear.

To ensure effective and sustainable management of the watershed all these agencies have, first, to work together, and second, to develop a participative culture if the programme has to be people driven.

To be truly participatory these departments must

- work at the people's pace and not in a hurry at the end of the year in order to achieve the targets they set – the latter course may require bulldozers where, if the work were properly scheduled, human labour would be adequate
- operate in a situation where the people are allowed to exercise a degree of

control not only in planning but also in controlling and monitoring the estimate of costs of the work done and flow of funds; wherever possible, village people should be allowed to do the work, if necessary, with some outside guidance
- accept people's initiatives in constructing structures for water and soil control (and not destroy them at the altar of an overall integrated plan prepared by experts).

An agreement was signed between the Swiss Development Co-operation (SDC) and the government under which the government's departments (Agriculture, Soil and Water Conservation, Forestry, etc.) would provide physical inputs and technical advice; another agreement between SDC and MYRADA gave the latter the role of providing the software or the dimension of people's participation. Appropriate structures were provided at project level with representation from government, SDC and MYRADA, and at state level under the chairmanship of the Secretary for Rural Development and Panchayat Raj to provide planning, management and monitoring of the project.

It soon became evident that the line departments could not respond to the demands of an integrated programme. Each department was constrained by its own programme pressure and schedules. The dimension of participatory planning and people's involvement could not find a place in this context. The three partners therefore decided to entrust the government's intervention to the Dryland Development Board which had an integrated approach to watersheds and to set up a separate project field unit to work with PIDOW. This new arrangement has worked for a year and has largely solved the problems caused by lack of an integrated approach. The participative dimension has also improved considerably with MYRADA introducing participatory methodologies through participatory rural appraisal exercises in which all actors (DLDB, MYRADA, SDC and people) are involved at every stage of the process. It will be enlightening to compare the earlier watersheds (where participatory methods were still being worked out) with the latter (where the degree of people's participation increased at every stage of the process in a systematic way) to see if this increased and organized participation improved the levels of sustainability and management of the resources in the watershed.

PARTICIPATORY RURAL APPRAISAL (PRA)/ PARTICIPATORY LEARNING METHOD (PALM)

MYRADA was introduced to participatory rural appraisal (PRA) in 1989 when five of its staff were trained in the techniques at its Gulbarga Microwatershed Project. It has subsequently become the country's leading exponent of PALM (as MYRADA prefers to term it), having by early 1991 trained over 3,000 persons in the techniques (Table 4.1).

Table 4.1 Taking stock of PRA in India: PRAs conducted, number of staff trained and number of trainers

Name of organization	*Number of PRAs conducted*	*PRAs no. of staff, field-workers and village animators trained [no. trained more than 3 occasions]*		*Number of trainers in organization*	
ActionAid[a]	10	50	[5]		3
Activists for Social Alternatives (ASA Trust)	2	40[b]	[4]		2
Aga Khan Rural Support Programme (AKRSP)	40	40	[10]	Staff	8
				Villagers	8
Xavier Institute of Social Service (XISS)	—	2			1
DRDA, Anantapur	5	160			1
Andhra Pradesh	15	410[b]	[20]		5
SPEECH	5	117[b]	[15]		8
Seva Bharati	1	34[b]			6
BAIF	6	8	[3]		2
Krishi Gram Vikas Kendra, Ranchi (KGVK)	6	26	[26]		7
MYRADA	—[c]	3,000[b]	[97]		31

Source: Compiled by Eva Robinson (see Mascarenhas *et al.* 1991)
Notes: [a] Does not include continuing PRAs in 75 villages
[b] Number trained does not include substantial numbers of villagers and sangha members also trained during PRAs
[c] No details for number conducted by MYRADA

A recent workshop in Bangalore (Mascarenhas *et al.* 1991) summarized the innovative techniques now being used in this approach in India. A particular emphasis is on maps and models produced by villagers themselves, of parts of villages, villages in their agro-ecological context and, in some cases, of entire watersheds. Some maps are designed to show socio-economic characteristics, such as the size and location of low-caste settlements, the incidence of nutritional disorders among children or the distribution of ownership and access rights to natural resources. These are complemented by what is now a well-known range of PRA methods (transfer, seasonal calendars, activity peoples, semi-structured interviewing, chain interviews, historical profiles, Venn diagrams, wealth ranking, resource inventories, key probes, 'futures possible', etc.).

However, training others in PRA methods is unlikely to succeed without significant change in attitudes, particularly on the part of government staff. The mnemonics REAL (*r*espect for the people; *e*ncourage people to share ideas; *a*sk questions; *l*isten carefully) and LEARN (*l*isten; *e*ncourage; *a*sk; *r*eview; *n*ote) summarize the attitudinal qualities required, which are often radically different from the authoritarian and self-important attitudes commonly found among government staff. A capacity for sharing is a third principal component of PRA, and its relationship with the methods and attitudes is encapsulated in Figure 4.3.

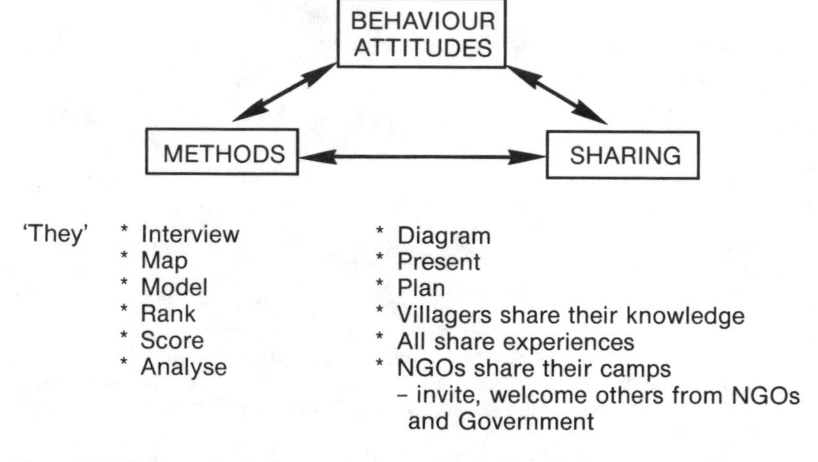

Figure 4.3 Three principal components of PRA: behaviour and attitudes, methods and sharing

Source: Adapted from Robert Chambers in Mascarenhas *et al.* (1991)

The proponents of PRA claim that it has already achieved successes in India in the following areas:

- raised awareness and self-confidence leads to greater willingness among villagers to act without outside financial support
- increased devolution of responsibility from implementing agencies to rural people themselves
- raised local awareness resulting in increased local monitoring
- improved capabilities in NGO and government staff
- breakdown of barriers within villages, between government and villages, and between government departments
- more sustainable watershed protection at lower cost

- improved monitoring through the use of villagers' maps.

The thirty-five participants in the Bangalore PRA workshop included five from government service, two of whom represented projects documented elsewhere in this volume (Kamal Kar from Seva Bharati KVK and D Satyamurty of the Karnataka State Watershed Development Cell - SWDC). A degree of interaction between MYRADA and government agencies is indicated by the fact that it has been actively involved with the KWDP and ran a PRA course at the KVK.

The then director of the Karnataka Watershed Development Project, Mr D Satyamurty, attended a PRA in early 1990. He trained SWDC staff on his return, and subsequently arranged training for other division-level staff. He and other government participants in the PRA workshop drew attention to problems faced by government staff in rural development planning, including:

- the difficulty of meeting deadlines with time-consuming conventional planning methods
- severe inadequacies in the maintenance of structures designed and constructed solely by government - people do not readily identify with them
- the arrival of government staff in a village no longer arouses curiosity
- villagers expect government officials to perform poorly, and to demand bribes

Substantial progress appears to have been achieved with the introduction of PRA. Satyamurty claimed that 'In one year our bureaucracy has become sensitive . . . PRA has totally changed our attitude to the planning process' Mascarenhas *et al.* 1991:27). Another participant, Somesh Kumar, claimed that PRA had helped to reduce disciplinary barriers between government staff (Mascarenhas *et al.* 1991: 28).

Attempts to make government staff's attitudes more modest and open have, in at least one case (Monnappa, ibid. p. 29), proven essential before PRA methods can adequately be used. Government staff have to 'learn to unlearn' in deep-seated biases, to search out villagers instead of expecting villagers to wait on them. All are required to stay overnight in the villages during PRA sessions, and senior officials are expected to show an example by refusing special comforts.

NGO LINKS WITH THE KARNATAKA STATE WATERSHED DEVELOPMENT CELL

MYRADA and the PIDOW project

K V Bhat and S Satish

This section describes the complexities of interaction between the State Watershed Development Cell (SWDC) of the State Government of Karnataka (GoK) and a local NGO (MYRADA).[7] This experience has to be seen against the wider political setting in Karnataka which has been conducive to pluralistic approaches to development, as demonstrated by the interaction which has evolved over time between the GoK and the Federation for Voluntary Organisations for Rural Development in Karnataka (FEVORD-K).

Since its registration in 1982, membership of FEVORD-K has grown to over 100 NGOs. Its aims include identification of potential areas for mutual help and promotion of better understanding among NGOs; design of appropriate programmes for training NGOs; facilitation of access to outside information; development of linkages with external agencies - government and non-government - and analysis and awareness creation in policy-related issues in the NGO sector.

The possibilities of influencing policy and decision-making have encouraged FEVORD-K to put substantial efforts into developing links with government, and the GoK response has been favourable. In 1982, for instance, a circular was issued to senior civil servants directing them to make all orders, circulars and policy papers available to FEVORD-K. A state-level committee for consultation (SLCC) with NGOs was set up in 1984. Following political and administrative changes made to promote decentralization, GoK issued an order in 1989 requiring the establishment of district-level consultative committees (DLCCs) with up to eight NGO members. The SLCC was widened in January 1990 to include bankers, industrial associations, Lion and Rotary clubs and chairmen of various voluntary agency sub-committees.

Requests by consultative groups for practical interaction between NGOs and GOs in the development programme were examined by the meeting of the GoK working group on NGOs in July 1989. Joint sub-committees were subsequently established for some areas of concern, including rural development; social welfare; health and family welfare; education; forests; and watershed development. The sub-committees present their suggestions to the government or SLCC for action. Their special focus at the time of writing is on the involvement of NGOs in development programmes during the Eighth Five-Year Plan (1990–95).

The GoK aims to work with at least one NGO in each of its nineteen District Watershed Development Programmes and, at FEVORD-K's suggestion, agreed to hold a workshop for twenty-five to thirty NGOs to promote awareness of technical issues, identify potential areas for collaboration and prepare a plan of action. The workshop was held in March 1991. The twenty participating NGOs and SWDC staff identified issues on which responsibilities needed to be clarified, or joint NGO–GO initiatives taken; within four broad areas of activity (Table 4.2).

However, poor organization of the workshop, including delays in sending out invitations, illustrate the lack of agility in GOs which constrains the pace at which they can interact with NGOs. It may also have reflected SWDC's growing awareness – partly as a result of working with MYRADA in the PIDOW project – of the complexities and challenges that working with NGOs generates. The PIDOW/MYRADA experience is now examined in detail.

GOVERNMENT OF KARNATAKA'S WATERSHED DEVELOPMENT PROGRAMMES (WDPs)

More than 75 per cent of the cultivable area in Karnataka State is under rainfed farming, of which 60,000 sq. km with a population of over 10 million are drought-prone. In an effort to minimize the risks of rainfed farming, GoK decided in the late 1970s to promote watershed development on a pilot scale in all the agroclimatic zones of the state. A watershed is a topographically delineated area drained by a stream system. Given rainfall, soil, vegetation and gradient, estimates can be made of water receivable, the amount retained and runoff for any given piece of land. This permits 'water budgeting' estimates which, in turn, provide guidelines for land utilization.

These pilot efforts envisaged WDPs of about 25,000–30,000 ha in each of the nineteen districts over a period of seven years. In seeking to reverse resource depletion and achieve sustainable increases in the production of crops, forage, fruit, fuel and timber, the WDPs bring together the activities of public sector institutions in agriculture, forestry and horticulture under a single umbrella. This aims to facilitate the integrated design of technologies and the collaborative treatment of differing land types, ranging from forest in the upper reaches to agricultural in the valley bottoms. It should be noted that these embrace several categories of ownership: forest lands are owned by government; the more productive lands in valley bottoms often belong to larger, wealthier farmers, and lower income farmers tend to own or obtain access to poorer hillside lands, with (often) some 'encroachment' into forest lands.

Table 4.2. Issues identified for further clarification during the GO-NGO workshop on Watershed Development in Karnataka, March 1991

OBJECTIVES	*INITIATIVES*
1 Mobilization of people for the formation of micro-watershed *sanghas*	(a) Awareness creation campaign (b) Direct access to the administration (c) Opportunities to develop common property resources (d) Timely supply of agriculture-related inputs
2 Breaking the barrier between government officials and the people	(a) Better working relation between NGO and GOs (b) Sharing of information (c) Setting up of project-level committee consisting of NGOs and the field level staff (d) NGO representation in the Karnataka Development Programme monthly meetings at district and taluka levels
3 Common property resources management	(a) How to maintain dialogue between GOs, NGOs and *Sanghas* (b) Determining rights to productive assets (c) Where local level NGOs do not exist, GOs themselves should undertake wider responsibilities; NGOs from outside should not be imposed (d) NGOs have their own agenda, hence they are at best facilitators; they should not be construed as an extended arm of government. The ultimate responsibility for programme planning and implementation rests with GOs
4 Training and education	(a) GOs to be given training in people's participation (b) NGOs require training in technical as well as the economic aspects of projects for help in discussing their feasibility with farmers (c) A network of NGOs participating in WDP, together with a newsletter, should be set up to enable cross-fertilization of ideas and experiences

The broad objectives of the WDPs include:

- large-scale verification of the available technology
- selection of technical components and criteria

- identification of investment limits and cost structures
- development of an organizational framework for rapid replication of project activities on large areas.

The specific components envisaged in preparation of WDPs include the following:

- *Rainwater management* measures to enhance moisture conservation and biomass production on non-arable lands, and management practices and structures to enhance soil and water conservation in arable areas.
- *Production management* more intensive farming permitted by rainwater management measures.
- *Use of marginal lands* for subsidiary occupations such as small livestock, fibre extraction, carpentry, etc.
- *Local-level institution-building*: watershed groups need to be formed for managing and sustaining watershed activities.
- *Credit management* to widen the range of (especially) non-farming options available to watershed communities and so ease the pressure on land.

PLANNING AND IMPLEMENTATION

The GoK's approach to WDP was innovative in two ways: first, it created an institutional 'umbrella' to bring together agriculture, horticulture and forestry departments in both planning and implementation. This was manifest at district (project/watershed development team), division (Dry Land Development Board with a multidisciplinary cell) and state levels (State Watershed Development Cell). This collaboration has led to the formulation of an agreed integrated plan for each micro-watershed. Second, these plans are finalized after conducting a series of meetings with local groups (*sanghas*). Work with these groups sought planning and implementation according to people's needs; better access to credit; and improved post-project maintenance. SWDC felt that people's views on, and participation in, watershed development could be more effective if they were organized by NGOs.

WATERSHED DEVELOPMENT PROGRAMME AND NGOs

The checkdams and bunds needed to retain water on slopes are necessarily small and of low individual cost. This makes it difficult to engage (and monitor the work of) contractors. Local materials can, in many cases, be used effectively in their construction, making it largely unnecessary to engage contractors. Local communities' inclination to undertake construction and maintenance increases if they are involved in planning from the beginning and feel confident that the various structures would become their assets and help to raise incomes over time. Achievement of the necessary levels of community organization is a task that SWDC envisaged as appropriate for NGOs, as the PIDOW project demonstrates.

PARTICIPATORY AND INTEGRATED DEVELOPMENT OF WATERSHEDS (PIDOW)

Origins

PIDOW is a bilateral co-operative endeavour between the governments of India and Switzerland with participation of a third partner, MYRADA, a Bangalore-based NGO, for which initial proposals were made in 1979.

Although envisaging participation by farmers (especially by women's groups) at the implementation stage, the early draft proposals were essentially 'top-down'. SDC responded by fielding teams to assist in the preparation of revised drafts that would, first, draw on people's participation at the early stages of design, and second, involve a local institution that would stimulate participation on a sustained basis. Since there was no appropriate institution in Gulbarga itself, after some discussion, MYRADA, a Bangalore-based NGO, was invited to participate in a planning mission in February 1984. The findings of the mission were agreed by the Rural Development Department in July 1984.

Given the project's objective of producing a replicable approach to participatory watershed development, MYRADA was requested to devise a means of testing whether the interventions planned to date corresponded with people's needs and opportunities. The district administration (District Rural Development Society – DRDS) would then identify whether the proposal should be modified.

During this phase (September 1984 to July 1985), MYRADA examined the procedures for selecting watersheds from the overall inventory; examined the socio-economic situation and undertook a resource inventory; and identified opportunities for action-research. A final draft was prepared and submitted in April 1985. The final agreement with SDC, however, was not signed until September 1986. This delay encouraged SDC to enter into a separate agreement with MYRADA in March 1986. Phase I of the project ended in March 1988 but was evaluated in October 1987 and a decision taken in favour of a Phase II for 3 years from April 1988.

Objectives of PIDOW

The overall objective of the project is to support farmers and village communities in their own efforts at improving their livelihoods in semi-arid rural areas. The project aimed to develop a replicable strategy for sustainable participatory and integrated development of small watersheds by seeking to identify how people's needs and capabilities could be brought to bear on proposals for watershed development and its respective components: soil and water conservation, horticulture, fodder, agroforestry and afforestation, and dryland agriculture.

ORGANIZATIONAL STRUCTURE AND IMPLEMENTATION (1986-90)

The project initially was based on a three-tier system:

- Joint Project Committee (JPC), a policy-making body at state level
- Standing Committee (SNC) to co-ordinate the activities of line departments at district level and
- Project Management Committee (PMC) to look after planning and implementation at the field level.

The project director for the District Rural Development Society was initially designated overall head of the PIDOW project. However, legislative changes at national level aimed at decentralized administration were introduced in 1988 and resulted in the creation of the *zilla parishad* (ZP) at district level. As part of this change, ZP took over the responsibility for PIDOW from the DRDS. But the ZP's range of responsibilities was too wide to permit adequate attention to PIDOW. The decision was therefore taken to entrust this responsibility to a special deputy commissioner, the executive director (ED) of the Water Technology Mission.

While the three-tier structure (JPC, SNC, PMC) ensured proper co-ordination among the three partners (GoK, SDC and MYRADA) on matters of policy, management and monitoring, the co-ordination of field-level activities was poor. Only some 25 per cent of physical targets were met during this period, and conflict arose between those in overall charge of the project and implementing agencies at field-level over responsibility for this poor performance. The project chief faced severe staffing constraints which were compounded by cash flow problems: the GoK's contribution to the project was channelled to line departments independently, which resulted in poor co-ordination and delayed disbursements. Frustration over the inadequacies of existing mechanisms for field-level co-ordination led MYRADA and SDC to propose a review of alternatives.

From this emerged a proposal in the latter half of 1989 that implementation of PIDOW should be transferred to the District Watershed Development Team (functioning under DLDB/SWDC) thus giving the DLDB a pivotal role in the management of PIDOW. Though acceptable in principle, formal agreement to these changes was delayed as a result of numerous shortcomings identified during an SWDC technical audit, including: inadequacies in non-arable land treatments and in some erosion control structures; excessive influence of revenue department boundaries (instead of watershed criteria) in definition of the project area; delays in planning; inadequate co-ordination of activities, both among line departments, and with the results of MYRADA's 'needs survey'. Doubts were also voiced over whether the *sanghas* formed by MYRADA could continue on a sustained basis. Furthermore, while information provided by the participatory rapid appraisals (PRA) conducted by MYRADA enhanced outsiders' understanding of local situations, there remained uncertainty among some government staff over whether the maps and diagrams prepared by villagers could provide insights useful to watershed management.

In the light of these observations, the transfer of responsibility for PIDOW to the DLDB/SWDC was formalized in October 1990 on condition that SWDC be granted overall responsibility for the planning process from micro-watershed upwards, and that a new relationship be worked out between technical planning and the community participation and organizational inputs that fell within the mandate of MYRADA.

PIDOW WITH DLDB/SWDC (1991 ONWARDS)

A new agreement among DLDB, MYRADA and SDC was formally reached in February 1991, the principal objectives remaining the same. The proposed collaboration is expected to draw on the skills and resources of project partners in a more co-ordinated way, and to bring PIDOW into the mainstream of WDP in the state to enable cross-fertilization of ideas, skills and methods. The overall division of responsibilities is as follows:

- *DLDB* is assigned the responsibility of all technological aspects of integrated watershed management to ensure technical soundness.
- *MYRADA* is assigned the responsibility of all 'non-technological' people-oriented aspects such as PRA, and formation of groups to facilitate people's participation. The testing and developing of new or alternative approaches and technologies for participatory watershed development and sustained resource management is included in its agenda.
- *SDC* is to play the role of co-ordinator and facilitator.

Activities for 1990–2 envisage the development of 1,649 ha at a cost of almost Rs10,500 per hectare. The fact that this is double that allowed by the National Watershed Development Programme for Rainfed Agriculture (NWDPRA) is partly explained by the additional overheads of engaging MYRADA in methodology development.

There are also other important differences between the PIDOW and NWDPRA approaches: the former includes the costs of group formation by MYRADA. Similarly, MYRADA has been responsible for the socio-economic aspects of other project components such as roads, health, education and community infrastructure. MYRADA's work also has a human resource development function. To date, some forty development professionals have gained experience in managing the socio-economic complexities of watershed technology, institutions, credit and participatory methods. These factors account for the fact that some 60 per cent of the per-hectare costs have been allocated to MYRADA for socio-economic work of this kind. Many see this as a valuable investment, given the spread of socio-economic perspectives and PRA methods which this work has helped to stimulate. Examples include the Western Ghats forestry development project in Karnataka, wasteland development programmes in Maharashtra, Kerala, Tamil Nadu and Gujarat and, significantly, a replication of the PIDOW approach in Andhra Pradesh.

The SWDC has also prepared proposals for some 'scaling-up' of many of the aspects of the PIDOW experience: a project valued at some Rs86 million from 1991/2 to 1995/6 has been prepared in collaboration with a local NGO, India Development Service, to develop watersheds over 28,000 ha involving 8,000 families in 27 villages.

CONCLUSIONS

The GoK has pioneered watershed-based programmes of rainfed agricultural development. Much of this initiative has been concerned – as e.g. in the district watershed development programme – with setting up institutional arrangements necessary to bring together the skills and knowledge available in specific line departments, so facilitating co-ordinated development efforts that integrate agriculture with the wider management of soil, water and trees.

GoK, through consultations with NGO umbrella organizations such as FEVORD-K, began to develop an appreciation in the early 1980s of the need for change to be 'owned' by local communities if it is to be sustained, and of the potential offered by NGOs for community organization. However, attempts to implement NGO–GO collaboration of this kind in the PIDOW project encountered a number of difficulties. These include

- problems in co-ordinating activities among line departments, and consequent difficulty in making technical interventions respond coherently to needs identified by NGOs
- senior staff discontinuity, especially in government departments, which not only contributes to co-ordination problems among departments, but also costs NGOs time and effort to re-establish their credentials and familiarize new staff with key issues
- staffing shortages and delayed disbursements on the government side, which again cause co-ordination problems
- uncertainty over which side is to take credit for success in particular initiatives
- some continuing restrictions by government on access to details of future policies, programmes and projects.

Two fundamental difficulties deserve particular emphasis since they may lie at the root of the apparently increasing unease among some SWDC staff over the role accorded to the NGO partner in the PIDOW project.

First, it is not clear to engineers how far the maps and estimates of water flow produced by PRA are compatible with their concepts of water budgeting and land use planning. Thorough research is needed to establish the levels of accuracy that different kinds of PRA conducted under different conditions are capable of achieving. Only then will it be possible to identify whether and how the resource-intensive data collection methods used by engineers can be replaced partly by PRA.

Second, and more intractable, are the challenges that PRA and community-based 'ownership' of interventions pose to the conventional operating procedures of government technical staff. At a broad level, they demand unaccustomed flexibility and responsiveness, but, more particularly, require changes in attitudes from government 'control' of a development process and from the top-down provision of concrete and steel structures (and the handling of large contracts with suppliers) towards processes and interventions which are community-led.

We leave this project at a time of uncertainty: initial enthusiasm, within GoK and specifically within SWDC, for NGO–GO collaboration has been tempered by experience that has demonstrated the potential for confusion over responsibilities, for uncertainty over how to interpret the results of PRA and for conflicting claims of success.

Realization among GO staff of how far their administrative procedures and attitudes will have to be reformed if they are to respond adequately to these new modes of project implementation has, in some cases, remained suppressed, leading to undercurrents of antagonism towards the PIDOW experience and towards further 'experiments' of this kind; in other cases, it has been faced boldly, but with the sobering realization that there is a long way to go before the full potential of NGO–GO collaboration can be realized.

ARE NGOs MORE COST-EFFECTIVE THAN GOVERNMENT IN LIVESTOCK SERVICE DELIVERY?

A study of artificial insemination in India

S Satish and N Prem Kumar

INTRODUCTION

A common concern of donor agencies is whether NGOs perform more cost-effectively than government in delivering services to the rural poor. In many instances, this question is irrelevant – indeed, misleading – since, as is demonstrated throughout this book, NGOs and government have different objectives and seek to achieve them by different means. In one case study, however, the efforts were described of an Indian NGO (BAIF: see pp. 136–44) to provide an artificial insemination service to produce cross-bred dairy cattle for small rural milk producers. The objectives of this activity are similar to those of a number of government agencies in India, so that some comparison of cost-effectiveness can be made. It should be noted, however, that cross-bred dairy cattle require higher levels of nutritional and veterinary care than nondescript cattle, and so the farmers catered for by BAIF and by government programmes, although small-scale, are not among the poorest rural households.

As the case study notes, BAIF pioneered frozen semen technology in India, and has set up a delivery system for artificial insemination (AI) and veterinary care comprising 453 village-level cattle development centres in six states of India. Overall BAIF has been responsible for producing around 10 per cent of the entire India cross-bred dairy herd.

BAIF's costs are compared with those of a government AI programme operated by the state of Tamil Nadu (TN). This state was chosen for several reasons: it is not one of those in which BAIF operates; it appears typical in size and level of efficiency of state-operated AI services (though resources did not permit systematic verification of this), and, at a practical level, staff were willing to make data available.

METHOD

A full economic assessment would have required consideration not only of the costs of providing services but also of programmes of cost recovery (which both BAIF and TN State have) through charges to farmers and, leading from this, some assessment of how far the service can be made financially self-sustaining in the long term. Given data limitations and the need to rely on several assumptions, as outlined below, the present study sets for itself the limited objective of comparing the full costs of service provision, and interpreting these figures in the light of detailed differences in types of service provided between the two institutions.

Costs are grouped into two broad categories for each organization: semen production, and AI-related services and supervision.

Semen production

These include the annualized costs of land and buildings,[8] and of the capital cost of equipping laboratories, generator plant, and cryogenic plant to the standards observed. Recurrent costs include AI materials, bulls, staff and the operating costs of plant and equipment. Comparative data are provided in Table 4.3. Clearly, TN State is the more expensive producer of semen straws, the average BAIF costs being only 66 per cent of those of TN State.

Table 4.3 Comparative cost of semen production, BAIF and TN State, 1989–90 data (Rs)

	BAIF	*TN State*
Annualized fixed costs	1,445,000	310,450
Total staff and operating costs	1,192,000	800,000
Total	2,637,000	1,110,450
No. of semen 'straws' produced per year	1,000,000	279,155
Average cost of producing one straw	2.64	3.98

This is surprising, given that BAIF's laboratory and equipment costs include a substantial research component, whereas TN State's laboratories are simply geared to semen production along well-defined guidelines.[9] Much of the difference can be explained by the fact that operating costs – particularly staff costs – are much higher in relation to the number of straws produced at TN State (Rs2.87 per straw) than at BAIF (Rs1.19 per straw), reflecting low labour productivity in the state sector.

Given the assumptions that have to be made regarding the proportion of land, buildings and equipment that should be allocated to the AI programme, these figures should be regarded more as indicative than conclusive. An important additional comment at this stage is that BAIF's research has generated considerable external benefits to other organizations. In a perfect economic system, the users of these benefits would be made to contribute to the research costs, thereby reducing BAIF's overall costs.

AI-related services and supervision

Differences between the two organizational and administrative structures in which BAIF and TN State are located require consideration of different types and levels of costs (Table 4.4). In the BAIF case, for instance, 5 per cent of the annualized resources of the information resource centre and 10 per cent of the services of the Kamadhenu office (i.e. head office – literally 'cow') are assumed to have been allocated in support of the AI effort. BAIF's principal costs, however, are those of operating the 453 cattle development centres through which local-level AI services are provided. These average Rs74,400 per centre per year, but provide health-care services and advice on animal nutrition as well as AI services. Our estimate is that, as a maximum, 75 per cent of their costs are allocated to AI (Table 4.4).

The situation is more complex in TN State. Officials at state, district and divisional level have responsibility for the AI programme as part of their duties, and a corresponding portion (20 per cent at state level, 22.5 per cent at district level and 25 per cent at division level) of total recurrent costs (including the rental cost of buildings) is charged to the AI programme. The overall supervisory and administrative costs per calf estimated in this way are remarkably similar between the two institutions (Table 4.4).

Table 4.5 brings together the costs estimated in previous tables. The low number of straws required by BAIF to achieve a successful conception results in greater efficiency in semen utilization, so that TN State's overall costs are some 15 per cent higher than those of BAIF.

CONCLUSION

Uncertainties over data quality and over some of the assumptions that had to be made mean that these estimates should be treated with some care. However, what is remarkable is that BAIF should be able to deliver AI services for a lower cost than a government department given that first BAIF conducts research into e.g. semen quality, and maintains an extensive programme of progeny testing, all of which add to its costs, and second, BAIF staff provide an *on-farm* AI service which enhances the prospects of successful conception (cows required to walk an average of, say, 10 km to an AI station as in TN State tend to be stressed and less likely to conceive). This service also reduces farmers' costs. A farmer who has to make an average of four visits to an AI station to achieve successful conception will lose an aggregate of two days' work, which, even at modest rates, will incur an opportunity cost of Rs20. This

Table 4.4 Comparative administrative and supervisory costs of BAIF and TN State AI programme (Rs)

	BAIF	*TN State*
State level administration (per district basis)	—	85,565
District level administration	—	64,800
Division level administration (per district basis)	—	63,237
Information resource centre services	27,000	—
Kamadhenu office services	13,000	—
Total annual cost of operating cattle development centres	25,277,400	—
Veterinary dispensary	—	44,529
Veterinary hospital	—	55,779
Clinician centre	—	79,091
Main office annual costs (average/district)[a]	—	621,986
Sub-centre costs (average/district)[b]	—	2,455,900
Total costs	25,317,400	3,470,887
No. of calves born	200,000	25,600[c]
Costs per calf	126.59	135.58

Notes: [a] These are an average of main office costs between those offices and service centres falling within the intensive cattle development programme (ICDP), and those falling outside the programme. The heavier overheads of the ICDP result in average costs per calf born some 50 per cent higher than in non-ICDP areas.
[b] Fifty sub-centres for each main office.
[c] Average per district.

Table 4.5 Costs of semen production and administrative/supervisory costs per calf (Rs) (1989–90)

	BAIF	*TN State*
Administrative and supervisory costs per calf	126.59	136.00
1 Semen production costs per straw	2.64	3.98
2 No. of inseminations required to achieve successful conception	2.5	4.6
3 Semen production costs per calf[a]	6.6	18.31
Total costs per calf	133.19	154.31

Note: [a] A successful conception in (2) is assumed to translate into a live calf born in (3). While some miscarriage and stillbirth occurs, its incidence is low and unlikely to differ systematically between the two case studies.

means that, to the farmer, TN State services cost 30 per cent more than do those provided by BAIF.

Finally, it would be unwise to conclude from this case study that NGOs are generally more cost-effective in rural service delivery than government. BAIF is an exceptionally large and highly organized NGO. It is also exceptional in seeking to provide some of the same services as does government. In the majority of cases, NGOs' objectives differ from those of government but the case study alerts us to the need – given the increasing trend among donors and governments to 'contract out' some service delivery to NGOs – to monitor and evaluate the cost-effectiveness of their performance.

GOVERNMENT INITIATIVES TO COLLABORATE WITH NGOs IN INDIA

Three distinct types of initiatives taken by government to collaborate with NGOs in the broad sphere of agricultural technology development can be identified:

- efforts to involve NGOs into schemes designed by task forces for watershed management
- efforts under the Eighth Five-Year Plan to hand over a range of extension functions to NGOs in specific geographical areas
- the creation of farm science centres (Krishi Vignan Kendras – KVKs) by the Indian Council for Agricultural Research at the field sites where selected NGOs operate.

In what follows, summaries are presented of proposals under the first two initiatives (since activities have not yet started) and three brief case studies are presented of activities to date under the third.

PROPOSED LINKS BETWEEN TASK FORCES FOR WATERSHED MANAGEMENT AND NGOs[10]

Strong evidence in the mid-1980s that rainfed areas (comprising 70 per cent of India's cultivated land) had been largely neglected by the Green Revolution led to increased government commitment to address their technology needs through flexible, multi-disciplinary approaches hitherto uncharacteristic of Indian research and extension services. A rainfed farming systems cell was established in the Ministry of Agriculture in 1987, many of the guidelines which it was to follow being promulgated by the national watershed development project for rainfed areas (NWDPRA) in 1990. These guidelines commit the government to 'sustainable farming systems' approaches based on the holistic concept of integrated watershed management. Implicitly, it recognizes the inadequacies of over-specialized, centrally managed efforts in research and extension which have tended to pursue narrow commodity- or discipline-based approaches.

Explicitly, it recognizes that more needs to be done to cater for the specific needs of women, and to promote collaboration with organizations outside the public sector, including NGOs.

This initiative follows a long history of efforts to generate more co-ordination among the government agencies responsible for the various aspects of watershed management. These started with efforts to promote sectoral co-ordination between the Department of Agriculture and that of Forestry at state level in the 1950s. Subsequent efforts in the 1970s sought closer co-ordination by allocating development funds to a single district-level agency, the District Rural Development Agency, which were to be used to draw in the services of such departments as Agriculture, Forestry, Horticulture, Animal Husbandry, Minor Irrigation, Dairy Development, Fisheries and Sericulture. An alternative model, developed in collaboration with the World Bank under the Uttar Pradesh Himalayan watershed management project, generated a central plan through the directorate of watershed management which provided the framework for resource allocations to the respective departments. However, the plan's co-ordination objectives were thwarted by the refusal of several departments to accept the costing norms under which funds were to be allocated to them.

Perhaps the highest degree of success has been achieved by a recently developed task force model, in which staff from the relevant departments are seconded into a multidisciplinary task force for the planning and implementation of watershed-based activities. Much of the experience on which this model is based was obtained in watershed management in Karnataka State (see the section by Bhat and Satish pp. 160–68). A refined version of the task force model is now scheduled for application to watershed development, its secretariat being located in a newly created division of the Ministry of Agriculture responsible for rainfed farming systems. The broad doctrine of this initiative is that watershed management should comprise a full range of renewable natural resources, production and conservation activities, unhindered by administrative boundaries, and embracing also the provision of wage-labour and household-based income-generating activities. To prevent jealousies among departments, at least one staff member of each department is made a watershed team leader at field level in a specific micro-watershed, and the chairing of technical meetings at state level is rotated among departments. A comprehensive organizational structure reaches down from national to district level: a national watershed development policy committee sets a policy framework, decides on funding allocations among states, provides a means of inter-departmental conflict resolution and monitors progress; a national watershed development implementation committee approves individual projects and monitors the utilization of funds. Technical support in resource-inventorization, mapping, production systems and socio-economic systems is provided by the Division of Rainfed Farming Systems. These three national-level tiers – policy, implementation and technical secretariat – are mirrored by a similar hierarchy within states. A co-ordination team operates at district level, a development team at watershed level, and a watershed development supervisory committee at block level.

The significance of this complex hierarchy – and the justification for allocating space to a description of it – is that, according to Seth and Axinn (1991), planners have located NGOs at the bottom of the hierarchy, designating as their functions:

- creation of awareness
- training in systems diagnostic approaches
- social impact evaluation.

Government-sponsored local-level activities in which NGOs would participate include the formation of groups based on the panchayats for joint purchase of inputs and shared use of equipment, the formation of self-help groups based on watershed-level nurseries, and the formation of savings and credit groups ('sanghas').

Much of the scheme described above remains at the level of proposals. Three features give cause for concern in terms of NGO–GO relations:

- The fact that the hierarchy of responsibilities is highly structured, and that NGOs and local groups are placed at the bottom of the hierarchy. Grassroots influence on policy and strategy issues is likely to be extremely difficult to achieve.
- The assumption that equitable groups capable of representing the interests of the poor will emerge under the aegis of existing, frequently highly inequitable, institutions such as panchayats.
- The fact that NGOs do not appear to have been consulted in the design of the hierarchies and institutional forms to whose operation they are expected to contribute.

A pilot project has been proposed by the GoI to UNDP–FAO for implementation of these approaches in forty watersheds, each of approximately 1,000 ha, over a five-year period. Funding requirements are estimated at some US$5.25 million of aid funds, with an approximately equivalent local contribution. However, in view of the above shortcomings, it would be surprising if, at the end of the pilot project, significant numbers of NGOs were found to have committed themselves to the programme.

A radically different approach has been taken jointly by an NGO, the Ahmednagar Social Centre (ASC) and a donor (GTZ) towards the area-based co-ordination of activities among government departments, NGOs and local organizations of various kinds. Under its 'natural resources management by self-help promotion programme', GTZ has indicated willingness to provide DM12 million over an initial five years in support of integrated watershed management in part of Maharashtra State. While much of this money will be channelled through a GO (the National Bank for Agricultural and Rural Development), project preparation has been largely in the hands of an NGO

(ASC) and much effort has been made to define GO and NGO roles and responsibilities to allow the needs of the rural poor to be incorporated into the development 'process' and to build in checks and safeguards to ensure that expenditures are geared towards these needs. The complex arrangements are summarized in Box 4.2.

PROPOSED DELIVERY OF AGRICULTURAL EXTENSION SERVICES THROUGH NGOs[11]

The Department of Agriculture and Co-operation of the Ministry of Agriculture has, for the Eighth Plan, proposed a pilot scheme for five states in which extension would be contracted out to NGOs. The proposals are based on the results of a survey conducted by the Delhi-based Participation and Development Centre,[12] which found that medium/large-scale NGOs (i.e. having a minimum of fifty employees) saw a role for themselves in agricultural extension, principally in the areas of

- non-formal education and training
- natural resource conservation
- group formation and group-based activities
- nutrition, especially of women and children
- sericulture, horticulture, fruits and vegetables
- animal husbandry and aquaculture
- demonstrations of new technologies
- identification of the needs of the rural poor.

They expressed willingness to collaborate with government providing that technical support and training were given to their staff, and that programmes were planned comprehensively to reflect their own requirements and those of the rural poor.

The report concluded that NGOs' perceptions largely coincided with those of government officers involved in agriculture. It envisaged a supporting role for government in which it would enhance the technical and project management skills of NGOs, and assist in the procurement of inputs.

GoI proposals developed on the basis of this study envisage a stronger role for NGOs in technology transfer and feedback in selected blocks in Gujarat, Rajasthan, Orissa, Kerala and West Bengal. The proposals envisaged technical support, both before and during the project to NGOs from the State Departments of Agriculture and from the State Agricultural Universities. Specific activities to be undertaken by the NGOs include the organizing of farmer groups, the procurement and distribution of literature, the organization of farmer group meetings at demonstration sites, and of visits to research farmers. An annual outlay by GoI of some US$0.5 million for NGO activity in these areas is envisaged, up to a limit of US$600 per NGO per year in administrative expenses, and US$1,400 in project-related expenditure.

Box 4.2 The Indo-German Watershed Development Programme - roles and responsibilities of NGOs and GOs

1 *The objective*
To develop micro-watersheds in a comprehensive manner through participatory village group initiatives so as to enhance the ecological basis of production and sustenance systems leading to mitigation of the impact of drought and the creation of adequate and sustainable livelihood opportunities for all.

2 *Programme organization*
(a) Project level
- The Gram Sabha (Village Self-Help Group) is the project implementor. The Gram Sabha nominates the Watershed Development Committee which executes the project.
- The NGO motivates and organizes the Gram Sabha and assists the Watershed Committee in executing the project.

(b) Programme level
- Financial support NABARD (the National Bank for Agricultural and Rural Development) assisted by the KfW (German Bank for Reconstruction and Development).
- Technical support Technical Support Organizations in the private sector, Government Institutions (Agricultural Universities, Training Institutions), Government Departments (Water Conservation, Agriculture and Forests).

(c) Linkage building/networking The Co-ordinator (based at Social Centre, Ahmednagar) is responsible.

3 *Programme operation*
(a) Reputed and selected NGOs will prepare a project proposal together with the Village Self-Help Group and the assisting Technical Support Organization (where applicable) and forward it to NABARD for technical scrutiny.
(b) The project proposal will then be approved by a Project Sanctioning Committee.
(c) Sanctioned Funds will then be forwarded by NABARD to a Bank Account jointly operated by the NGO and the Watershed Committee.
(d) Monitoring/supervision will be done by NABARD and ongoing support provided by the Co-ordinator.

4 *Approach*
(a) Watershed development projects, to be implemented by village communities mobilized by NGOs in watersheds having a minimum gross areas of 1,000 ha.
(b) Training of educated unemployed youth (Human Resource Development) to enable them to mobilize participation of village communities so as to create a people's movement for Watershed Development.

Source: Lobo (personal communication); see also Lobo and Kochendörfer-Lucius (1992).

The survey commissioned by GoI from the Participation and Development Centre represents a serious attempt to obtain the views of GO and NGO staff on the potential involvement in agricultural extension. However, the resulting proposals would have benefited from more detailed efforts to identify the comparative advantage of NGOs in specific tasks. As they currently stand, the proposals envisage that NGOs will carry out a full range of extension functions, effectively replacing government extension services in specified geographical areas. While NGOs may have some advantage in group formation, it is unlikely that they would in responding, for example, to technical questions that farmers might raise during demonstrations. Two further difficulties under the proposed arrangements are likely to present themselves: the first is that NGOs are heavily dependent on the quality and relevance of the technical recommendations made by government researchers. Given long-standing concerns over the weakness of systems-based approaches in the Indian agricultural research establishment, there is some danger that current technical recommendations may not be relevant to small farmers working under difficult conditions. Second, the channels through which NGOs might articulate the needs of the rural poor – either those which they have just identified, or those arising as feedback on the use of existing technologies – are not clearly defined. Reward systems in the Indian agricultural research establishment do not encourage receptivity by scientists to feedback on the impact of the technologies deriving from their work. Unless deliberate steps are taken to stimulate their interest in monitoring and evaluation, it seems unlikely that any attempts by NGOs to bring grassroots pressure to bear on the definition of research agenda will have much success.

Overall, this initiative is an imaginative expression of interest in collaborating with NGOs at field level, and as such is, in principle, to be welcomed. However, lack of prior definition of the respective roles of NGOs and government, and of the steps necessary to facilitate performance of those roles by each side, mean that the project is unlikely to generate clearly identifiable advantages over and above existing systems.

NGO LINKS WITH GOVERNMENT FARM SCIENCE CENTRES (KVKs)

Farm science centres (Krishi Vignan Kendras – KVKs) are under the authority of the Indian Council of Agricultural Research (ICAR) and serve as centres for demonstration and training in 'scientific farming'. The ICAR's policy is to increase the number of KVKs to one per district. Since they are an officially designated mechanism for the dissemination of research findings from the forty-four ICAR central research institutes and in many cases have formed links with State Agricultural (and other) Universities, they offer, in principle at least, the advantage of ready access to the public sector. Such access may be particularly attractive to NGOs, assuming, first, that these public sector institutes have technologies and methods to offer which are relevant to NGOs' clientele and, second, that the linkage can go beyond the 'delivery of packages' or 'top-down' training courses. Instead, NGOs are likely to be interested in securing flexible responses from research institutes to their needs – whether in the form of a range of technologies or methods instead of a single package, or in the form of willingness to *listen* to the needs and aspirations of the rural poor and modify future research agenda to accommodate these. Three cases in which KVKs have been located at NGOs are discussed below, and their methods of operation and achievements reviewed.

KVKs

KVKs are a central component of the ICAR's national plan for linking scientific progress at research institutes and universities with village life. They are intended to provide need-based skill-oriented training to rural people to enhance their capacity to take up technologies developed by the public sector. Their philosophy is grounded in teaching and learning by doing; a focus on the weaker sections of the community; and self-betterment. KVKs have a standard professional staff complement of twelve persons, the disciplines or areas of competence being determined according to local requirements.

NIMPITH KVK

S Satish and Dipankar Saha

The Sunderbans region in the southern part of West Bengal, with a population of 2.2 million, comprises almost 4,500 sq. km of land characterized by poor infrastructure, heavy rainfall (1,750 mm per year), high salinity, and dissected by innumerable small streams and inlets. In 1962 a monk from Ramakrishna Mission in Calcutta began to introduce education and development programmes in health, education and agriculture in both Nimpith and the island of Kaikhali. Early efforts to draw on the advisory services of government institutes attracted little response, and so, in 1970, an agrodevelopment centre was established as a focal point for applied scientists capable of developing appropriate technologies by building on local knowledge. Following a visit by the director-general of ICAR, a KVK was attached to the centre in 1979. This allowed an expansion of the centre to its present complement of seven scientific and nine technical professionals, with a research/demonstration farm of 40 ha, a dairy farm, soil testing laboratory and a hostel and training facilities for farmers.

Training is the major function of the centre, with over 150 courses, held almost entirely on campus, and 4,000 farmer participants annually; a number of specialist government institutes and universities are drawn into the programme. Research in support of the training programmes has centred on two principal problems: poor drainage and salinity in the wet season (June–December) and inadequate access to fresh ground water in the dry season, underground water being saline to a depth of some 7 m. These basic characteristics led to a classification of land according to drainage characteristics:

- Type A: free of standing water in October–November
- Type B: free of standing water in December
- Type C: free of standing water in January.

Research for Type A lands is focusing on horticultural crops. The donation of plastic sheeting by a commercial company keen to diversify demand for its products has stimulated research into 'poly-mulch' techniques, but so far has not led to their adoption by farmers.

On Type B lands, researchers have concluded that cotton is one of the few crops that will grow under the high prevailing salinity conditions. Variety trials have been conducted, and the centre has developed a system of growing seedlings for one to two months in nurseries prior to transplanting, in order to reduce time required in the field to the six months available. This research has been in progress since the mid-1980s, and an action research project sponsored by the ICAR began in 1990 in an effort to popularize cotton production. Hitherto, however, there has been very limited uptake by farmers.

On Type C lands, the recommended solution is re-profiling, to create some areas of land that remain permanently above water, and others that remain below. While inherently logical, structural changes of this kind require high levels of investment and so far there has been virtually no adoption by farmers.

Complementary research has been conducted into aquaculture, an area in which there appears to be scope for improved technologies, given the high density of ponds in the area (12 villages surveyed recently were found to have over 1,200 ponds) and the frequent mixing of incompatible species. While research into stocking density and supplementary feeding does not yet appear to have stimulated adoption by farmers of these technologies, there has been high demand for seed-fish produced by the centre: some 2 million carp spawn and 380 kg of fry were supplied to farmers in 1990–1.

While the Nimpith KVK has achieved some success, it has been plagued by problems of irregular release of funds from ICAR. It has also proven extremely difficult to retain research staff, given the poor working conditions at the centre, and its isolation from urban areas and from the wider scientific community – an isolation that can be only partly reduced by access to some Indian institutes' publications and by infrequent exchanges with them.

SEVA BHARATI KVK

S Satish and Kamal Kar

Seva Bharati is a non-political, non-profit NGO established in 1947 at Kapgari, Midnapore District of West Bengal by an horticulturist (P K Sen) working at

Calcutta University, with a focus on agricultural research, extension and training. The almost complete absence of irrigation has meant that farm incomes, relying on one crop per year in most areas, have been extremely low. In addressing these issues, Seva Bharati's farm for research and training – both formal and non-formal – has grown from 18 ha in 1953 to its current 50 ha. Located at the farm are centres for social studies and for agrarian research. The latter enjoys strong links with the College of Agriculture, Calcutta University, of which P K Sen was principal in its formative years, and has hosted sixteen PhD students and a number of MSc students in recent years. It has defined its mandate as follows:

- improving rainfed rice yields
- increasing the extent of off-season cropping
- introducing relay-cropping into areas with some irrigation
- organizing farmers into co-operation action.

Seva Bharati has experienced a number of difficulties analogous to those at Nimpith KVK, which have been compounded by excessively rigid approaches to staff selection taken by its now ageing founder.

The KVK's mandate was defined on the basis of a survey in 1976–7 revealing that 80 per cent of families were below the poverty line, more than 50 per cent of the population in Kapgari and surrounding villages comprised scheduled castes and tribes, and literacy was only 9 per cent. The mean annual rainfall of 1,460 mm falls mainly in July–September, but its distribution is erratic, and little dry-season irrigation is found.

Over 1,200 training programmes have been held from 1976 to 1990, with some 13,000 trainees, comprising mainly farmers, but also local branch staff and the clientele of NGOs (Catholic Relief Service; Women in Social Action; Forum for Rural Development). Trainers and resource persons have been drawn from state government departments, the universities and the commercial sector (e.g. Kirloskar Cummins on diesel engines).

To be relevant and effective, training programmes have to build on practical field experience. The KVK's main research effort has lain in the development of water resources for off-season irrigation, which is currently available to only 17 per cent of cultivated land in and around Kapgari. Observations of local farmer practice revealed that shallow ditches often dug at the foot of low hills were recharged by seepage through the dry season and so a useful source of irrigation water. Experimentation at several of these sites indicated that these could profitably be increased in size to rectangular tanks 60 ft x 30 ft by 15 ft deep. These were found to recharge, on average, in 7–8 hours, allowing 1 acre of paddy, 7 acres of wheat or 10–15 acres of vegetables or oilseeds to be irrigated from each tank.

Other research has included vegetable and field crop variety trails, and the introduction of sweet varieties of ber, of which some 15,000 new plants have been produced since the mid-1980s.

UPASI KVK

P Swaminathan and S Satish

The United Planters' Association of Southern India (UPASI), founded in 1893, is the apex organization of the growers of plantation crops (tea, coffee, rubber, spices) in South India. UPASI's affiliates include the State Planters' Associations of Tamil Nadu, Kerala and Karnataka, and the many district-level planters' associations in these states. It is financed by an area-based levy on its members. UPASI's primary mandate is to represent growers' interests in national and international policy fora. It is also engaged in research, information exchange, training and estate supplies (including planting material).

The UPASI Tea Research Institute (TRI) is staffed by ten PhD-level scientists and a further forty at MSc level, and disseminates technologies via seven advisory centres located in the main planting areas. UPASI provides research and advisory inputs on tea to national-level bodies, at the same time drawing on other institutes for advice on other components of South Indian production systems (coffee, cardamom, horticulture, forestry, livestock).

UPASI's funding structure has meant that its research responds primarily to the requirements of large-scale producers. Concern within UPASI began to arise in the early 1970s over its neglect of smallholders' requirements, despite the fact that 50 per cent of the cultivated area in the Nilgiris hills (where UPASI is situated) is owned by smallholders. This led to an approach to the ICAR to enhance the level of resources available at UPASI for smallholder-related work, and a KVK was established in 1982.

UPASI's philosophy in establishing training courses with the KVK has paralleled its approach to large-scale planters: it has drawn on the expertise of a number of other government agencies for advice on products other than tea. However, while training courses have included issues relevant to diversification, the main focus has been on tea improvement.

Tea yields among the 43,000 smallholders in the Nilgiris were low in the early 1980s, averaging 800 kg per ha per year against 200 kg per ha per year among large-scale producers. A baseline survey conducted by the KVK suggested six principal reasons for low yields: widespread poor management; extensive gaps in plant stands; poor genetic material; difficulties in acquiring inputs in small quantities; marketing difficulties; problems in obtaining institutional finance. In response, the UPASI-KVK, drawing largely on existing UPASI research, worked out an investment strategy for tea improvement incorporating three levels of options.

- *Level 1* changes in cultural practice involving little or no investment, including the adoption of optimum pruning cycles, timings and heights; increasing the number of plucking rounds and relating plucking standards to cropping patterns; optimum stand of shade trees and appropriate pollarding and lopping practices.

- *Level 2* low-cost improvements, including correction of zinc deficiency and soil acidity; selection of correct type of fertilizer and split dose application; timely adoption of plant protection; burial of prunings to increase soil organic matter.
- *Level 3* improvements in involving higher investment; including adoption of soil and water conservation measures; filling in plant vacancies; replanting with improved genetic material.

From 1979 to 1990 the KVK has conducted 714 training programmes on aspects of these strategies, involving 15,000 trainees. A thirteen-part radio series incorporating the principal items from these training courses has been broadcast for those unable to attend courses. Further use of the radio is made to broadcast warnings of pest outbreaks and appropriate preservation or control measures, and to transmit quiz programmes on aspects of tea production.

To assist in gap-filling, UPASI's nurseries are providing up to 25,000 improved tea plants per smallholder at a subsidized price of Rs0.50 per plant; 120,000 clonal tea plants have been supplied to date, together with 500 shade tree seedings (*Grevillea* sp.)

New tea plantings have been introduced to 500 small farmers on 180 ha of land previously allocated to potato production, which has become unprofitable owing to increased nematode problems and rising input costs. Detailed advice is given on contour-planting and on management of the new crop. Other development agencies in the Nilgiris are now considering the introduction of similar schemes.

The increasing demand for new clonal material by smallholders has generated substantial employment opportunities in newly established nurseries. UPASI has trained educated, unemployed youths in nursery management since the early 1980s, resulting in the establishment of 340 nurseries which to date have produced over 96 million plants.

Overall, the UPASI programme has achieved considerable success: large quantities of improved clonal material have been planted, and some improvement in management practices has been achieved. Smallholders participating in the various training and improvement programmes have been able to raise average tea yields from 800 kg per ha to 1,575 kg per ha. This has been achieved by drawing on existing UPASI research which has been geared to large-scale producers' requirements. UPASI scientists see further improvement towards the large-farm average of 2,290 kg per ha as possible, providing that further research on small farmers' fields is carried out. The prospects for carrying out such work will depend largely on the ability of the KVK's local management committee – comprising scientists, extensionists, development administrators, local NGOs and smallholders – will be able to pull resources from both ICAR and UPASI towards their interests in the face of likely opposition from UPASI's long-established plantation members.

NOTES

1 For a convenient overview, see Robinson (1991).
2 For a review, see Unia (1991).
3 Examples of such organizations are PRADAN (see pp. 102–9), PRIA (Society for Participatory Research in Asia), based in Delhi, and another NGO – SEARCH – in Bangalore.
4 The Federation of Voluntary Organisations for Rural Development in Karnataka (FEVORD-K) and the Association of Voluntary Agencies (AVA) in Tamil Nadu are two examples of such networks.
5 'NGOs, natural resources management and links with the public sector' held as part of the study on which this book is based at Hyderabad, India, 16–20 September 1991.
6 This was extremely complex, since DoA could provide funds only to villages registered under the oilseeds programme, and arrangements had to be made by NGOs to provide the remainder.
7 For a summary of MYRADA's philosophy and objectives, see the section by Fernandez and Mascarenhas elsewhere in this Chapter.
8 The assumed life of land and buildings is fifty years, of equipment fifteen years and of bulls eight years,
9 Guidelines which BAIF's research helped to develop in the first place.s.
10 This section draws heavily on Seth and Axinn (1991). However, the opinions expressed here are those of the editors. The editors are indebted to the authors for permission to draw on their material.
11 Information in this section is based on interviews with GoI staff. However, the opinions expressed are those of the editors alone.
12 Participation and Development Centre (1990) 'GoI sponsored special study on the scope for enlisting the support of voluntary agencies in agricultural extension, with suggested linkages', mimeo, New Delhi.

Chapter 5

NGO–GOVERNMENT INTERACTION IN NEPAL

OVERVIEW

Narayan Kaaji Shrestha and John Farrington

ECONOMIC AND SOCIAL CONDITIONS

Real growth in Gross Domestic Product (GDP) has fluctuated widely in Nepal in recent years (Table 5.1). Its first development plan (1954–5 to 1960–1) produced no appreciable increase in per capita income which, by 1987/8, had reached only US$170, one of the lowest in the world. After fiscal expansion in the early 1980s which proved inflationary, a period of stabilization was followed by a World Bank-supported structural adjustment programme in 1987. This was producing promising results until the March 1989 breakdown in the negotiations for renewal of trade and transit treaties with India. Although the resulting physical disruptions to trade were short-lived, increases in the tariffs applied to goods imported through India contributed to inflation and to a severe deterioration in public sector finances.

Nepal is heavily aid-dependent: aid receipts accounted for 15.5 per cent of GDP in 1989–90, by far the highest in Asia, with Bangladesh in second place at 10.1 per cent, and most other countries in the 1–3 per cent range. Its external debt currently amounts to 54 per cent of GDP.

Agriculture accounted for 52 per cent of GDP in 1987–8, down from 66 per cent in 1975–6. Agricultural production has achieved positive growth since the mid-1980s, but, as with GDP, exhibiting wide inter-year fluctuations. Agriculture remains one of the few sources of employment, still providing livelihoods for 93 per cent of the labour force in the late 1980s.

Table 5.1 Nepal: recent economic indicators

	1986–7	*1987–8*	*1988–9*	*1989–90*	*1990–1*
Real GDP growth %	2.7	9.7	1.5	-2.2	3.0
Population growth %[a]	2.6	2.5	2.5	2.6	2.6
Growth in GDP per capita %	0.1	7.2	-1.0	-4.8	0.4

[a]Based on Central Bureau of Statistics, Population Projections of Nepal 1985–2000

Major constraints to development in Nepal are the weak fiscal base of the public sector, its cumbersome procedures, large size (approximately 90,000 civil servants in 1988–9), lack of performance incentives, compressed pay differentials and career prospects seen by middle-level civil servants as extremely constrained. Nepal government allowances for field-station postings have recently been increased, but many civil servants still have little incentive to work outside the Kathmandu Valley unless their emoluments are topped up by donor funds.

Social indicators in Nepal are among the weakest in the region, but have improved substantially since 1970: life expectancy at birth has risen from 42 years in 1970 to 51 years in 1989. Secondary school enrolment rose from 10 per cent in 1970 to almost 30 per cent in 1988, with primary school enrolment rising from 26 per cent to 85 per cent over the same period.

NGO-STATE RELATIONS

Formal NGOs do not have a long tradition in Nepal. However, the Nepalese have long responded to their geographical isolation by forming groups for community work and social or religious activities. Many bridges, trails, schools, temples, irrigation canals, rest houses, resting places (*chautara*) etc. are the legacy of those self-reliant actions. Much of the physical and social infrastructure owes its existence to the labour, capital and indigenous knowledge of those groups (Adhikary and Suelzer 1985). Guthi,[1] parma,[2] dhikur,[3] neighbourhood organizations and resource creation and management groups were prevalent, but they were informally organized for specific purposes. These institutions are still widespread, particularly in more remote areas.

Until recently, autocratic forces sought to control NGOs and subservience became a requirement for recognition. It was almost impossible to operate smoothly without recognition. Especially during the Rana regime (1847–1950) any kind of formal organization was perceived as subversive. Two prominent NGOs, the Nepal Charkha Pracharak Gandhi Smarak Mahaguthi (1927)[4] and Paropakar (initiated in 1945 and recognized in 1947),[5] established themselves against the wishes of the ruling elite after a long struggle.

The 'partyless' system introduced after the 1951 revolution looked unfavourably upon any local-level organizing, since this was perceived as undermining the unity that defined 'partylessness'. However, government-sponsored NGOs (GONGOs) were initiated, partly in order to generate employment opportunities for the elites who supported the political system. The Nepal Family Planning Association (1959), Nepal Red Cross Society (1960) and Nepal Children's Organization (1964) emerged as potential welfare motivated NGOs. During the 1960s and 1970s there was an upsurge of politically approved clubs and NGOs of this kind.

The Social Service National Co-ordination Council (SSNCC) was formed in 1975 to co-ordinate the activities of NGOs. The council was headed by the

Queen, and a further six committees were headed by royalty.[6] Membership of the council and of the committees was accorded to elites loyal to the system. Membership accorded a privileged position in the society due to the royal connection. All NGOs had to be affiliated with the council in order to be officially recognized and 219 NGOs had done so by 1990. The Social Service Act 1978 required a two-stage procedure for NGO registration: at local level the founding members are required to submit an application specifying objectives, plan of action, and background of the members to the chief district officer (CDO). The CDO takes particular care to examine the background of proposed members, and, once approved at district level, an application for registration with the SSNCC at national level may be made.

The 1990 Constitution drawn up in response to pressures for multi-party democracy recognized the formation of organizations as a basic human right. It abolished the six committees of the SSNCC and made it necessary for an NGO to register with the SSNCC only if it sought government or foreign financial support. As a result, in the eighteen months from April 1990, almost 200 new NGOs became affiliated with the SSNCC. There are also numerous registered but unaffiliated NGOs.

The future role of the SSNCC remains unclear. Some, but not all, of its powers of patronage and control have been removed. While many NGOs seek the liberty to operate as they wish, the Ministry of Finance in 1991 made it clear that permission from His Majesty's Government of Nepal (HMGN) is required before local NGOs can receive resources from foreign donors, either directly or via international NGOs (INGOs), and there is some indication that NGOs would accept the SSNCC in a new role of this kind, providing it operated objectively and acted as a 'one-stop' decision-taking unit, i.e. without having to forward papers to HMGN line departments for decisions.

Recently expressed views see NGOs increasingly as implementers of development projects. The Finance Minister for instance noted in a recent budget speech that 'By making the projects undertaken for poor and disadvantaged groups by the NGOs as well as the government complementary to each other, an environment will be created to motivate the NGOs to provide as much benefit as possible to the target groups'.[7] Indirectly, it is assumed that the NGOs are the better means to reach the poor and work with them.

How far NGOs in Nepal can contribute to rural poverty alleviation, and what modes of operation and planning mechanisms will be necessary, are central questions in current policy debates in Nepal.

Two items of legislation being debated suggest a strengthening of the SSNCC, and of local government institutions. While the former may prove excessively restrictive, and is open to political abuse, the latter may result in improved compatibility between NGOs' work and local development efforts in ways which are not necessarily threatening to NGOs.

First, a draft 1992 directive from the Ministry of Labour and Social Welfare outlines policies for the 'running, coordination and regularisation of national and international NGOs'. The document envisages the prohibition of direct project implementation by foreign NGOs, severe curtailment of the number of visas issued for expatriates to work on NGO projects, control over the salaries and benefits payable to both local and foreign staff, and numerous other financial regulations. The document proposes significant strengthening of the SSNCC: it will assess all NGOs' proposed activities and their previous record, in addition to ensuring that NGOs active in Nepal adhere to the new regulations outlined above. SSNCC will also ensure that no NGO 'disrupts religious harmony'. Its regulatory powers will be substantially increased by the requirement that all foreign funds for NGOs in Nepal should be transferred through the SSNCC.

Second, three Bills have recently been passed which aim to strengthen local self-government: the Village Development Committee Bill, the Municipality Bill, and the District Development Committee Bill. The purpose of these is to define the structure, obligations and rights of local government organizations, and their relationship with the national government. The Bills provide enhanced capacity for local revenues to be raised, and for development plans to be drawn up through wide-ranging consultative processes. For instance, the chairman of each village development committee and municipality is required to set up an eleven-member advisory committee, on which NGOs and local groups will be represented, and which will meet twice-yearly to discuss development proposals and budgets. The Bills envisage greater involvement of NGOs at village, municipality and district level, and the committees established under the Bills are required to formulate and implement development projects with the participation of NGOs. NGOs may also be contracted to carry out development activities by the committees, and may, in agreement with the committees, carry out development projects with their own funds. Local government bodies are given powers to ensure that the activities of NGOs are consistent with the agreed local development framework, to insist on co-ordination of activities among NGOs, and to audit their accounts.

From all sides, expressions are to be found of the desirability of closer working relations between NGOs and government departments. Fears are increasing among NGOs, however, that government sees NGOs only as a means of implementing government programmes. Maniates (1990: 156) concludes: 'the systematic integration of local NGOs into state-dominated systems of rural resource development will not necessarily foster greater program effectiveness'. Nepal's complicated and time-consuming procedures required by the government and the aid agencies encourage NGOs to collaborate.

Many government staff fear NGOs as competitors for donor funds. Distrust is compounded by the gap in organizational culture and ethos between NGOs and GOs, and many have begun to realize that the establishment of joint task forces in which a common view of issues, necessary actions and respective roles can be achieved is one of the few ways of reducing mutual suspicion.

Many feel that the main priority at this stage in Nepal is to lobby for 'policy, rules and regulations' that will facilitate process and activities of the NGOs by retaining their salient features – flexible, people-oriented, service-oriented, empowering, participatory, consciousness-raising, consensus-building, and humanistic.

NGOs IN NEPAL: TYPES, ACTIVITIES AND GEOGRAPHICAL DISTRIBUTION[8]

A study conducted in 1988 indicated that the staff of service-type NGOs, and the membership of nationally-registered membership organizations, were highly educated (65 per cent with tertiary education), middle class and predominantly (90 per cent) male.[9] NGOs' governing bodies – as undoubtedly in many other countries – rarely have members drawn from the clientele (especially women and the poor) whom the NGOs are intended to serve. The resultant dangers that services provided by NGOs may be irrelevant to clients' needs are self-evident.

Apart from the traditional grassroots organizations described above, three further types of NGOs can usefully be distinguished in Nepal: village-based NGOs, international NGOs, and private research and development organizations.

Village-based NGOs

Village-based NGOs were established in response to administrative and legal requirements: central government budgets are increasingly allocated on condition that the projects for which they are intended are managed jointly with local groups. Local officials in many cases therefore create such groups in order to acquire the funds allocated. Forestry user groups, for instance, have to be formed according to the guidelines of the 1988 forestry sector master plan. Prominent among this category of NGO are user groups in forestry, drinking water and irrigation, groups involved in credit or saving and women in development, and service delivery groups in livestock. The initial support provided by local officials tends to be withdrawn once funds are obtained, and various INGOs and bilateral projects have, in some cases, provided continuing support to the groups formed in this way.

International non-governmental organizations (INGOs)

Involvement of the international NGOs in social and welfare-related activities in Nepal started after the downfall of the Rana regime in 1951. By 1990 fifty-four INGOs were affiliated with the SSNCC. However, the undertones mentioned above of political restriction and patronage in the SSNCC's operations discouraged several INGOs from becoming affiliated, among them the United Mission to Nepal (UMN), Save the Children Fund (SCF)/UK,

CARE/Nepal, and various volunteer organizations. These organizations have direct agreements with the Ministry of Finance, with concerned line ministries, or as in the case of the International Union for the Conservation of Nature (IUCN), even with the Planning Commission. The activities of these organizations have ranged from a few volunteers operating individually in health-care and teaching in village schools to sizeable organizations with well-equipped offices and substantial numbers of expatriate and local staff (Affal 1988). Unconfirmed sources report that there are several INGOs which are supporting local NGOs without any affiliation to any of these legal entities.

INGOs in Nepal are operating through three modes: implementing activities themselves (ActionAid, SCF-USA, UMN, CARE, Redd Barna, etc.), providing financial and technical assistance to local groups (Oxfam, World Neighbors, SAP, Helvetas, etc.), and a combination of both (USC, Plan International, Lutheran Missions, CECI, etc.). INGOs are mostly involved in relief and rehabilitation, health-care, education and training, community development and income generation. A few are oriented towards agriculture or forestry development although generally within the wider context of community development or income-generating activities.

Private Research and Development Organizations (PRDOs)

A further group of organizations is worth noting in Nepal, although they do not operate on a strictly non-profit basis. Several PRDOs created in the 1970s served to absorb skilled labour which either could not or did not wish to be absorbed by the public sector. Independent of the SSNCC, these focused on the preparation and implementation of action-oriented projects catering for the poor and disadvantaged.

According to one PRDO, the failure of conventional development models propelled the PRDOs to explore effective alternative approaches to development and to undertake even those activities that normally lie in the realm of NGOs, that is motivating the poor farmers and other local groups at the grassroots level through action research oriented programmes not only to meet their felt needs but also to assist them in articulating those needs and enabling them to take control of the decision-making processes which affect their lives. The PRDOs therefore seek to assist these groups to undertake small self-help and self-managed village-level activities. The groups are also encouraged to link their activities with government programmes so as to maximize the benefits that are likely to accrue to them (New Era 1989: 5).

Some 200 PRDOs exist, but the majority are small and ill-equipped and many are more directly profit-oriented than the above ideals imply. However, they were perceived as enough of a threat to orthodoxy for the SSNCC in 1988 to attempt (unsuccessfully) to force them to register.

AREAS AND SCOPE OF ACTIVITIES

More than 75 per cent of the 390 NGOs affiliated to the SSNCC in 1990 are either based in the Kathmandu Valley or the district centres. Far-western and mid-western regions are almost totally neglected (Table 5.2), yet these are the areas where people are poorest and still under feudal forms of oppression. NGOs' geographical expansion paths are closely tied to the construction of new roads.

Table 5.2 Classification of NGOs affiliated with the SSNCC, Nepal, according to Development Regions ($n = 390$)

Committee	*Central*	*Eastern*	*Western*	*Mid-western*	*Far-western*
Health service	88.9	11.1	00.0	00.0	00.0
Child welfare	100	00.0	00.0	00.0	00.0
Women service	55.5	16.7	16.7	11.1	00.0
Youth activity	73.0	6.7	14.2	6.1	00.0
Community service	73.6	4.8	15.2	2.4	4.0
Hindu religion	59.1	20.4	13.6	2.3	4.6
Total	73.3	8.5	12.6	3.8	1.8

The SAP study suggests that most NGOs are small, have few professional staff, and have informal managerial and administrative procedures. It notes that in 1988:

> Out of the total number of projects implemented or currently being implemented by the NGOs, 32 per cent were found to be health and sanitation projects (most of these were actually drinking water supply schemes). Off-farm income generation constitutes 28.5 per cent of the projects while income generation in the agriculture sector represents 10.5 per cent. . . . Adult education, women's development, environment protection and building construction are the other areas of the activity in which the NGOs are involved.
>
> (SAP-Nepal 1988: 12)

NGOs' PERCEPTIONS OF FUTURE ROLES AND INTERRELATIONS

In response to the 1990 political changes in Nepal, forty local NGOs and sixteen INGOs with observer status met in June 1991 in Kathmandu to discuss the implications of the changes and the opportunities they offered. Partly in response to the SSNCC's previous political influence, the meeting began by defining the characteristics of non-profit social welfare and community-oriented NGOs. Nepali NGOs resolved that in principle INGOs should over a five-year period withdraw from direct implementation and restrict themselves to working in support of local organizations.

Nepali NGOs also resolved to press HMGN for streamlined procedures for approval of NGO projects and of their financing, whether from domestic or foreign sources. They also pressed for greater ease in access to government support through a 'single window' instead of through individual contacts with a plethora of government bodies as had happened hitherto. The meeting requested that clear legal distinction be made between NGOs and largely commercial organizations such as PRDOs.

A concrete result of the meeting was the creation of a national NGO Federation charged with pressing government to respond to the above proposals, and to seek the support of NGOs not represented at the meeting.

GOVERNMENT POLICIES AFFECTING AGRICULTURE

General efforts to improve incentives and out-station allowances in the civil service are likely to have a particularly strong impact on agriculture. The sector will also benefit from efforts under the SAP to streamline procedures for the release of government funds for 'core' or high return projects. Thirty-five of these are to be guaranteed funds even if there is an unexpected shortfall in government revenue. Particular efforts to improve the management of public resources in agriculture relate to irrigation works, found almost exclusively in the lowland Terai areas. Farmers' associations on irrigated areas have now been granted legal status, and there is increasing emphasis on their participation in water management.

Other policy reforms have focused on fertilizer supply. Subsidies on fertilizer to keep its farm-gate price roughly at parity with that in India (and so minimize smuggling) mean that the Agricultural Inputs Corporation is the sole importer. Improvements in its management and the encouragement of fertilizer retailing by private dealers through reduction of trading zone restrictions and increased margins for the hill areas are also likely to be of benefit, but the great majority of agrochemical consumption is in the Terai, implying relative discrimination against the hill areas.

PUBLIC SECTOR AGRICULTURAL RESEARCH[10]

In public sector agricultural and forestry research in Nepal, the emphasis from the early part of the twentieth century to the late 1960s was on the acquisition of technology from abroad and its adaptation to local conditions.[11] Most attention during this period was focused on the lowland Terai districts of high potential, and on the Kathmandu valley. In 1965–6, for instance, USAID began to fund a co-ordinated, intensive agriculture development programme that was eventually to cover sixteen districts in these areas.

It was only in the mid-1970s that a major agricultural research effort was made for the mid-hills when the mandates of Lumle (mid-western) and Pakhribas (eastern hills) agriculture centres were broadened from extension and training to research.

HMGN expenditure on agricultural research in the early 1970s was Rs15 million per year (approximately US$1.5 million) rising to Rs19 million (US$1.7 million) by the end of the decade. (More recent data are unavailable.) This amounted to some 19 per cent of the agriculture sector budget, equivalent to 0.22 per cent of agriculture GDP over the period, which is well below the 0.6 per cent average for fifty-one less developed countries (ldcs) surveyed by Oram and Bindlish (1981). Recent substantial expansion of funding especially from donor sources has raised research expenditure to approximately US$4 million, thus bringing Nepal close to this average.

Only approximately 40 per cent of the national research budget is allocated to the hill and mountain areas, which contain 60 per cent of the population. The substantial external funding of research for the hill areas (especially through Pakhribas and Lumle) has raised this figure considerably, but it should be noted that agro-ecological conditions are much more diverse in the hill areas than on the plains, requiring higher expenditure per hectare of cultivated land to achieve comparable rates of technical change.

RECENT INSTITUTIONAL AND POLICY CHANGES IN AGRICULTURE AND FORESTRY

The remaining set of influences governing NGO–GO relations in renewable natural resources management find their roots in recent attempts to arrest environmental decline. Since several of the Nepal case studies reported below deal with forestry- or environment-related issues, the background to recent policies is described in some detail here.

Two broad characteristics of the public sector in Nepal have had (and may continue to have) a strong bearing on its relations with NGOs in agricultural development. First, the Ministry of Agriculture was organized by crop and discipline until the mid-1980s. Some reduction in the barriers that these imply was achieved under the 1985 reorganization, and a research co-ordination committee to prioritize research was established with support from the USAID-funded agricultural research and production project (ARPP). However, the ARPP has faced a number of difficulties. A symptom of these is the low proportion (ten from nineteen by 1987) of the local posts in ARPP-supported programmes that had been filled by permanent staff. The underlying malaise was the defence of vested interests by senior civil servants against change, and it is not clear that even the political changes that started in 1990 have reduced these interests to any significant degree.

Second, while there appears to be no shortage of skilled personnel in the ministry (over 1,400 graduates by 1987; over 60 with doctorates by 1990), the capacity of staff to respond to local needs is constrained by:

- The need for new appointees to spend three to six years initially in temporary positions, and the 'urban bias' caused by their unwillingness to leave Kathmandu, where the best prospects for elevation to a permanent post lie.
- The fact that promotions are usually made only within the disciplinary 'faculty' in which an officer serves. This inhibits staff movement among the twenty-two faculties.
- The fact that overall emoluments, and field allowances in particular, have lagged behind inflation for long periods means that many officers are reluctant to take on field assignments and many take on supplementary employment which, again, is best found in Kathmandu.

Nepal's environmental conditions reflect the problems of widespread poverty, low labour productivity, rapid rates of population growth (2.6 per cent per year), low levels of off-farm employment, a long history of settlement and the intensive use of natural resources.

The expansion of agriculture into more fragile lands with poor terracing not only poses an environmental threat, but also imposes additional demands on livestock manure which is a primary source of soil nutrient particularly in the hills where it contributes more than 80 per cent of fertilizer used. This in turn places heavier demands on forests since some 40 per cent of animal fodder is derived from them. Heavy demands for timber and fuelwood further increase depletion rates.

Forest land covers 38 per cent of the country. From 1979 to 1985 overall forest area was reduced by only 3.3 per cent but the lowland tropical forest of the Terai by 25 per cent. These trends were made worse by the 1990 trade dispute with India, and by recent political uncertainties, which have diminished forest officers' motivation to maintain strict control over felling. Widely quoted estimates suggest that in the Terai for short periods during 1990 and 1991 commercial contractors successfully put pressure on forest officers to allow them to fell at rates five times higher than previous felling levels. Reduction in forest quality is of particular concern throughout the country: all but one-fifth of Nepal's forests have been reduced to less than 70 per cent crown cover and can be considered as overused and depleted. Several reports (e.g. Carew-Reid and Oli 1991) have drawn attention to the increasingly endangered status of plant and animal species.

It is widely recognized that the nationalization of Nepal's forests in 1957 (the Private Nationalization Act 1957) was a failure. Responsibility moved from villagers to the authorities who were unable to protect, manage and utilize the forests. The process resulted in many villagers clearing their private forests to retain ownership of their land.

Steps were subsequently taken to promote wider participation in management through the Panchayat Act 1961 and the Panchayat Forest Rules and Panchayat Protected Forest Rules of 1978. As a result of these changes, local communities found it easier to request HMG to assign forest areas to be managed by and for their own purposes. Subsequent amendments to the Forest Act 1961 formed the basis for community forestry projects in numerous hill districts and the terai with World Bank and bilateral donor assistance.

Results from these large community forestry projects were mixed. While the projects managed to establish new plantations, little progress was made concerning people's participation and engagement, largely because the 'Panchayat' often proved to be too large a unit to mobilize general interest. The smaller unit, 'user group', is now frequently being used to address community forestry efforts. The basic concept is that user groups themselves should direct the establishment and sustained management of their local forest for their own benefit.

These approaches and concepts represent such a drastic change from the past that the Forest Department field staff have to switch from being 'forest police' to supporters of villagers in their efforts to manage forests for their own use. Despite the problems in this process of generating villagers' confidence in officials, many NGOs, GO staff and donors, see it as the route most likely to lead to success in community forestry activities. Since the early 1980s there have been a great number of multilateral and bilateral forestry and watershed management projects initiated in Nepal as a result of growing international concern for environmental issues.

Participants at a joint HMG–donors meeting on forestry in 1984 decided that an in-depth analysis of forestry programmes and policies was required to both rationalize and prioritize investments in forestry on a nation-wide basis over a twenty-five year time frame. The resulting forestry master plan (FMP) is a multi-donor effort developed in 1986–8 to better co-ordinate and focus the estimated sixty projects and forty organizations then active in the forestry sector. The plan represents a concerted approach over the next two decades.

Among the key components of the plan is development of a national community forestry programme involving substantial policy and institutional change. An important focus is on the promotion of people's participation in forestry resource development, management and conservation, and, although provision was made for strengthening the framework for development of institutions working in the forestry sector, no detailed proposals were made regarding the role of NGOs in developing and promoting new technologies and management practices, or in facilitating people's participation. Several of the Nepal case studies presented below indicate how NGOs, despite the shortcomings of the master plan, have made substantial progress in these areas.

AN OVERVIEW OF NGOs' AGRICULTURAL ACTIVITIES IN NEPAL

Narayan Kaaji Shrestha

This study was undertaken in 1991 through a combination of (largely unpublished) literature reviews and semi-structured interviews with twenty-four NGOs, whose head offices are located in or near Kathmandu.[12] Inadequate prior documentation of NGOs' activities prevented rigorous sampling, and so those interviewed are best regarded as 'key informants'.

NOTEWORTHY NGO INITIATIVES IN THE AGRICULTURAL SECTOR

For the present study, twenty-four NGOs (thirteen INGOs and eleven NGOs) were contacted and interviewed. Out of thirteen INGOs, six directly implement their projects in the villages, four provide training and financial support to local NGOs, and the rest do both. Agriculture is one of the main components of the community development efforts of all INGOs. Out of eleven local NGOs, none except the INSAN (Institute for Sustainable Agriculture Nepal, which is involved in low-cost housing, environment-related and energy-saving projects) has agriculture as an important component. However, agriculture features in the awareness-raising activities of all, and three are coalitions mainly facilitating other NGOs and working as pressure groups.

All twenty-four NGOs are involved in awareness-raising activities. Nineteen implement agricultural extension by direct involvement with farmers or through written materials. Fifteen have initiated on-farm trials and eleven are involved in seed improvement and multiplication activities, with only four involved in marketing. Half the sample explicitly mentioned group formation activities as a means of providing local resource mobilization.

Analysis of progress reports and discussion with the sample made it clear that some (especially INGOs) have initiated innovative processes characterized by integrated 'systems' approaches analogous to farmers' own perspectives. By contrast, the sectoral approaches pursued by HMGN have been fragmented and more oriented towards production than to the needs of people. There is a similar contrast between the group formation activities of NGOs and those arising from government legislative and administrative provisions – the latter tend to be dominated by elites whereas NGOs seek representation from all strata, including women. One NGO (ActionAid) focuses specifically on those below the poverty line.

Technical innovations successfully introduced in agriculture include new varieties of existing crops and improved animal species, the introduction of entirely new crops (e.g. herbs and spices), and the promotion of agrochemicals. Improved management practices have focused on intercropping, improved stall feeding and the more efficient production and use of compost. The great majority of trials have been conducted informally, and results impressionistic. A number of NGOs (ActionAid; SCF-US) have introduced group-based credit in support of agriculture, often with a savings component. Extension efforts have centred on farm-based trials and demonstrations from which farmers can select components according to their conditions.

FUTURE PERSPECTIVES

The right to organize established by the new government of April 1990 has led to a proliferation of NGOs. Some 200 additional NGOs registered with the SSNCC in the following eighteen months. While in many respects likely to make a positive contribution to the livelihoods of the poor, this trend has generated concern over the number of NGOs that simply appear to be aiming to acquire whatever resources are on offer.

A number of NGO networks have recently emerged. Some (e.g. NGO Federation) act as a pro-NGO pressure group; others (NGO Forum) encourage the sharing of experiences. Nepal Agroforestry Foundation (see pp. 210–13) is perhaps unique in the range and depth of technical information it assembles, processes and makes available.

Redd Barna has agricultural projects (including new practices such as rice-fish culture, hybrid goats, mushroom and ginger cultivation, and seed banks) within three rural community development projects aiming ultimately to improve the conditions of children. It facilitates the dissemination of new ideas through on-farm demonstrations, trials, discussion groups and study tours. It provides imported seed (which has to be repaid in kind), seed storage bins, other planning material, seed fish, veterinary services and technical advice.

Redd Barna has good links with local and international NGOs and with GOs at both local and national levels. It sees the slow development of local organizations and skills to take over many of the above functions and so allow Redd Barna to phase out its activities as a wider problem. Nevertheless, to develop local organizations remains a central future objective.

ActionAid (Nepal) operates a large integrated community development project in the west-central mid-hills. A principal objective is to increase the area's agricultural productivity to its maximum sustainable level within a six to eight year period. It is concerned with production improvements in a wide range of cereals, vegetables, fodder and livestock. A particular focus is on seed improvement and multiplication. Strong community involvement in project planning, partly through the official village development committees, is a particular feature of ActionAid's work. Agricultural innovations include crop trials and demonstrations, soil fertility improvement, zero tillage and organic means of enhancing fertility and controlling pests and diseases.

Dissemination is organized through demonstration plots, audio-visual media, agricultural fairs and study tours. ActionAid provides planting material, lime for soil acidity correction, improved study animals, and animal health-care services.

ActionAid maintains close links with local NGOs and GOs, including the Lumle Agriculture Centre, and government research stations. Future plans include further strengthening of grassroot groups and village development committees, and the drawing up of strategic plans for its own activities and for staff development.

Agricultural development strategies among the NGOs interviewed vary widely. Brief summaries of some activities illustrate the similarities and differences in strategies and activities. Agriculture is a component of five community development projects implemented by the United Services Committee. USC sees community organization (e.g. into groups dealing with income generation, savings and credit, women's issues) as a precondition for sustainable community development. Its focus has been on the low external input production of fruit, fodder, seed, and livestock using compost-making techniques that it has introduced. USC provides saplings and seedlings, veterinary services and technical advice. 'Local agricultural workers' are trained to provide extension advice, and training is provided in agriculture and animal health, literacy, and the operation of savings schemes.

USC enjoys close working relations with other INGOs (including UMN) but has virtually no links with GOs. It sees the lack of skilled Nepali staff willing to work in rural areas as a major constraint and, through future links with local organizations, plans to develop skills within them.

MAJOR AGROFORESTRY ACTIVITIES OF NGOs IN NEPAL

Jeannette Denholm and Min Bahadur Rayachhetry

INTRODUCTION

Background

Agroforestry here is defined as a land-use system which intentionally combines annual agricultural crops with trees or shrubs and/or livestock within the same land management unit to achieve diversified, increased production. The purpose of the present survey was to learn of the activities, successes, constraints and needs of NGOs actively engaged in agroforestry and related fields. While Nepal contains a plethora of forestry and agroforestry projects sponsored by HMG in collaboration with bilateral and multilateral donors, the activities of smaller, independent organizations are largely unrecognized in the development community. As forestry development turns increasingly to the concept of working through user groups to establish and protect forest resources, it becomes imperative to examine the experiences of those projects which have combined integrated approaches with a commitment to community participatory development to involve users of resources.

Methodology

This survey of agroforestry activities was limited to the Kathmandu-based head offices of the NGOs in question. The method of information-gathering included the use of a semi-structured questionnaire and the review of available project documents. The questionnaire was used to interview key persons of the organizations involved in planning and implementing agricultural, forestry or agroforestry activities in different parts of Nepal. Information obtained through such in-depth discussions, in addition to that obtained in project background papers, was used to complete the checklist of topics to be covered. Once completed, each NGO was sent a print-out of its respective section for corrections and revisions. Locations of project activities were marked on maps of Nepal.

SUCCESSFUL NGO INITIATIVES

During the survey, it became clear that some NGOs have achieved successes in activities which have posed difficulties to larger organizations with more formal, sectoral-based styles of working. These are noted here to draw attention to their achievements. It is believed by many that NGOs, due to their flexible nature, can take the lead in demonstrating new methods of working with communities to improve living standards. The integrated strategy followed by most NGOs interviewed presents a style that differs significantly from that used by most agencies promoting agriculture and forestry programmes in the country; it allows an approach that may be more suited to the world view of farmers managing their family members, homes, livestock, and private and communal lands as an interlinked system. We shall review some of the areas where NGOs have been particularly successful.

Involvement of women

Many of the NGOs have recognized the key role that women play in activities related to cropping systems, forest use, and livestock maintenance. ActionAid, CECI and the 'women in forest resource management' project have made women the focus group of agroforestry programmes. CARE, UMN, Redd Barna and SCF-US all consider the participation of women to be very important to achieve their goals. ActionAid, SCF-US and some UMN projects (Lalitpur and Andhikhola) have been able to draw women into the planning and implementation of agroforestry projects through their participation in literacy classes. These gatherings of women (over 90 per cent of SCF-US's class members are female) provide informal fora in which members can discuss and motivate one another to participate in such activities. Redd Barna and ActionAid have the structure of child-care centres to use as entry points for participation. In UMN, women are active as *naikes* and in user groups; ActionAid has organized women to form livestock groups that encourage the planting of fodder trees and grasses, and has community nurseries managed by groups of women.

Smokeless stoves

A few NGOs have achieved successes in the field of smokeless stoves (*chulos*). This is another topic that has proved problematic to planners of forestry programmes who had initial high hopes that the promotion of such devices could reduce the consumption of fuelwood. Attempts to date by various projects have not been able to generate acceptance of these improved stoves by large numbers of hill women due to inappropriate designs and technologies, unavailability of stove parts, and ineffective extension methods. NGOs such as SCF-US and UMN (Andhikhola) brought to attention the existence of their programmes in this area, and expressed their opinions that such efforts have proven successful. Further study on this topic would be needed to reveal more details of NGO activities in this field.

Traditional agroforestry practices

The question related to the staff's awareness of farmers' traditional knowledge of agroforestry practices did not elicit much information, except in one case (ActionAid). One reason for this is that staff who would have these answers are most likely to be found in the field, not in Kathmandu headquarters offices. However, several NGO staff stated that they would like to integrate a knowledge of indigenous management practices into their agriculture and forestry programmes. CECI staff noted that their programmes would begin with a survey of such knowledge before project planning would commence.

ONGOING CONSTRAINTS

From the answers provided to the questions posed to NGO representatives, it is clear that many of these organizations have made substantial contributions to the research and development of agroforestry in Nepal. However, it also became evident during these discussions that all face similar kinds of constraints to improving or expanding such activities, which are described in the survey. Some organizations lack technical training and advice; for others, the problem is a shortage of seeds and seedlings. But one interesting and encouraging point that emerges is that many of the stated needs could potentially be addressed from resources found within the NGO community itself, through the sharing of knowledge, experience, and other resources. The breadth of experience within the NGO sphere is such that one organization's weakness is another's strength. This is demonstrated below, according to expressed areas of need.

Seeds, seedlings and inocula

The procurement of quality seed has been expressed by many as a crucial basic requirement for the functioning of agroforestry programmes. ActionAid and World Neighbors obtain grass and tree seeds from the National Tree Seed Unit and the forest research project, sources perhaps unknown to the other NGOs in need. Still, though, quantities obtained are insufficient, particularly of grass and fodder species.

CARE/Nepal has been discouraged in its efforts to promote *ipil-ipil* (Leucaena spp.) by a lack of the inoculum needed to propagate seedlings, yet other NGOs are supplied with the critical ingredient by World Neighbors and contacts at the Agriculture Department at Kumaltar, Lalitpur. Tree seedlings always seem to be in short supply at the time of planting, leaving NGOs without their own adequate nurseries dependent on the favour of DFOs or project managers who can spare some of their leftovers (which are often not of good quality nor are desired species). This is an area of possible collaboration among NGOs; SCF-US, for instance, is planning to construct a very large nursery that will produce 100,000 seedlings per year for the market.

EXTENSION MATERIALS

NGOs require extension materials such as films, flip charts, etc., to raise awareness of resource issues and motivate farmers to become involved in activities that will improve the resource condition of their croplands and forests/pastures. Most of the NGOs interviewed expressed a need for simple, relevant materials, but were not aware of the sources and types of such materials that do exist. The scarcity of materials was noted by Redd Barna, and the need for them stated by CECI and SCF-US. ActionAid is in possession of filmstrips and solar projectors developed and sold by World Neighbors, but does not make full use of them due to the staff's unfamiliarity with the equipment. Bauddha Bahunipati Project (BBP) staff continue to design their own materials, as do the staff of UMN's Andhikhola Project (dramas, flip charts) and Lalitpur Project (a video on forestry). Difficulties experienced in borrowing popular films such as *The Fragile Mountain* from the HMG/DSCWM office were noted by ActionAid staff.

TRAINING

Training needs for NGO staff fall into two categories: technical training and training in methods of community development. Training in technical subjects such as silvicultural methods, lopping techniques, and methods of nursery propagation for species difficult to produce, were mentioned by ActionAid, SCF-US, CARE and the women in forest resource management project staff. The UMN and World Neighbors (BBP), which both have more experience than most NGOs in research related to technical aspects of agroforestry, may be in a position to assist on the training side. The rural development centre of UMN will be expanding soon to provide training to project staff other than their own. BBP staff have already provided training to staff and villagers of other projects such as ActionAid. Short sessions on specific subjects are available from the research staff of the forest research project. Lumle Agricultural Centre has assisted many NGOs with their agroforestry training needs, including CARE, ActionAid, UMN and Redd Barna. INSAN also offers short-term courses related to sustainable agriculture.

A need for training related to community development methods was expressed by both CECI and CARE. The staff of women in forest resource management project and CECI have used the trainers of South Asia Partnership to teach these skills. Other projects have sent their staff and some villagers on tours to see how some initiatives have succeeded through the active participation of local farmers; the Bauddha Bahunipati Project has been a popular destination for NGO and government staff visits for several years now, and has reportedly sparked the interest of many farmers who have been motivated to try to replicate the results on their own farms (as were women involved with the women in forest resource management project). NGOs which have achieved some success in forming user groups and eliciting the participation of women and poorer groups of the society in agroforestry activities could easily share their experiences more thoroughly through more visits of this nature. CECI also mentioned the need for training in methods of rapid rural appraisal.

Research

Research needs were not expressed by all NGOs, but were limited to those which have had more focus on technical aspects of agroforestry, and which have specialized agroforestry staff, such as UMN, World Neighbors, CARE and ActionAid. These organizations seek additional information on fodder tree management techniques, shade effects of trees on crops, species combinations, effects of agroforestry combinations on nutrient recycling, etc. It is generally felt that too little attention is being paid to these issues; only the forest research project, Institute of Forestry, and Nepal coppice reforestation project are known to be pursuing answers to this type of question. More on-site research could be conducted by NGOs in their projects, as World Neighbors and UMN (Butwal) are doing.

Monitoring

Those NGOs which are without their own technically trained staff have mentioned the desirability of having occasional site visits made by experts to assist in the planning and monitoring of agroforestry programmes. CECI and ActionAid have both specifically stated that such people could analyse each programme on an individual project basis and offer advice on techniques, species selection, etc., and link the NGO to appropriate agencies which could offer additional help.

Networking

The need for more sharing of information and experience among NGOs and other service organizations was mentioned by all NGOs. The extent of linkages that each NGO interviewed has developed differs significantly, and probably has some impact on the degree of success experienced in agroforestry activities. With fewer financial and technical resources available to them than to large projects or government agencies, NGOs depend to a much greater extent on their links and contacts with other organizations. The International Union for the Conservation of Nature (IUCN) has requested that an NGO-co-ordinating body be established to promote the activities of NGOs in the fields of environment and resource management; they themselves are unique in their linkages to communication organizations which could operate as focal points for information sharing. From evidence generated by this survey of the wealth of experience and knowledge that exists among NGOs in the field of agroforestry, it seems that a formal system of information sharing could be developed at low cost to solve some of the problems stated above, and magnify the success of these community development organizations in their efforts to improve the access of poorer sectors of society to crop, forest and livestock resources to meet their basic needs.

EXPRESSED NEEDS OF NGOs

The areas in which NGOs involved in agroforestry feel their efforts need further strengthening can be summarized as follows:

Research

- optimal agroforestry species combination
- development of fodder tree management techniques for use in the farmer's field
- screening and establishing source of appropriate rhizobial inocula for nitrogen fixing trees
- total biomass production in farmer's field through agroforestryactivities
- shade effects to agricultural crops in agroforestry activities
- digestibility of fodder by species
- lean period for fodder and its effect on livestock
- rooting characteristics of agroforestry trees
- information on nutrient recycling in agroforestry and its long-term impact on soil fertility
- development/screening of shade-tolerant poplar strains for intercropping with agricultural crops.

Training

Many NGOs feel that there is a need for development and strengthening of a training institution targeted at the following:

- field-level workers of different NGOs in agroforestry
- community development aspects
- providing appropriate diagnostic tools for rapid identification of local problems

- skill development of the nursery *naikes*
- Environmental extension work.

Professional services

There was an expressed need for professional services to be made available in:

- planning and monitoring field sites
- specific technical aspects (i.e. silviculture).

Extension

- Development of extension materials on environmental awareness, and resource conservation and utilization for use in workshops/training arranged in villages.
- Development of demonstration sites to enhance resource conservation activities.

Seeds

Insurance of quality seed sources for grasses, tree fodder, fuelwood and timber species.

Tree tenure

Many felt there is a need for clear-cut, simplified government policy with regard to private tree tenure.

Networking

Sharing of information and experiences of different professionals and organizations working in the field of agroforestry.

FOLLOW-UP ACTIVITIES INDICATED

From the discussions with NGO staff, it became clear that increased networking among NGOs was a commonly felt need, and one that could fairly easily be addressed. With this in mind, it was decided to provide each NGO interviewed with several copies of this paper to allow them to learn of activities and experiences of the others surveyed, with the hope that they would form links, on their own, with other NGOs able to offer assistance. In this sense, the wide distribution of this paper is intended only as a first step in a process to increase the sharing of information among NGOs in order to meet their needs for resources in the field of agroforestry. It is hoped that NGOs in addition to those represented here, including national and local NGOs, will be able to make use of this report and contribute their experiences as well.

NEPAL AGROFORESTRY FOUNDATION (NAF)

Nurseries, training, demonstrations and networking

B H Pandit

This section explains the contributions of NAF to other NGOs, government organizations and private institutions of Nepal in promoting agroforestry activities.

WHAT IS NAF?

NAF is a recently registered and locally managed organization devoted to the promotion of agroforestry action research and training. It aims to achieve balanced and sustainable development through appropriate ways of managing complex interactions among various components of the ecosystem, that is agriculture, livestock and forestry. Previously, similar activities were being carried out by World Neighbors' supported Agroforestry Advisory Service for agroforestry development in Nepal. NAF board members represent several NGOs whose focus is on small farmers who fall within the economically and socially disadvantaged groups of society and operate in areas which lack such infrastructure as irrigation. NAF's main purpose is to expand the initial work started through World Neighbors (WN) and to develop a local self-supporting institution with the capacity to support and replicate proven agroforestry extension work in Nepal.

WHY NAF?

The degradation of forest in the hills of Nepal has been intensified by the fodder, fuelwood and timber needs of the rapidly growing population. In this context, agroforestry support programmes focusing on fodder and green manure cover crops can assist farmers by increasing soil fertility, sustaining increased numbers of livestock, meeting needs for fuelwood and timber, and reducing pressure on common forests.

The Government of Nepal has a long-standing forestry programme, but, as discussed below, has faced numerous difficulties in extending this to embrace agroforestry. However, NGOs working at the community level have a long track record in identifying and promoting agroforestry technologies that have been taken up by farmers. This experience can supplement government programmes, but, at the same time, there is a lack of research on the management of various fodder trees in agroforestry systems on terrace faces. Information of this type needs to be generated and shared in support of NGO work. NAF was established to meet some of these needs and to supplement the national programme of the government.

The Nepal Agroforestry Foundation's objectives are as follows:

- To promote throughout Nepal the demonstration and extension of agroforestry components such as trees for fodder, fuelwood, horticulture and timber, and also farm crops, grasses and livestock. To promote the 'systems' integration of these, with particular attention to soil and water interaction and to the role of green manures.
- To strengthen skills, especially among NGOs, in programme planning and support, needs surveys, monitoring, evaluation, action research and the conduct of trials and demonstrations on farmers' fields. To strengthen skills in the identification of appropriate technologies and management practices, and in extension processes.
- To provide these services to and establish networking with NGOs and GOs and with farmers, focusing especially on NAF members' projects and programmes with small and marginal farmers.
- To produce seeds and planting materials of promising species for trials and distribution to farmers and projects.
- To co-ordinate research and trials in agroforestry species with development and extension programmes in the field.
- To become self-supporting as a non-profit organization.

ACTIVITIES

Sixteen project/community nurseries and twenty-eight home nurseries in six NGO sites in the central hill districts of Nepal are being supported by NAF during the current year. A total of 110,000 saplings of fodder trees, fuelwood/ timber and fruit trees (excluding grasses) have been distributed to farmers from these nurseries in which fodder trees represent 60 per cent of the total.

Two demonstrator farmers at each nursery site plant an average of 300 (surviving) fodder trees each per year. Farmers usually plant 200–400 seedlings a year and add trees each year to reach the total of 1,000, providing that land-holdings are sufficient. NAF produces and collects appropriate seeds and planting materials to meet the demands of both farmers and NGOs for agroforestry species on a self-sustaining basis, using small farmers as seed producers. Moreover, small farmers are encouraged to establish private nurseries of horticulture and grass species and to become self-supporting from their sales. NAF charges for services to cover seed production, administration and training costs in order to achieve financial self-sufficiency. NAF seeks to scale up successful agroforestry demonstrations conducted by NGOs by having teams of trained farmers replicate them. In addition, NAF provides seeds and advice to twenty NGO projects interested in promoting agroforestry activities.

Training and exposure workshops are among the key activities that NAF conducts throughout the year at times most suitable for dealing with specific field problems. In 1990–1 a total of 400 participants from different NGOs, government organizations and private institutions participated in NAF's training programmes. Demand was particularly high for training in agroforestry and basic nursery skill (almost 60 per cent). Participant organizations included International NGOs, such as CARE/Nepal, Redd Barna (Norwegian Save the Children Fund), ActionAid/Nepal, CECI/Post Earth Quake Project, Swiss Development Co-operation (Dolakha), Lutheran World Service, and United Mission to Nepal (all Nepal). Local NGOs included Baudha Bahunipati Family Welfare Project/Family Planning Association of Nepal (FPAN), Tamakoshi Sewa Samiti (Ramechhap), Non-Formal Education Service Centre (Dhading), ABC-Nepal, and Integrated Development Systems (Kathmandu). Finally, government and semi-government projects included the Lumle Agriculture Centre, Kaski (research), the Watershed Management Project, Kathmandu (watershed protection), the Annapurna Conservation Area Project (ACAP) Pokhara (mountain conservation), and the Integrated Watershed Management Project, Kulekhani.

Future priorities for exposure and training in agroforestry include:

- exposure of NGO staff and farmers to successful farmer demonstrations of agroforestry
- training of trainers for the establishment of agroforestry programmes, forming community user groups, nurseries, extension of agroforestry, and the methodology of farmers' training farmers
- home nursery training methodology, and training of instructors who work with farmers
- basic nursery skills training
- livestock user groups: upgrading of livestock management and health
- citrus propagation and citrus pest and disease control.

As a general training process, NGO project staff and farmers are exposed to the potential of agroforestry through a one-day course. This is followed by specific training for field staff or farmers and 'cross-visits' from one agroforestry demonstration area to another in order to increase the sharing of ideas among farmers and the rate of adoption.

INTER-AGENCY NETWORKING

NAF has become one of the members of a field implementation working group under the Forestry Sector Co-ordination Committee of HMG Nepal.

This group consists of twenty-seven NGOs and GOs and has formed an NGO co-ordination committee of which NAF is also a member. Particularly strong links have been developed between NAF, other NGOs and the Department of Livestock of the Ministry of Agriculture. A sub-group formed in co-ordination with the Department of Livestock is preparing guidelines on the basic steps of co-ordinated livestock development for promoting livestock and fodder development activities in their working areas.

A one-day workshop in 1991 on the impact assessment of Bauddha Bahunipati Project (BBP), one of the integrated rural development projects supported by World Neighbors, was organized. The study was done by New Era, a private research organization with financial support from the Ford Foundation and World Neighbors itself. The main purpose of the workshop was to disseminate the findings of the study on such aspects as bio-mass generation, crop yields, income from livestock and labour allocations. Twenty-five NGO and GO projects of Nepal participated in this workshop. The list of participants has provided the basis of a networking group in agroforestry, including government and non-government projects.

Co-ordination between the Division of Entomology, Ministry of Agriculture, to release psyllid predators *Curinus coeruleus* and *Olla abdominalis*, the Coccinellid beetles, into Nepal from Thailand, was made possible when the government selected field project sites for releasing and testing predators. Sites are also visited by government-sponsored consultants.

THE INTERNATIONAL UNION FOR THE CONSERVATION OF NATURE (IUCN) AND NGOs IN ENVIRONMENTAL PLANNING IN NEPAL

J Carew-Reid and K P Oli

INTRODUCTION

Since 1948, IUCN, the world-wide conservation union, has been actively involved in conservation of natural and cultural resources in many countries around the world. By the close of 1990, it was operating in some 119 countries with 627 institutional members including states, government agencies and leading conservation organizations.

Nepal joined the IUCN as a state member in 1973 and after the endorsement of the World Conservation Strategy (WCS) by His Majesty's Government of Nepal (HMGN) in 1980, IUCN is currently assisting in the implementation of the National Conservation Strategy (NCS), through the National Planning Commission (NPC).

As part of this programme, NPC/IUCN is working with seventeen government agencies and over twenty local NGOs in Nepal. The major programme areas include environmental planning and assessment, environmental education and public information. The development of integrated policies and procedures for planning and environmental impact assessment is co-ordinated by the intersectoral National Environmental Planning and Assessment Steering Committee chaired by the National Planning Commission.

An environmental core group of forty senior government officials from some seventeen ministries and the NPC has been established to facilitate the consideration of environmental factors in policy formulation. This core group is participating in policy development workshops, and testing national and sectoral planning guidelines and procedures for Nepal. Environmental policies are therefore being formulated and tried by local experts, according to local needs and in practical decision-making situations.

This policy work is complemented by an intensive field programme at village and district levels which seeks to develop decentralized methods and structure for planning which can be implemented through a national system.

Existing short-term, centralized and sectoral planning has not promoted the sustainable use of resources. An integrated, cross-sectoral approach to environmental planning in which local communities control and manage their environment with technical assistance from government is being tested under the NCS implementation programme. This section introduces the concept of environmental planning and the approach taken by IUCN to develop model environmental planning at local level for sustainable resource management in Nepal.

ENVIRONMENTAL PROBLEMS AND NEED FOR PLANNING IN NEPAL

Effective environmental planning requires collaboration, among the various sectors (forest, agriculture, livestock, and so on) of the system, a participatory approach decentralized decision-making and genuine devolution of power over resource management to local communities.

THE ROLE OF NON-GOVERNMENTAL ORGANIZATIONS (NGOs) IN ENVIRONMENTAL PLANNING

In Nepal, IUCN has been working to promote people's participation through local NGOs. The value of NGOs to local environmental planning derives in part from well-known characteristics: their size, proximity to rural dwellers, poverty focus, innovativeness, flexibility and interdisciplinarity. Of particular value in Nepal are their enthusiasm for collaborating with other organizations and for creating community networks of expertise and resources, and the fact they are associated with the people's movement, with democratic ideals and with the new government's emphasis on local self-sufficiency.

For developing model village environmental plans, two NGOs were deployed in eight villages of Lamjung and Arghakhachi. At the same time, development of environmental planning guidelines for all four levels of government (village, district, regional and national) was initiated using the environment core group members from different ministries at national level. Both exercises were facilitated by a multidisciplinary NPC/IUCN planning team.

PLANNING AT LOCAL LEVEL: VILLAGE DATA COLLECTION

In the IUCN programme, the NGOs decided on a participatory rural appraisal (PRA) methodology for data gathering combined with longer-term participatory research. The intensive interaction with local people, site visits and collection of written material encouraged in PRA allowed the interests of all affected groups to be incorporated into the plan in a culturally sensitive way.

The NGOs worked with their communities in their assigned villages to form village environmental planning committees responsible for overseeing the entire environmental planning process. The committees included women and representatives of lower socio-economic groups. One local supervisor, usually a school teacher living in the area, was trained in PRA techniques and stationed at each village to facilitate the process of information collection.

In all, eight village committees and seventy-two sub-committees – the latter based on settlement patterns and resource distribution – were established in eight villages. A checklist of information which needed to be gathered was prepared by the NGOs following discussions with the village committees, and included:

- the condition of access to and use of, existing natural and cultural resources; socio-economic conditions and felt development needs
- relevant central government directives, guidelines, by-laws and services; specific environmental problems; ongoing major programmes and projects.

PRA was then conducted over a number of months by committees and sub-committee members and the facilitators in order to prepare comprehensive village profiles.

VILLAGE ENVIRONMENTAL PLANS

From these data, village environmental profiles are prepared jointly by village committees and NGOs, and discussed at plenary village meetings. The profiles are submitted to district-level environmental steering committees which integrate the local process with district planning. They are also used in village plan formulation, which usually requires visits to specific sites by the committee, sub-committee members, the facilitator and the village-based line agency workers. The NGOs provided this group with cadastral maps, aerial photographs and records of land-ownership of the area. Participation by people from the area itself allows areas to be blocked out on maps according to common features of land use, cropping systems and altitude. Options for future land-use systems are discussed on the basis of 'zoning' which allows limited resources available for development and conservation to be applied to those areas of most immediate concern, and then agreed by all concerned.

A high level of local participation has so far been forthcoming, and allows activities to be divided into three categories: those which the local community can undertake without outside help; those which would require a minimum of technical support through line agencies of government; and those which would require substantial outside resources. Environmental indicators and monitoring programmes were also defined to be undertaken by farmers and other users' groups to ensure that unwanted environment impacts are quickly identified and remedial measures taken.

This participatory approach, although time consuming, integrated modern approaches with local knowledge and involved the local community in establishing a village database. Following the on-site visits and community gatherings a draft plan was prepared by the facilitator assisted by the NGO and made public for comment. A series of community meetings were conducted by the village committee and comments fully incorporated in a final draft plan. This draft was formally endorsed by the village assembly which then presented it to the appropriate district environmental planning committee. In both districts, line agencies agreed to incorporate the plans into their own strategies and work together on a co-ordinated basis in facilitating their implementation.

INTEGRATION OF VILLAGE ENVIRONMENTAL PLANS WITH GOVERNMENT SECTORAL PLANS

Despite the effort to integrate government line agencies into the village planning process, a gap remains between the focus of planning at the local level and the vertical process retained within departments and driven by the annual budgetary cycle.

However, some success has been achieved in reflecting sectoral objectives in village level plans. The delivery of sectoral services can be integrated simply enough into those plans which can be implemented by the village committees themselves. Village environmental plans demanding greater technical and other resources or covering more than one village are integrated with the respective sectoral plans at the district level, which has a key role particularly in case of shared natural or constructed resources.

This experience at local level is feeding into the environment core group's guidelines for a national system of environmental planning. Legislation is also being drafted with the intention of instituting the environmental planning process and the concepts it embraces of devolution, integration and long-term perspective in resource use planning and management.

Sensitive approaches to local environmental planning have succeeded in incorporating elements of farmers' traditional practice into village plans, including dependence on wide biological diversity at the farm level as a hedge against risk; the selection of locally adapted cultivars and animal species; the application of manure to increase soil organic matter, and the interplanting of legumes with principal crops such as maize, sorghum and rice.

DISCUSSIONS AND CONCLUSIONS

In Nepal national economic development planning has been in practice since the late 1950s. Effective control and management of local resources by 'user' communities is the only way to arrest the steady depletion of natural and cultural resources that orthodox planning has permitted.

The IUCN programme, carried out since the early 1980s in eight villages in central Nepal, found local NGOs more efficient than government agencies in discussion and dialogue with local communities, and in creating awareness about environmental issues of concern to local people and in supplementing scientific methods of assessing e.g. water quality, erosion and forest loss with local knowledge.

Villages – through village committees and sub-committees – and sectoral agency personnel are heavily involved in the planning process. Many disputes over drinking water, irrigation and grazing rights were discussed and even settled during the plan preparation process. NGOs' role in forming effective committees and sub-committees is therefore a vital contribution to institutional sustainability in environment plan preparation and implementation.

Probably the most important feature of environmental planning at local level is the emphasis it gives to traditional management systems. In one of the model villages, for example, people continued to manage their forest in the traditional way despite government intervention. They have set rules for harvesting, have raised funds for forest guards and introduced special controls for seasons when there is a danger of fire. They do not allow outsiders to enter or use the protected forest area. This village also has developed rights and controls on drinking and irrigation water and animal grazing. The IUCN approach to environment planning builds upon such successful indigenous environment management practices, seeks to incorporate the lessons from them into inter-sectoral planning at higher levels, and so make it possible for other villages to manage local resources by their own criteria in the same way.

UNITED MISSION TO NEPAL's INVOLVEMENT IN FORESTRY ACTIVITIES WITH SPECIAL REFERENCE TO THE NEPAL COPPICE REFORESTATION PROGRAMME

Duman Singh Thapa

INTRODUCTION

The United Mission to Nepal (UMN), established in 1954, is an NGO comprising over forty Christian organizations from eighteen nations which have agreed to work in unity under a single administration with a common purpose. The UMN supports various programmes to strengthen the capability of local communities to manage their own development by providing human and financial resource and training on the basis of agreements with His Majesty's Government of Nepal (HMGN).

UMN's programmes are administered under its four sectoral departments: health, engineering and industry, education, and rural development. UMN's own relations with government have evolved since its establishment. In the 1950s and 1960s UMN developed and operated its programmes almost completely in isolation from a government whose own programmes were still largely non-existent in rural areas. By 1970, however, national government had developed its own policies and sought to have more direct control over certain areas of UMN activity, even to the extent of taking over or closing some programmes.

UMN responded by planning its programmes in ways compatible with those of government. However, government's incapacity to provide even basic services in many areas, together with recent political changes in Nepal, have made it more open to collaboration with NGOs. At the same time, UMN faces the choice of whether to move completely into supporting the voluntary sector (increasingly including local NGOs) or to continue its dual approach to supporting both public and private sector programmes.

RURAL DEVELOPMENT DEPARTMENT (RDD)

The RDD was initiated in November 1987 to give a greater focus to this sector and to consolidate the diverse RDD work being carried out by UMN. The RDD implements community development support programmes, projects and activities, and represents the needs and concerns of these efforts within UMN, to UMN's supporters and to the authorities concerned. The department also supports activities being undertaken by other departments within the United Mission and by Nepalese organizations who are independent from it. Most of UMN's forestry and agriculture work falls under the RDD.

SUMMARY OF UMN FORESTRY WORK

UMN forestry work takes place in the context of wider rural development efforts at eight locations, at which four broad programmes (community and private forestry; natural and leasehold forestry; medicinal, aromatic and minor forest products; soil conservation and watershed management) are pursued in varying combinations.

Briefly, these activities include the following:

- Support to Karnali technical school (Jumla); training forestry guards and forestry motivation in three village development committees (VIDCOs).
- Awareness creation in three VIDCOs in Surkhet.
- Species trials in forestry and agroforestry systems to enhance material supply for Butwal plywood factory.
- Awareness creation and non-formal education in five VIDCOs in Andhikhola as part of a resource conservation programme in support of hydroelectric and irrigation work.
- A link with the Pokhara rural development centre provides technical assistance to UMN projects. UMN assists in following up trainees' work in establishing nurseries and setting up forest committees; particular attention has been given to bamboo propagation techniques.
- UMN's community development and health programme in Lalithpur was started in six VIDCOs in 1983, and has resulted in wide establishment of private and community plantings and, more importantly, in the reversal of habits that had led to ecological deterioration. Seedlings are provided through twenty contract nurseries.
- In Okhaldhunga, UMN started to work with three VIDCOs in 1987, principally motivating tree planting in private land for which several private nurseries have been established. A forest committee has been established to maintain a community nursery.

UMN's work with the Nepal coppice reforestation project (NCRP) since 1990 has focused on Naubise village, 26 km west of the Kathmandu Valley in Dhading District. NCRP is the only UMN forestry project under a formal agreement with the Ministry of Forest and Environment. Under this agreement NCRP is the facilitator between HMG and community user groups to transfer government forests to villagers with community ownership.

Prior to 1990, Phase I of the NCRP, funded by USAID and implemented by a US contractor, comprised:

- A permanent multipurpose tree nursery primarily for the production of planting stock, but also as facility to conduct applied experimental work on plant propagation, nutrient and water requirements, nursery practices, outplanting techniques and species provenance and clone trials.
- A research and demonstration programme that would evaluate and demonstrate several high-yield production systems for tree products, methods for the maintenance and management of these systems, and their economic and cost effectiveness.
- A programme of technology transfer and technical assistance aimed at increasing employment and income in the forestry sector through the training of technicians, farmers and students in the use and management of multipurpose tree production system.

Only the first of these objectives was achieved under Phase I. Participation with local people to understand their forest management practices and constraints, to incorporate their priorities into the research agenda, and to adapt technologies to their requirements had been negligible, as had contacts with local forestry officials. UMN's approach has been to develop community-based forestry activities in parallel with the transfer of land from the government to user groups. Following a UMN survey of user groups in the area, land has already been transferred to one, and requests from five more have been forwarded to the District Forest Office for approval. Other groups from neighbouring areas have approacl ed UMN to take advantage of the support it provides to group formation through education and motivation.

In this programme, UMN staff provide information and stimulate dialogue only around certain issues that have been raised in order to foster a learning process around a specific idea or problem, but the final decisions to implement or to abandon an idea rest with rural people themselves. All this enables the people to analyse critically their total situation, to gain the power to change that situation and to see themselves as capable of changing it. The development of user groups in this way is seen as the most viable way of achieving long-term institutional sustainability.

As part of the drive towards institutional sustainability, it is planned to transfer the project's research activity to the Department of Forestry and Plant Research (DFPR), and seedling production will be ensured through contracts both with local user groups and with DFPR. Research currently focuses on the coppicing qualities of thirty multipurpose tree species, and, as a further component of institutional sustainability, a research guidance committee comprising representatives of programme staff, of DFPR and of donor agencies, has been established.

GOVERNMENT AND NON-GOVERNMENT RELATIONSHIP USING THE NCRP EXPERIENCE

UMN has found it difficult to convince villagers that forestry projects are essential in order to prevent deforestation and soil erosion since they tend to see forests as the property and responsibility of government. They identify other needs like health, drinking water, electricity, roads, bridges, etc. and suggest that these should be given priority. These attitudinal problems have compelled UMN to raise villagers' awareness of recent forest legislation and to integrate various forestry into other development activities so that these communities feel they have control over community forestry and so have the confidence to carry out forestry activities.

UMN's approach has been cross-sectoral, integrating forestry work with, for example, projects to improve drinking water and to enhance the status of women. This has the advantage of allowing UMN to act as a bridge between the narrowly defined activities of government departments. This involves some time cost to the NGO – certain proposals to conduct non-formal education in forestry have, for instance, had to be approved by the district Education Office – but it has the potential advantage of enhancing one Department's appreciation of the other's work. Another time-consuming aspect of UMN's integrated approach is that activities conducted under an agreement with the Forestry Department must be justified in terms of their benefit to forestry activities. For instance, a drinking water system being constructed by UMN in Naubise village had to be modified to take into account the water needs of local nurseries. One of the greatest strengths that UMN has been able to share with government departments is the capacity to work with local farmers, to encourage experimentation among them, and to draw both on their knowledge and experience and on that of technical scientists in prioritizing more formal research agenda.

NOTES

1 Two types of guthi are prevalent. One type is a caste or kinship group formed to ensure continuity and security of social, cultural, and religious activities of its members. The second type relates to land-ownership which was endowed for religious and charitable purposes.
2 Parma are reciprocal labour exchange groups.
3 Dhikur is a voluntary credit association usually initiated to raise capital for investment, and is prevalent among Thakali and Gurung ethnic groups.
4 'The Mahaguthi, originally created to perform some religious rites, later expanded its activity to develop cottage industries through the hand spinning wheel, the "Charkha" ' (New Era 1989). The Mahaguthi runs schools, literacy and health-care

centres, provides training for poor and destitute women, and runs shops to sell Nepali handicrafts.

5 Originally started to provide medical services in Kathmandu for people suffering from diarrhoea and dysentery. Paropakar now runs a school, orphanages, and provides health and maternity care services.

6 Health Service Coordination Committee, Hindu Religion Service Coordination Committee, Child Welfare Coordination Committee, Youth Activity Coordination Committee, Community Service Coordination Committee, and Women Service Coordination Committee.

7 Ministry of Finance (1991b: 20).

8 This section draws heavily on Shrestha (1991).

9 SAP (South Asia Partnership) – Nepal, 1988, 'Strengthening Nepalese Non-Governmental Organisations: Human Resource Development Needs Assessment', Kathmandu, Nepal.

10 This section draws on Farrington and Mathema (1991).

11 The Department of Agriculture was established in 1924. The Forest Service was established in 1942, but conducted little formal research until the 1970s.

12 Resource constraints did not permit wider coverage.

Chapter 6

NGO–GOVERNMENT INTERACTION IN THE PHILIPPINES

OVERVIEW

Aurea G Miclat-Teves and David J Lewis

The Philippines, a country consisting of approximately 7,100 islands, covers a total land area of 30 million hectares. Agriculture accounts for more than one-quarter of the country's GNP, contributed 22 per cent of total exports in 1989 and is therefore a priority in the country's development planning. Along with Bangladesh, the Philippines is now the poorest country in the region. Almost half of the labour force are engaged in agriculture, although only a small minority actually own the land they work. A national survey carried out by the social weather stations in 1987 (quoted by Liamzon and Salinas 1989) reported that 71 per cent of Filipino households own no agricultural land. There are over 2 million sharecroppers and 1.5 million people farming public lands without title. Two out of every three families live below the poverty threshold of P2,531 per year for a family of six; 85 per cent of all school children in the country are deficient in protein and calorie intake and up to 40 per cent of all fatalities are the result of malnutrition (Steinberg 1986).

Foreign capital maintained a position of control over the Philippines elite during the Marcos period. Control of large tracts of land and other resources have passed in recent decades into the hands of large corporations (both local and foreign MNCs) through government leases, contracts and concessions. In Mindanao, for example, over one-third of the prime agricultural land is owned by thirty-one corporations, half of which are foreign controlled. Land-ownership patterns are highly skewed and benefits from technological innovations have tended to reinforce these inequalities. A World Bank (1988) report argued that the

> growing number of the poor and the persistence of income inequality in the Philippines is a result of three structural factors: unequal asset ownership, particularly land; population growth; and the lack of productive growth.

At the same time, the country has a US$29 billion foreign debt and fast deteriorating forest and marine resources. The population, which currently stands at 62 million, is growing at a rate of around 2.5 per cent per year.

The rural natural resource base of the Philippines is vital to the country's economy, and agriculture, forestry and fisheries considered together employ more than 50 per cent of the labour force, earn two-fifths of export revenues and contribute 25 per cent of GDP (World Bank 1989a). As a result of inequitable land-ownership and the reduction in available lowland arable land, migration to semi-cleared uplands areas has increased in recent years so that in the early 1990s one-third of the population live in upland areas, 8 million to 10 million of whom are farming forest land. The uplands have become severely degraded through insecure tenure (which promotes short-sighted and ecologically damaging agricultural practices) and commercial logging. For similar reasons of poverty and commercial interest, the coastal areas have also been subjected to unsustainable extraction of natural resources. This section explores the background relevant to gaining an understanding of the relationship between a sub-set of the NGOs and government agencies working in agricultural development, and argues that a certain complementarity exists in the sphere of technology development and delivery. However, in common with other countries, it would be naive to assume that a simple functional compatibility can transform the structural tensions which strongly influence GO–NGO relationships in the Philippines. We therefore compare the relatively successful partnerships undertaken in upland agriculture with the failure of the Aquino government to confront the urgent need for agrarian reform.

GOVERNMENT AGRICULTURAL POLICY

In the 1960s and 1970s, the government's agricultural strategy was primarily externally directed and based on the new 'high-yielding varieties' (HYVs) of rice being developed with donor support at the International Rice Research Institute (IRRI) at Los Baños. At this time the Marcos administration faced deteriorating levels of poverty, falling employment levels and the exhaustion of reserves of new cultivable land. The prospect of new 'miracle seeds', which could produce up to three times the yield of existing varieties, was readily grasped by an administration faced with the growing threat of rural unrest. While there was much unorganized activity originating from a variety of ideological perspectives, the unrest became increasingly co-ordinated by the New People's Army (NPA), the military wing of the Philippines Communist Party.

Despite the widespread increases in paddy yields, the Green Revolution constituted a technological solution to rural underdevelopment and did not address the need for redistributive reform (Feder 1983). The gap between rich and poor continued to widen. This approach which tended to devalue indigenous agricultural methods, leading to a reduction in bio-diversity and the damaging ecological effects of fertilizer and pesticide dependence (Shiva 1991). Soil erosion and leaching became a serious threat to continued expansion of HYVs (Steinberg 1986). In comparison with the effort with which programmes were developed for lowland agriculture, the uplands received little attention. They were regarded as backward and less important for national food production objectives, although the export of timber increased substantially. There were other political reasons for this, since the more isolated uplands became strongholds of the NPA, which found support among areas of the most extreme local poverty.

During the Marcos years, rural development was therefore left largely within an unregulated context, except for a number of *ad hoc* agricultural programmes designed and implemented primarily by outsiders. As Krinks (1983) points out, the workings of the market economy – set within an international context in which foreign companies had been able to insert themselves into positions of considerable economic and political power in the Philippines – were allowed to continue relatively unhindered in agriculture, fisheries and forestry leading to massive depletion of the natural resource base in each of these sectors. Krinks (1983: 119) presents an overview of this period which in many ways represents a low point in Asian rural development:

> A laissez-faire policy might be thought to imply a 'trickle down' approach to welfare, but looking at the needs of the tenants and labourers on plantations or the squatters in Mindanao and setting them against policies, it is hard to believe that decision-makers had any serious concern for even such residual spreading of the benefits of growth. Undoubtedly there were and are sincere people in the bureaucracy, among foreign advisors and even for a time in Congress, but the record of attempts and achievements conveys the impression that the measures taken either were intended to enhance the dominance of the privileged or were grudging concessions to reduce pressures for redistribution.

NGOs IN THE PHILIPPINES

Against this background, the need for popular development alternatives was keenly felt by people in the rural areas, although non-government organizations have traditionally faced an extremely difficult political context for their activities. The history of NGOs in the Philippines is linked to the evolution of peasant organizations. The National Union of Peasants in 1919 directly addressed problems faced by the rural poor and forced the government to deal with them. After the Second World War, agrarian unrest grew into what became known as the Huk Rebellion. Organized groups of peasant guerrillas (who had refused to lay down their arms) forced the newly independent government of Manuel Roxas to focus on the plight of the peasantry in Luzon, the continued victims of insecure tenancy and poverty (Steinberg 1986).

In the 1950s other peasant and workers' federations were founded, some by the Catholic Church, such as the Federation of Free Farmers (FFF) and the

Federation of Free Workers (FFW). At the same time the community development approach evolved from the work of Dr James Yen, who founded the International Rural Reconstruction Movement. The Philippine Rural Reconstruction Movement (PRRM) was also established, with its four-point programme on health, education, livelihood and self-government. The Catholic Bishops' Conference of the Philippines (CBCP) formed its National Secretariat for Social Action (NASSA) which in 1974 redefined social action as justice for total human development. This led to an emphasis on human rights and targeted assistance to the landless. In the late 1960s and early 1970s growing conflicts within Philippines society and crises brought about by natural disaster led to the formation of NGOs such as the Philippine Business for Social Progress (PBSP), which sought broad-based NGO strategies for disadvantaged people.

While the imposition of martial law by President Marcos in 1972 prevented traditional institutions such as political parties and business centres from functioning effectively, the number of NGOs working in pro-democracy, pro-poor advocacy and organizing work mushroomed (Garilao 1987). A sharper political analysis of social problems developed among NGOs as students, workers and farmers organized more effectively during the early 1970s. However, martial law brought many more practical NGO programmes to an effective halt. Some NGOs learned methods of working under highly repressive conditions. Others became highly politicized in their rejection of the Marcos government's authoritarian development model and NGO grassroots work was often suspected by government of being part of insurgency operations. NGO politicization was reflected in many NGOs' increasing involvement in political movements, both reformist and revolutionary. Many NGOs began to work in specific areas such as policy advocacy, community organizing, education and consciousness-raising.

The Aquino assassination in 1983 and the political and economic crisis – which by this time had alienated even the middle and upper classes from the Marcos regime – crystallized the growth of popular movements, pressure groups, coalitions and networks that, with the framework provided by the NGO sector, became the mass movement of 'people power' which eventually brought down Marcos (Villegas 1990). NGOs and other cause-oriented groups provided organizational support to the huge, largely spontaneous crowds of opponents to the Marcos dictatorship defending the Manila barricades.

There are now around 18,000 organizations registered as NGOs in the Philippines, of which two-thirds are 'voluntary membership organizations'. According to Constantino-David (1992), most of these fall into two major groups:

1 primary groups (or people's organizations) such as unions, community associations and co-operatives, operating on a mainly voluntary basis

2 civic and professional organizations, which tend not to have a grassroots membership.

The remainder are generally the organizations normally referred to as the development NGOs: intermediate agencies which aim to provide a wide range of services and programmes to and facilitate the emergence of self-sustaining local people's organizations (POs). The NGOs we are concerned with in this study are indigenous or foreign non-profit-making development agencies concerned with seeking the sustainable development of the livelihoods of the poor.

THE AQUINO GOVERNMENT

When the new government took power in 1986 after the peaceful 'people power' revolution, it gave recognition to the contribution of NGOs to development and responded to pressures from them to improve NGO–GO relations through a number of formal actions:

- The 1987 Constitution stated that 'The state shall encourage non-governmental organizations, community-based or sectoral organizations that promote the welfare of the nation' (Article II).
- The Medium Term Development Plan 1987–92 called for greater popular involvement in decision-making, planning and implementation of programmes along with a greater scope for private initiatives in pursuing economic growth.
- The government signed several international conventions, such as the FAO-World Conference on Agrarian Reform and Rural Development's 'Peasant Charter', and the ILO's Convention 141 which recognized active NGO participation at the national and international levels.

These changes were achieved with the help of a number of NGO activists who took up important positions within the new government, including the new NGO Outreach Desks in the Department of Agriculture (DA) and the Department of Environment and Natural Resources (DENR) (see Boxes 6.1 and 6.2). For some NGO workers, the mutual expectations proved difficult to reconcile and after a few years they returned to the NGO camp. For other NGO activists and government personnel, the new conciliatory approach began to bear fruit. Some of the collaborations which resulted from these changes, such as the adoption of the NGO-developed Sloping Agricultural Land Technology (SALT) in the DA's South Mindanao Agricultural Programme (SMAP) and the DENR's work with NGOs in implementing the Debt-for-Nature Swap scheme, are described in detail in later sections of this book. But while the DA NGO Desk, for example, sought to maximize the democratic 'space' for NGOs to carry out local level organizing, many NGOs failed to see the strategic value of taking up this space, or failed to understand how to work with government (which they continued to see as monolithic), or simply did not want to collaborate with government at all.

In many ways, the prediction made in 1987 by the Filipino NGO writer and activist Garilao has proved to be the case: 'NGOs will develop Third World leaders who will leave the NGO sector and move to government. . . . [They] will begin to enter into public service and politics because, in addressing the structural problems of poverty, it is the logical next step' (Garilao 1987: 119). This development highlights an important issue: that NGOs and governments are often interrelated in quite complex ways and cannot be conceptualized in terms of a simple 'opposition' between the two.

The resulting collaborations between NGOs and government agencies are an expression of the increased accountability of government, and a desire to encourage participation from wider sections of the community in government and development. The role of donors such as the Ford Foundation has often been an important catalyst. However, the potential for 'co-optation' by government remains an issue. As the initial expectations of the Aquino government faded, some former NGO activists found themselves stifled within the confines of routinized government agencies.[1] A functional conceptualization of the compatibility between NGOs and government roles has proved difficult to translate into action. As development NGOs became of interest to government, so government has tried to determine the type of role that it wishes NGOs to play.[2]

Many of the resulting initiatives are based largely on a vision of government working with NGOs as delivery systems, in sectors where government lacks the resources, organization or credibility to perform adequately. NGOs are seen as being able to 'reach' grassroots associations and communities where government has little expertise. One project which exhibits these tendencies is the Ecosystems Research and Development Bureau (ERDB) agri-livestock project. This is primarily an extension project aimed at delivering inputs to upland areas, where government, on account of its previous neglect or top down interventions, has little credibility. However, many NGOs had moved away from agricultural development work during the 1960s and 1970s, and according to ANGOC (1984), an emphasis on education, advocacy and consciousness-raising reduced the NGOs' potential ability to help deliver the development hardware and services under the new administration.

The tendency of organizations to appear as repositories for available funds remains a problem. The phenomenon of GRINGOs (Government-formed NGOs) tends to undermine efforts by the government to work sincerely with existing organizations. At the same time, GRINGOs are often a convenient way for government to gain access to foreign funds intended for bona fide NGOs.[3] As Constantino-David (1992) points out, there has been a spectacular increase in the number of social development NGOs since 1987 due to the increase in local and foreign funding and 'the ferment that developed from the anti-dictatorship struggle, the victory of the people-power revolution

Box 6.1 Problems and prospects arising from early activities of the Department of Agriculture NGO Desk

In 1989 in Camarines Sur, the DA's NGO Outreach Desk persuaded the DA to host and finance a provincial congress of farmers' associations and to allow NGOs and peasant organizations to participate in the design and the consultation. The design was approved after much heated debate and some disagreement. The DA felt that it had been inadequately consulted by the NGOs in the preparatory stages and that the design was too open and flexible. This reflected the process orientation of many of the NGOs on the one hand and the task/output-oriented mentality of some government officials on the other. During the conference, the DA received some criticism from NGOs and farmers' associations, to which some government staff reacted badly. While they accepted some of the criticisms, they were offended that such a response showed a lack of gratitude for their role in arranging and financing the whole initiative. When the DA complained to the Outreach Team, members responded that participants should be encouraged to comment freely since the DA was after all a public service paid for with taxpayers' money. Despite the intra-government tensions, the conference succeeded in securing representation of NGOs and peasant associations on the DA's Agriculture and Fisheries Councils (AFCs) at the municipal and provincial levels. Previously these councils wre dominated by large business interests, landlords and co-opted peasant organizations. The new independent presence on the AFCs has brought significant gains for local people, such as reduced costs for using DA training centre facilities.

Source: Personal communication from George L Villegas, former head of DA NGO Outreach Desk

and the consequent frustration over the obviously failed promises of EDSA' (1992: 2).

The Aquino government also took power with a promise that a far-reaching agrarian reform would be undertaken. Many NGOs and peasant organizations saw themselves as potential partners in this process. However, both the government's commitment to NGO participation in the planning and implementation of such a reform, and its wider ability to deliver a substantial agrarian reform programme have come to very little in practice.

Agricultural development NGOs have worked with some success in income-generation projects in upland agriculture. In the absence of government services (which particularly in the Marcos era has had very little credibility in many of these areas), many NGOs have worked with marginal farmers to develop technologies appropriate to marginal farmer needs and aimed at reversing the increasingly serious environmental degradation brought about by increased population, organized illegal logging and rural poverty. Some religious NGOs have been drawn to these areas by their 'indigenous' populations, while others have focused exclusively upon environmental concerns. Some are not new technologies as such, but combinations of existing

Box 6.2 The NGO Desk in the Department of Environment and Natural Resources (DENR)

The NGO Desk of the DENR consists of an advisory committee and the executive officer and staff. The advisory committee is composed of two assistant secretaries and a planning and policy division chief. The advisory committee is thus high up in the bureaucratic hierarchy. This is important as the NGO Desk requires top-level support for many of its activities and decisions. Its executive officer is an assistant director of the DENR Special Concerns Office. Later on, it was found out that the presence of top officials was important as the NGOs saw their presence as an indication of the importance given by DENR to GO–NGO interactions. NGOs wanted dialogues and meetings where key decisions can be made. These DENR officials can make key decisions and are known to NGOs as being able to do so.

It is also important to note that all of those in the NGO Desk, from the advisory committee members, the executive officer and the staff, came from a strong NGO background or at least possessed a strong belief in the importance of NGO participation. The chairman of the advisory committee was formerly a co-ordinator of a nation-wide networking of environmental NGOs while the executive officer was a member of NGOs working on tribal concerns. The DENR NGO Desk was thus seen as an ally whenever NGOs were requesting support from DENR or making complaints over its actions.

The DENR has expanded its NGO Desk into the regional offices. Initially, however, the regional NGO desks were not as effective as desired. One key reason was that the selected desk officers were at the lower middle rank of the bureaucracy and therefore lacked decision-making powers and consequently also the capability to draw resources into organizing a strong regional NGO desk. The latest decision of the DENR is to strengthen the regional NGO desks by making the regional technical directors for environment as direct supervisors. The DENR NGO desks will also receive a boost as the DENR built into its upcoming major programmes the strong participation of NGOs, such participation to be co-ordinated by the central and regional NGO desks.

methods, practices and materials assembled into packages which can be selected to meet individual farmer needs in different ecological contexts.

THE POTENTIAL CONTRIBUTION OF NGOs TO DEVELOPMENT

One striking lesson from the Philippines is that NGOs have helped to bring critical development issues and concerns into open public debate and to the attention of policy-makers. A particular feature of the NGO sector in the Philippines are the highly developed and extensive NGO networks and coalitions that exist in the various sectors of education, agriculture, health, land reform and ecological campaigning and are active in lobbying at the local, national and international levels, of which the Asian Non-Governmental Organizations Coalition for Agrarian Reform and Development (ANGOC) is an influential example. More recently, ten networks of development NGOs within the country combined to form the Caucus of Development NGO Networks (CODE-NGO) which seeks to focus NGO efforts on issues relating to national life, international linkages and inter-NGO relations (Constantino-David 1992).

Many NGOs have also provided a focus for study, documentation, debate and exchange of views and experiences on a more specialized range of issues. This is illustrated by case study material presented below, where the participatory development of sustainable, upland agricultural technology has been pioneered by NGOs such as Mag-uugmad Foundation (MFI) and Mindanao Baptist Rural Life Centre (MBRLC). Training in these techniques has been provided by the NGOs to farmers and to both government and non-government personnel. Innovations in adult education, literacy, primary health-care and co-operative development have also been developed on the basis of successful small-scale, localized experimentation by NGOs.

At a conceptual level, it is possible to envisage a productive partnership between government and NGOs based on certain 'comparative advantages'. As in Bangladesh, NGOs have highlighted the needs of sections of Philippine society which have tended to remain powerless and invisible to mainstream policy-makers: the indigenous communities, landless rural people and women workers. NGO success at the micro-level has been largely based on the ability of many NGOs to address the particular problems of a community and local target groups:

> Given their accessibility and acceptability to grassroots groups and communities, and their adeptness at utilising innovative approaches to development work, NGOs have been regarded as vital links between the government and the people. NGOs are seen to play a vital role in mobilising not only beneficiaries, but also government agencies towards ensuring timely, responsive delivery and utilisation of knowledge, technologies, resources and services
>
> (Quizon 1989: 1)

More success has been achieved with the short-term objectives of evolving models than with the attempt to replicate successes on a wider scale or at the national level. The government with its national constituency, on the other hand, can disseminate information and technology far more effectively through widely networked institutions and superior resource base. Nevertheless, government has also become associated with large-scale, untargeted and essentially 'top-down' approaches to the provision of development services. (For a notable exception, see Box 6.3.) For Philippine GOs and NGOs, the main issue in considering collaborative approaches is whether and how a joint approach can reach the poor and the marginalized sectors of the community more effectively, especially under worsening economic conditions.

While there is evidence that this view of NGO–GO complementarities can

Box 6.3 The Philippines National Irrigation Administration

The Philippines National Irrigation Administration (NIA) is an example of a large government agency carrying out a major programme which has empowered local people to take an active role in their own development. By 1976 the NIA faced contradictions between its strategy and its structure, since while the nation's irrigation policy called for an increasing role for irrigation associations, the NIA had become accustomed to making decisions itself, informing farmers of the decisions later. The need for transformation of the NIA under a new policy, financial arrangements and approach to farmers' own efforts was then recognized.

The NIA has therefore given assistance to small-scale 'communal' irrigation systems throughout the country, and has sought to involve farmers from the start of the project – in planning the project's scope, determining the layout of the proposed system and carrying out the construction of the required dams, canals and canal structures. Drawing on farmers' centuries-old experience of organizing local irrigation systems, the NIA sought to simulate the establishment of irrigation associations by placing community organizers to 'catalyse' local organization and generate farmers' irrigation associations. These organizers represented a new kind of personnel within the NIA, which had previously been primarily technical in character.

The organizers lived in the village and assisted with – but did not implement – tasks which farmers could carry out themselves. Farmers became involved in committees responsible for topographical surveys, obtaining rights of way and assembling labour for construction. The associations were encouraged to develop skills in decision-making, resource mobilization and conflict resolution, which they would later need for operating and managing the irrigation system. After construction, the agency handed over full control and authority to the association which was required to pay at subsidized rates for the assistance received.

As recently as the 1970s, the NIA's methods had been non-participatory. Engineers planned irrigation infrastructure and constructed systems with only nominal consultation with the people presumed to benefit from the effort. Instead, the managerial focus of the projects tended to be on completion of engineering work within the time schedule agreed with the World Bank and the Asian Development Bank, which were the financing institutions. Now, both technical and social dimensions were now seen as integral parts of the system.

The NIA gave legal recognition to farmers' associations to strengthen collaboration with government and gave financial incentives for self-management; they did not pay a service fee to the government, but had to repay partially the construction costs. The willingness of farmers to pay for the service provided by the NIA was taken as a key performance indicator of client satisfaction. The key to the NIA's success therefore lies in its ability to transform its programme on the basis of a strong economic and cultural tradition of farmer irrigation management.

Source: Korten and Siy (1988)

be made to work in a number of areas, it nevertheless omits several key areas of importance to rural development initiatives, the most important of which is the issue of land reform.

THE AGRARIAN REFORM ISSUE

Few people today question the need for a genuine agrarian reform to break land monopoly and redistribute incomes in order to address the problems of poverty, instability and violence common in the Philippines countryside (Liamzon and Salinas 1989). The Aquino government had promised to make agrarian reform its centrepiece programme, but its limited and slow implementation led to widespread disillusionment and dwindling support for the government. Early attempts by NGOs and peasant organizations to assist with its design and implementation were not treated sympathetically by the government. In an optimistic article, Garilao (1987) describes the various tactical alliances which quickly emerged among NGOs across various points in the political spectrum to push for structural change at this time.

A few months after the new government took office, the Peasant Movement of the Philippines (KMP) offered a 'Programme for Genuine Land Reform' to the new government. This impressively detailed proposal predated the government's own attempts to draw up even a draft plan. The document included proposals for free distribution of land to agricultural tenants and marginal farmers; priority to female-headed households for the use and ownership of land; protection of ancestral land rights of indigenous minorities; selective compensation to former owners; and nationalization of lands controlled by foreign multinational corporations (Putzel and Cunnington 1989).

In 1987 pressure from farmers impatient with the government's inactivity led to the foundation of the Congress for People's Agrarian Reform (CPAR) by thirteen peasant organizations drawn from a variety of political positions. This represented a more 'moderate' version of the earlier proposal which removed the 'free' distribution stipulation. The CPAR went on to form the basis for discussions and congressional lobbying, and was itself used as the basis for House Bill 400, an early Bill presented to Congress and later rejected in favour of the final Comprehensive Agrarian Reform Programme (CARP), which was signed by President Aquino in 1988.

The CARP law has been heavily criticized by NGOs, peasant leaders and other activists since it contains an explicit bias towards the interests of large landowners, agribusiness corporations and traditionally powerful families (Putzel and Cunnington 1989). For example, landowners will be able to dictate the terms of receiving 'fair market value' rather than 'just compensation' for land, and have the right to contest the Department of Agrarian Reform's (DAR) decisions in the courts. There are facilities for redistributing land to the owner's children, which could remove as much as 70 per cent of land from CARP's coverage (Putzel and Cunnington 1989). Between 1988 and 1998 CARP intends to redistribute 10.3 million acres of land. By 1990, 293,427 acres had been distributed to 172,556 farmer beneficiaries.

Many NGOs have now rejected the government's agrarian reform strategies as 'too little, too late' and have concentrated their energies elsewhere. Others have continued to work with the CARP. According to one NGO worker from Kaisahan (an advocacy NGO), many NGOs, in their rejection of CARP, have failed to grasp the centrality of agrarian reform to their agriculture programmes and are content instead to work on income-generation projects at local level, without continuing to address the fundamental long-term issues of land tenure. Kaisahan, along with some other NGOs, therefore calls for more participation in this area. Others have preferred to oppose government efforts until such efforts approach the pressing need for real land reform more realistically.

CONCLUSION: AN UNCERTAIN FUTURE

The Philippines government, like many of the donors, can be said to have 'discovered' NGOs. Indeed, in a sense the government itself now contains elements of the NGOs and it is no longer possible to posit a simple structural 'opposition' between the institutions, objectives and procedures of government agencies and NGOs.

The Aquino government has undoubtedly been keen to work with the NGOs, but on whose terms? The rhetoric of NGO–GO partnership has become a high-profile feature of the development discourse in the Philippines. In 1988 staff at the DA NGO Outreach Desk initiated the formation of an informal group composed of other government agencies such as DAR, DENR and National Irrigation Administration (NIA). The group set about discussing and clarifying basic issues of NGO definition and NGO–GO collaboration, and following considerable advocacy work within both the NGO and GO communities, the group grew to become the Inter-NGO Liaison Offices Forum. During 1989 an NGO Liaison System (NLS) was formed to oversee NGO–GO co-ordination at all levels, and the Forum took a leading role in this initiative.[4] Two of the main outcomes were amendments to the 'Code of Conduct for NGO–GO Relationship' originally drafted by the DA Outreach Desk, and creation of a number of task forces in relevant areas.

In planning collaborative ventures, a number of issues have emerged as being important:[5]

- A recognition of distinctive areas of competence of GOs and NGOs.
- The need for an information and education programme on the respective activities of the NGOs and government agencies.
- Institutionalization of systems of procedures that are suitable and acceptable to both the GOs and the NGOs.
- Recognition of the invaluable assistance of the funding institutions in pursuing development work.
- Planning and even dialogues between GOs and NGOs often require intermediaries or facilitators, which have to be acceptable to both and should

have a good understanding of how government and NGOs think and operate.

The need for partnership has led government to develop a two-pronged approach to development, in which GOs are judged to have a capacity to work more effectively at the macro-level, addressing national problems and NGOs at working at the local level, addressing specific needs. The government believes that with continuous dialogue, areas of complementary activity can be worked out between GOs and NGOs, particularly in areas where NGOs have built up specific expertise. For example, the neglect by the government of upland development led to a niche into which NGOs moved and which now forms the basis for potentially constructive partnerships based on this model.

The political future of the Philippines is currently uncertain, since the elections of 1992. The extent to which collaborative initiatives take a firmer shape and prove constructive in addressing the urgent needs of the Philippines' poor depends on how far the government is prepared to accept the significant changes necessary at the policy level for bringing about the structural changes required by many of the NGOs, and on the success of the development NGOs to pursue their own poverty-focused agendas and safeguard their objectives in their relationships with the state.

THE MINDANAO BAPTIST RURAL LIFE CENTRE'S SLOPING AGRICULTURAL LAND TECHNOLOGY (SALT) RESEARCH AND EXTENSION IN THE PHILIPPINES

Harold R Watson and Warlito A Laquihon

INTRODUCTION

The Mindanao Baptist Rural Life Centre (MBRLC), founded in 1971, is located on the island of Mindanao among the rolling foothills of Mount Apo, the highest mountain in the Philippines. The NGO's target group is the poor upland farmer, whose income averages around $300 per year and farms 2–3 acres. Influenced by Schumacher, the MBRLC considers that farm people themselves are the most valuable resources in development.

The MBRLC has three main aims in its development strategy:

- to develop appropriate farm technology to enable subsistence farmers to produce sufficient food and provide adequate housing, clothing, education and health-care to their families
- to develop spiritual awareness in the farm household in a relationship with God, their community, their nation, and the earth from which their livelihood is derived
- to promote development in such a way as to conserve natural resources, and promote ecological balance.

MBRLC measures its progress directly in terms of its relevance to local farmers using four key criteria (which also form the acronym 'SALT'): *S*impleness, so that technology can be readily understood by farmers; *A*pplicability, such that training and development can benefit at least 50 per cent of the upland farmers of the area; *L*ow cost, based on in-country inputs; and *T*imely in terms of addressing the immediate problems faced by farmers.

In order to achieve the centre's three-fold purpose, the REDEEM programme was developed (Research, Extension, Development, Education, Evangelism and Mission) and this approach is integrated into all the centre's activities.

THE SALT TECHNOLOGY

MBRLC's approach to developing technologies begins with consultations with local farmers and agriculturalists, including those in government. The centre's highest priority project, which has become known as SALT, originated from a need expressed by a group of upland farmers who met with the NGO's staff and discussed informally their problems and aspirations. They had become alarmed by falling production and deteriorating soil conditions. During the discussions, farmers pointed out that:

- They had suffered a 75 per cent reduction in maize production, mainly due to soil erosion, over a ten-year period, which had drastically reduced farm incomes.
- They needed a better distribution of farm income throughout the year.
- They had no capital for buying fertilizers, insecticides and seeds of improved varieties of corn and other crops.

The dialogue revealed that soil erosion was the cause of their loss of production rather than deficiencies in techniques of cultivation or varieties. The conservation of topsoil and rehabilitation of subsoil became the priority. Working with the farmers, the MBRLC initially tried without success to build graded terrace lines with bungs. Consultations between the NGO and the government's Department of Agriculture yielded no better method of dealing with the problems. Most government research extension efforts were on the contrary directed primarily at lowland agriculture.

Instead, a new technology was developed based on farming between rows of *ipil ipil* (*Leucaena leucocephala*) and using the stems and leaves as a fertilizer for corn. Double rows of *ipil ipil* were planted on a contour line 4–5 metres apart. A system emerged in which soil erosion is controlled and the soil itself rebuilt, increasing incomes by 50–100 per cent. This became known as SALT (Sloping Agricultural Land Technology). Additional related technologies followed and training was given to farmers in livestock production (goats, sheep, ducks and swine). A close dialogue was maintained with the government Bureau of Animal Industry and many government officials therefore watched the development of SALT with interest.

SALT developed into an integrated, diversified hillside farming system that uses thick hedgerows of nitrogen-fixing species as soil binder, fertilizer generator and moisture reservoir. The development of other, related technologies followed. Simple Agro-Livestock Technology (SALT 2) is a system which integrates livestock raising (particularly dairy goats) into the SALT scheme. Sustainable Agroforest Land Technology (SALT 3) is a combination of food-wood production. Using Properly Lowland Integrated Farming Technology (UPLIFT) is a sustainable lowland farming system and Food Always In The Home (FAITH) garden – a 'refrigerator in the farm' – provides vegetables all the year round.

The centre has developed a training centre for local farmers, members of other NGOs and government staff interested in using the SALT technologies. Between 1980 and 1990 a total of 10,660 people in 440 groups were trained. A total of thirty-one organizations (both NGO and GO) have gone on to adopt SALT. MBRLC also became active in developing seed production so that local farmers could be provided with all the inputs they needed to begin tackling the pressing environmental and production problems.

In 1989, after a visit by the Governor of Davao to the centre to view its work, a request was made for the MBRLC to link up with local government and the Department of Agriculture to develop a tripartite approach to upland community development. Batches of farmers were trained in SALT 1, UPLIFT and FAITH gardening. The DA selected the farmers in batches of twenty to twenty-five and supplied seeds, the local government provided food and travel to training sessions and the centre provided the training itself along with major planting materials.

SALT is therefore a *S*imple, *A*pplicable, *L*ow-cost and *T*imely method of farming hilly lands. It is a technology developed for farmers with few tools, little capital and little training in agriculture. It can be said that it is a humanized technology in that the farmers' culture, resources and abilities were considered in its development.

THE SOUTH MINDANAO AGRICULTURAL PROGRAMME (SMAP)

The government acknowledged the success of the MBRLC's work with the SALT technology and invited the centre's participation (along with other local NGOs) in the planning of a larger upland agricultural development for the area. Through the NGO Outreach Desk at the DA, which was established in 1986, a partnership was institutionalized between government and NGOs working in the uplands to help define policy. One of the fruits of this partnership was the Southern Mindanao Agricultural Project (SMAP), funded by the European Community (EC) in which three local NGOs were invited to contribute to planning, training and community organizing in the early stages of SMAP.

The Resources Ecology Foundation for the Regeneration of Mindanao (REFORM) was invited to identify project areas and conducted surveys. The Santa Cruz Mission worked with the government to develop strategies for sustainable use of Lake Sebu's resources. The MBRLC had by this time come to be regarded by government as the best training organization for sustainable upland technologies and SALT was adopted as the centrepiece of the SMAP approach in the uplands. MBRLC was invited to provide the training component and several sessions have already been held for farmers participating in the programme in the Baranguay Balutakay located on the slopes of the Southern Mount Apo range in Davao sel Sur. In September 1990 the MBRLC was the venue of the launching of the SMAP.

In the SMAP contract, the government will select the farmers and provide transport for the programme, while the MBRLC will provide training and materials, charging the government for services rendered. Although the programme is still in its infancy, the MBRLC has shown its comparative advantage in developing relevant needs-based technology for upland farmers in the absence of government work in this area. Government has recognized this and taken up the technology for replication.

CONCLUSION

The donors have played an important role in prompting the shifts towards the issue to NGO–GO collaboration, brought about by the growing need to reach the grassroots more effectively and to gain 'better value for money'. In the MBRLC case, technology has flowed from the farmers to the NGO, and from the NGO to government agencies. There was previously no existing technology known to either party which addressed the specific needs of these farmers, and it was the NGO which addressed its efforts towards filling this gap. For the MBRLC, this fits with Schumacher's point that development must begin with people, through training and education. Although the credit for developing the technology is due to the farmers and NGO, it is sometimes helpful in collaborations such as this for the NGO to 'swallow its pride' and allow the government to be seen to be the originator of the idea.

However, there are a number of tensions which relate to NGO–GO collaboration which need to be considered:

- *Funding jealousies* the government is wary of the ability of NGOs to secure impressive grants from donors.
- *Bureaucracy* the government agencies are often extremely slow moving and this can be frustrating for NGOs.
- *Politics* in an area such as South Mindanao there are volatile political issues centring on Islamic separatism and the NPA guerrilla movement. The MBRLC has a policy of staying out of politics.
- *The issue of fake NGOs* (which exist only for gaining funding and are insincere in their efforts), and non-credible GOs (which sometimes make little effort to develop local relationships with people and emphasize their status in relation to the local community) can contribute to poor NGO–GO relations.

In order to guard against many of these potential pitfalls, the MBRLC takes no funding from the government or from donors and maintains a friendly, open policy towards all parties involved in local community development issues.

THE INTERNATIONAL INSTITUTE FOR RURAL RECONSTRUCTION (IIRR)

Developing an agroforestry kit

Julian Gonsalves and Aurea G Miclat-Teves

INTRODUCTION

The International Institute for Rural Reconstruction (IIRR) is an international NGO committed to changing the quality of life of the rural poor in the developing countries of Asia, Africa and Latin America. IIRR evolved from the revolutionary, grassroots development project initiated in China by Dr Y C Yen in the mid-1920s, integrating education, livelihood, health and self-government.

Under the tenet of 'release not relief', IIRR is concerned with promoting sustainable development strategies which enable rural people – the principal agents of rural reconstruction – to release their own potential and transform their own lives.

Technology development has tended not to address the needs of poor farmers. For this reason, IIRR introduced the concept of regenerative agriculture in the early 1980s, promoting technologies that could help arrest environmental degradation, attain food security for the farm household, increase farm income and regenerate its resource base.

Technologies were developed with the participation of poor rural people in a number of sectors: bio-intensive gardening, agroforestry, freshwater fish culture, livestock production, plant genetic conservation, low external input rice production and non-land-based micro-enterprises.

IIRR has promoted its regenerative technologies nationally and internationally through short courses, workshops and materials production. Alumni from IIRR's courses are active in eight Asian countries, five Latin American countries and eight African countries.

THE AGROFORESTRY KIT

Information kits have been produced in co-operation with both NGO and GO partner agencies and are now in use in twenty-five countries. The agroforestry kit developed by IIRR successfully combines experiences from a number of different agencies in the Philippines. Environmental degradation is proceeding at an alarming rate in the Philippines: more than 119,000 hectares of forest were depleted yearly from 1976 to 1986, affecting both uplands and lowlands according to data from the Department of Environment and Natural Resources (DENR). Soil erosion has increased dramatically. Declining tree cover has increased the level of floods and droughts; land productivity has declined.

The DENR's integrated social forestry programme (ISFP) aims to give poor upland farmers access to a maximum of 25 hectares of forest lands for a tenure of at least twenty-five years. However, the programme lacked a comprehensive resource book of appropriate technologies, despite the existing reserves of knowledge and experiences resting with NGOs, GOs and universities in the Philippines. DENR therefore requested IIRR to co-ordinate production of an interdisciplinary agroforestry information kit, drawing on these diverse sources.

IIRR was known for pioneering an approach to developing educational materials using participatory workshop methods. Two technology information kits had already been produced in this way ('bio-intensive approach to small-scale household food production' and 'regenerative agriculture') and were then simplified for use by the development worker. The single-concept sheet approach of producing science-based materials is a special feature of the IIRR kits. Loose-leaf format ensures easy revision and cheap production ensures maximum distribution.

Both of these kits were shared widely, especially with government agencies such as the Department of Agriculture (DA). The DA's undersecretary for operations has even ordered the distribution of the two kits (with UNICEF support) to all DA municipal officers to serve as guidelines in technology adaptation in connection with its household food production projects. The Department of Education, Culture and Sports has also distributed this kit nationally.

The basic premise of the IIRR approach is that experts – academic and non-academic, GOs and NGOs – have engaged in specific or related endeavours in agroforestry. Some of these experiences and recommendations have been documented but often not adequately circulated. Few have appeared in print. Other materials may be written in complex scientific language or are cluttered with peripheral information. Some are oversimplified and lack scientific rigour.

IIRR organized a workshop in November 1989 to develop an agroforestry technology information kit with Ford Foundation support in the form of a grant to IIRR for the workshop costs, pre-production costs and eventual printing of copies for use by the DENR. A steering committee responsible for the overall production of the agroforestry information kit was formed of representatives from the GOs and NGOs, specifically from the DENR, IIRR, the University of the Philippines at Los Baños (UPLB) and the Ford Foundation. The first activity was to draw up a list of topic areas to be included in the kit. The DENR first came up with a long list of information needs defined by the ISFP technicians. The committee arrived at a list of themes and sub-themes and drew up a list of persons, many of whom were farmers, who were invited to contribute. The resource persons were asked to write drafts of single-concept sheets for presentation at the workshop.

THE WORKSHOP

At the workshop, the drafts were continuously refined through active discussions and criticisms by participants, whose different perspectives provided for critique and improvement of materials. This complicated the process but it also generated materials that eventually passed the scrutiny of both GO and non-government sectors. A pool of experienced artists and editor-writers complemented the technical experts. The final drafts of 'technology single sheets' were completed at the end of the workshop.

During the workshop, representatives of GOs, NGOs and academics worked together effectively regardless of their organizational origins, focusing their efforts solely on production of the kit. This methodology saved time because it ensured that feedback was obtained during the workshop itself and reduced the time needed for production of the final drafts. The kit was finished by the middle of January 1990 and combined the following:

- an integrated and inter-disciplinary approach to agroforestry, representative of a cross-section of agencies
- information on environmentally-oriented technologies on agroforestry originating from diverse sectors: GOs, NGOs and academics
- key concepts in a simplified form
- promotion of sustainable agricultural technologies without an infusion of external capital resources
- a selection of relevant topics chosen through participatory, needs-based discussion
- single-concept sheets, colour-coded to facilitate easy access
- a technology options approach
- multi-agency use to reduce confusion resulting from each agency having its 'own' technologies
- combined training hand-outs, reference materials and synthesis sheets for a wide range of possible users such as extension workers, trainers and educators.

The workshop approach shortened the process of developing information materials without sacrificing the technical and artistic quality of the output. The pre-workshop planning was crucial to its success. The kit contained a broader perspective in terms of sharing of experiences in both content and process: academics were provided with practical applications, while field-workers and farmers were provided with theoretical knowledge. Technical validation was given to field experience. The kit was tailored to field technicians as an easy reference and guide. It contained the primary information needed for basic agroforestry practices using a holistic household-focused approach.

CONCLUSIONS

Feedback from users has been generally positive. In one area, materials from the kit were repackaged into booklet form and into individual hand-outs at UPLB. Materials were repackaged and adapted and used in international newsletters; such material was also translated into local languages. There is a plan to bring together users from the DENR to assess how the materials were used and to suggest changes in future revisions/versions of the kit.

The workshop had some problems. It had been agreed beforehand that the resource persons must have their drafts prepared. Some unfortunately had only documents and reference materials and could not concentrate on critiquing other papers since they were still busy preparing their own materials.

During the discussion, the different perspectives of the participants became apparent. Some were too academic while others based their statements solely on experience. The group had often to be reminded of the objectives and purpose of the workshop and the information kit. However, considering the diverse characteristics of the participants, the project was able to show that NGOs and the government can work together.

The logistics of nation-wide distribution of information materials needs to be improved. It took time before the kits were distributed (three months) and some technicians who needed them did not receive them on time. The editing and verification of scientific names used also hindered the speedy printing of the workshop output. This again was the result of an attempt to ensure scientific accuracy, a dimension rightly emphasized by participants from the university sector.

NGOs operate on a scale, so the government's help in spreading sustainable and regenerative agriculture technology creates greater impact and helps influence future policies. The NGO experiences, now incorporated in the kit, benefit from the wide-scale distribution undertaken by the DENR.

The DENR has the mandate, the experience and technical expertise, yet it has recognized that it can benefit from working with NGOs. The IIRR has the capability and expertise to produce information kits in terms of content and production. Its in-house capacity is strong, having the facilities readily available for support. Various participating agencies which either do not have the time or expertise to produce their own materials were able to use the workshop resources to get their own materials produced professionally and with the benefit of peer reviews.

MAG-UUGMAD FOUNDATION'S (MFI) EXPERIENCE OF UPLAND TECHNOLOGY DEVELOPMENT IN THE PHILIPPINES

Soil and water conservation strategies

Lapu-Lapu Cerna and Aurea G Miclat-Teves

INTRODUCTION

The Mag-uugmad Foundation Inc (MFI) has its origins in 1981 in the Cebu soil and water conservation programme (CSWCP) established by the US NGO World Neighbors. Cebu is an island located in the Central Visayas region of the Philippines. It is predominantly hilly with 70 per cent of the island with slopes of greater than 18 per cent and severe deforestation has taken place. Recent satellite photos show that virgin forest cover has now disappeared. Population pressure on the remaining land is severe. The primarily rainfed farms average about 0.75 hectares each and income from these farms is barely enough to sustain an average six-person household.

The CSWCP aimed to address local problems of

- severe erosion and declining soil fertility
- low income levels of upland farmers
- flooding of lowlands and decrease of potable water sources in mountainous areas.

Initially there was no local counterpart for the programme but in June 1988 a broad federation of farmers and fisherfolk was established for programme activities, called the Mag-uugmad Foundation Inc. This was linked to another related World Neighbors programme in the area. 'Mag-uugmad' is a Cebuano term meaning 'tiller'. The board of directors consists of the implementing staff of both programmes and includes three farmers and one fisherman.

APPROACH

MFI has subsequently developed a farmer-based extension strategy for its soil and water conservation programme in the Guba, Argao and Pinamungajan areas of Cebu. This is based on three main principles:

- Each production system must be economically viable, with an adequate return on investment for the farmer.
- It must be environmentally sound, otherwise it cannot be sustained.
- It must be culturally and socially acceptable to the people implementing it.

The programme is based on a community development approach with six stages: (1) start where the people are, (2) discover the limiting factors, (3) choose a simple technology, (4) test the technology on a small scale, (5) evaluate the results, and (6) train trainers to spread the technology.

Initial meetings were held with the farmers and MFI explained that its aims were to facilitate learning and exchange rather than provision of financial assistance packages or 'hand-outs', unlike many other programmes. Instead, the intention was to work towards overcoming constraints to farm production and introduce farmers to technologies which might assist them with problem-solving. Participants were expected to judge for themselves whether the technologies and practices could bring them adequate returns on their investments in time and labour. The project works through part-time farmer instructors rather than relying fully on professionally hired extension workers with formal education.

The project has developed a range of appropriate farming technologies with the farmers and improved many existing practices:

- construction of contour canals with hedgerows, drainage canals to remove excess water from the fields.
- soil traps and dams for gully stabilization.
- bench terracing, contour ploughing, hedgerow planting and in-row tillage.

Other technologies applied by participating farmers include crop diversification, animal production and management, and soil fertility management.

Soil conservation measures require considerable labour input by farmers. In order to reduce this burden, the indignous 'alayon' work-group system, based on kinship or residence, is used by MFI to organize reciprocal work. Each week, a different farm is selected and a rotation cycle ensured that all farms received equal attention. This also allows for sharing of draught animals, which was one of the constraints in increasing farm production identified by the farmers.

As the project gathered momentum, the need for a more defined organization was realized in order to:

- provide a legal framework to the organization with authority to enter into contracts and transactions with other organizations
- gain access to resources and direct funding, which will broaden the project base
- develop autonomous decision-making programmes, making possible direct farmer participation (Since members of the board are also the staff, there will be more voices in the planning and decision-making processes.)
- secure better access to information.

MFI therefore registered as an NGO and presently MFI has fourteen full-time staff and twenty-five part-timers.

The 'alayon' system plays an important role in sustaining the gains made by the project. It will be the instrument towards the formalization of a people's organization (PO) whether objective of self-reliance has been met. At this point, MFI will phase out. Forming a PO has the added benefit of providing a legal entity for the people, through which resources and funds can be delivered effectively.

RESULTS

MFI's soil and water conservation technologies have gained widespread acceptance in the project sites and in neighbouring localities.

- Yields of the upland farmers have increased. Some farmers now have more stable incomes, but others still lack sufficient food for their families. Out-migration in Argao and Pinamungajan among the project participants has been reduced significantly.
- Diversified farming systems have gained widespread acceptance. Fertility of the soil has improved through the use of organic fertilizers. Where they are ecologically damaging, shifting cultivation practices have been minimized in the project areas.
- Environmental consciousness has been raised. The upland farmers participating in the project are conscious of and active in helping to protect the forests and the environment.

A spin-off of the project is the establishment of POs to sustain the gains made. An example is the KAMACA (Kahugpungan sa mga Mag-uuma sa Cabalauan) in Argao and ADUNA or (Alayon Dugukan sa Nasud) in Pinamungajan. Three other POs exist in Guba. MFI has extended its experience to a broader range of farmers and technicians throughout the Philippines. This outreach programme conducts regular monthly training sessions for farmers and field technicians from the Department of Environment and Natural Resources (DENR) and NGOs involved in upland development.

Farmer-instructors at programme sites serve as the primary training staff assisted by a small MFI support staff. Farm tours, small group discussion, slide presentations, farm analyses and planning sessions are included. MFI also serves as a frequent venue for national and international workshops.

WORKING WITH GOVERNMENT

MFI works with farmers with different land tenure conditions in different parts of the island. In Guba, land is privately owned or sharecropped by individual households. Crops grown include vegetables, flowers, fruits, root-crops and grains. Goats, hogs, chicken and cattle are raised as an important source of additional food and income and as a source of organic fertilizer. Firewood gathering and the production of charcoal are other major economic activities.

By contrast, MFI's projects in Argao and Pinamungajan are located on government forest lands. In addition to livestock, maize, cassava and sweet potato are the main crops grown. Contacts with government have therefore been established. The Bureau of Forest Development (BFD), now called the Forest Management Bureau (FMB) of the DENR, and the Department of Agrarian Reform (DAR) assisted with project site selection.

Since Argao is under BFD jurisdiction, it was decided there would be a joint NGO–GO programme with the BFD allowing the project access to the areas, and MFI assisting the BFD in environmental conservation. Since Guba consisted of privately owned and tenanted lands, activities were conducted directly with farmers without government intervention.

A general rationale for NGO–GO partnership therefore emerged. Upland farmers, especially those on government forest lands, tend to see declining soil fertility and deforestation as a natural phenomenon rather than a problem of poor resource management. Unstable land tenure status leads their daily subsistence needs to block their concern for the wider ecosystem.

The BFD is a branch of the DENR, the government agency responsible for the broad environmental concerns of the country. Given DENR's large scope of responsibilities, it became interested in assistance from NGOs working towards sustainable resource management.

With the assistance of World Neighbors, a Letter of Understanding was therefore drawn up between BFD and the Cebu soil and water conservation project. Training was made a joint activity for farmers and BFD staff. Successful farmers using the technologies become trainers. BFD technicians in turn trained farmers at the project site in nursery practices. Cross-visits of BFD nurseries and MFI sites were conducted.

CONFLICTS OVER CONTRACT REFORESTATION

Co-operation over land tenure however proved more difficult. In areas covered by the DENR's integrated social forestry programme (ISFP), land 'stewardship certificates' were issued to farmers. The DENR then supplied the seeds and seedlings for reforestation, the farmers provided the plastic bags to be used as seedling containers.

In late 1988 a community-based contract reforestation agreement was signed by the MFI and DENR. This programme provides extended woodlot lease agreements to farmers working in marginal designated lands. MFI helps to organize and co-ordinate these activities by establishing community nurseries, site preparation and formation of 'alayons', including monitoring and maintenance tasks. Under the scheme, 40 hectares have been replanted with trees by the farmers in Pinamungajan and Argao.

MFI's farmer-to-farmer extension approach has been effective. Successful farmers have become trainers, sharing their experiences and training other farmers. But land tenure status of the project participants remains a problem. Since the farmers do not have the security of tenure on the land, they cannot move beyond immediate subsistence needs towards long-term, environmentally sound farming practices.

These problems have emerged more clearly in MFI's experience with another joint project with DENR: the family-based reforestation contract, which also started in 1988. An agreement was reached to begin reforestation efforts by MFI farmers with DENR through the Cebu assistant regional director. MFI understood from this that farmers would receive ownership of the trees they planted and long-term rights in the land in return for their work. However, the agreement was verbal, since the DENR had not yet produced implementation guidelines. The community environment and natural resources officer (CENRO) – the DENR official at the municipality level – was not informed of the agreement leading to misunderstandings between the CENRO officials and MFI and severe delay in payments for project billings.

When the guidelines were finally produced, the terms of reference contradicted earlier verbal assurances over ownership of trees and land tenure and effectively made the farmers contract farm workers. Under the contract, the upland farmers would not own the entire produce of the land. After three years, the land including the trees will be returned to the government. If an agreement is reached, the farmers would receive a lease for the woodlot for twenty-five years under a sharing system with the government, the terms of which are not yet decided.

This also contradicts the DENR's ISFP which gives the upland households stewardship of up to 7 hectares of land, allows farmers the full benefit of their produce, exempts them from regular fees on forest products, allows access to land and credit programmes using the stewardship certificate as collateral and permits the option to renew the contract upon expiration.

CONCLUSIONS

The conflicts over contract reforestation are still to be settled. The lesson learnt by MFI is that GO–NGO relationships are often based on personal relationships, which works well as long as there is trust. However, it may often be better to secure specific terms of reference and clearly stated guidelines. This is particularly true if government staff are moved to other jobs. Terms of reference should be the product of consultation among all persons involved including the lower rank officials to avoid misunderstandings.

Many NGOs exist because there are gaps in the services of the government. NGOs complement the work of certain government offices but have no intention to replace them. Since many NGOs have established important local, farmer-centred initiatives, government and the NGOs should meet on a national level to assess potential for linkages and plan clearly defined partnerships.

THE PHILIPPINES DEPARTMENT OF AGRICULTURE'S NGO OUTREACH DESK

Carlos Fernandez and Tess del Rosario

INTRODUCTION

In response to pressure from NGOs in the wake of the 1986 revolution, the Aquino government set about exploring the potential for working with NGOs in rural development. This was expressed through the following actions:

- The Government Mid-Term Development Plan (1987–92) calling for greater involvement of people in the decision-making, planning and implementation of programmes, and also a greater scope for private initiatives in pursuing economic growth.
- The Philippine Constitution of 1987, specifically Article II, that 'The State shall encourage non-governmental organizations, community-based or sectoral organizations that promote the welfare of the nation'.
- The Philippine government signed several international covenants, e.g. the FAO World Conference on Agrarian Reform and Rural Development and its Peasant's Charter, the ILO Convention 141 recognizing the active participation of NGOs at the national and international levels.

Although the Aquino government sought to restore relations between NGOs and GOs after a long period of antagonism during the Marcos era, this was not a smooth process and distrust remains in some quarters on both sides. Government has had to learn how to deal with NGOs, and how to work with them, given such different working styles and ethics. NGOs are also just beginning to learn to relate effectively to the bureaucracy.[6]

GO–NGO collaboration continues to be a problem of defining the precise nature of a partnership between two diverse work partners. The following problem areas have affected relations in general:

- *Definitions* NGO is a catch-all term that has bred more confusion than clarity. A proliferation of organizations after the 1983 Aquino assassination brought NGOs to the public view, along with their conflicting goals, aims and definitions. Various terms, such as private sector, cause-oriented groups, private voluntary organizations and social development agencies, have been used interchangeably.

One direct consequence of this confusion in terminology is the domination of representatives from the private sector (the business community, bankers' associations, commercial fishermen, landowner's associations, etc.) in the various councils initiated by government.

Being better organized and better linked to the government's decision-making structures, these private sector groups referring to themselves as NGOs became the government's showcase in demonstrating GO partnership with the NGO sector. The smaller producers, farmers and fisherfolk were by and large left out of the government.

- *GRINGOs* Some government agencies prefer to relate with their own kind, i.e. to collaborate with units which they themselves initiated. The historical experience with the Samahang Nayon Programme – a nation-wide experiment by government to form co-operatives – is instructive of the dismal failure of such attempts. Yet GRINGOs (government-run and initiated NGOs) abound, and these are utilized by government agencies to take advantage of foreign assistance and/or to promote the political agenda of government functionaries who organized them.
- *Outlook* The distrust between GOs and NGOs was not eliminated in the last five years of the Aquino administration, despite an open and supportive policy to work with NGOs. At grassroots level, NGOs continue to face intimidation and harassment while engaging in organizing work even for government programmes. NGOs, on the other hand, view government as a monolithic bloc plagued by red-tape, lack of dedication and commitment, bureaucratic inertia and the like. This simplistic view of a rather complex phenomenon disenfranchises many NGOs from securing valuable resources that may be made available to them, had their attitudes remained more open.
- *Institutional constraints* Within government agencies, there were no formal mechanisms to facilitate GO–NGO collaboration. While the call to facilitate came at a very early stage in the Aquino government, the lack of a concrete model militated against more effective and speedier linkages with NGOs. Thus, the years 1987–92 may be considered as experiments – often clumsy – to discover a workable partnership which would endure beyond this administration. In the early 1990s there are still no clear-cut models and guidelines, although some directions seem to be emerging.
- *Operational bottlenecks* Experiments with partnership at project level are probably the most difficult aspect of the volatile relationships between GOs and NGOs, more because of the absence of any concrete examples of a project partnership that works.

Partnership is usually limited to a few elements of the project development cycle, like planning or implementation of specific project components. Attempts to propagate an enduring partnership in all stages of the project development cycle remain elusive. Besides, most NGOs are still developing their own technical competencies in the 'hardware' of project development, to complement their strength and skill in social technologies.

- *Legal impediments* Finally, a legal framework to guide GO–NGO collaboration is still absent. Basic principles, such as the respect for NGO autonomy and independence, form part of the conventional wisdom by now. But the operationalization of this principle is yet to be legally established, if only to settle the sensitive problem of NGO accreditation. Other problems potentially arising from the lack of a legal framework are corruption both in government and the NGO sectors; programme and project mismanagement; incentives and penalty systems; inter- and intra-GO–NGO relations, that is several government agencies relating to several NGO partners, and NGO partners relating to one another. There are others, but these seem to be the most immediate.

THE NGO OUTREACH DESK AT THE DA

The Outreach Desk at the DA started as a task force under the Special Projects Office (SPO) in April 1986. Its basic objective was to establish a working relationship with NGOs, then considered as one of the innovations or experiments which the SPO was willing to undertake. Although housed under the SPO, the Outreach Desk was meant to be operated as autonomously as possible, in order to avoid it becoming a 'company union'.

Through various surveys and consultations at the local, provincial and national levels, a draft corporate plan and an Interim Consultative Council (ICC) was created in late 1986. Also, a national secretariat was created as a result of the reorganization of the task force.

The draft corporate plan contained the mission statement of the OD:

> to contribute to the promotion of people's participation through the institutionalization of the partnership between an intergovernmental agency body in an inter-NGO/SPO network, particularly at the local level.

The plan targeted a period of three years to come up with an operational model for institutionalizing the partnership between the DA and the NGOs. The Interim Consultative Council which resulted from a series of consultations was composed of fifteen major national NGO networks.

The main task was to provide policy directions to the DA in matters affecting the NGOs. It also assists the national secretariat in implementing agreed upon plans and programmes. The national secretariat was primarily engaged in groundwork, data gathering and consultations. This was necessary during the early years of the DA to enable staff to become informed of what and who NGOs are, and to identify possible areas of coordination. What became evident during these consultations was the need to define and refine the term NGO, so that those using the term would be subscribing to a common meaning. The other was to pinpoint areas of collaboration as early experiments for a longer and more durable partnership. Some collaborative activities are as follows:

- Orientation and training programmes on community organization, emerging concepts and theories of development given to selected officials and extension workers of Regions One and Two.
- Joint conduct of five-day seminar workshop of local self-help groups in Central Visayas.
- Joint conduct of indicative planning on the Comprehensive Agrarian Reform Programme (CARP), participants coming from the different line agencies and from the NGOs sector in Negros Occidental and Batangas.
- Joint conduct of leadership training and network building for small farmers' organizations in Bukidnon and Misamis Oriental.
- Joint conduct of the First Farmers' Congress in Camarines Sur.
- Joint efforts in coming up with a Fisheries Code and a proposal on the piloting of resource management concepts in selected areas.
- Joint conduct of a seminar-workshop on project development planning in Camarines Sur, Albay, Nueva Ecija, Pampanga and Bataan.
- Training of DA technicians on mango propagation, production and processing techniques in Iloilo and Guimaras.

OUTCOME: SMAP[7]

The Outreach Desk has subsequently been responsible for co-ordinating the EC-funded five-year South Mindanao agricultural programme (SMAP) which is making extensive use of NGO skills and innovations in technology development and training provision. Originally conceived in 1990 with the DA as the head implementing agency, it seeks to benefit the whole region through a sustainable development strategy, and was planned to involve NGOs in a wide range of supporting activities.

The project was due to begin in 1992 after a transition stage in which agreements were being sought as to identifying the precise nature of collaboration between NGOs and government, selecting staff, procuring equipment and mobilizing project structures.

The programme will take place in South Mindanao and will engage the services of the following NGOs:

- The Mindanao Baptist Rural Life Centre (MBRLC), which has achieved a reputation for developing the SALT technology and training farmers and extension workers in its use (see pp. 240–43).
- Resources Ecology Foundation for the Regeneration of Mindanao (REFORM), which will participate in the planning and identification of the project areas, and socio-economic research.
- Santa Cruz Mission for focusing on the potential of using the resource of

Lake Sebu in more productive and sustainable ways for the benefit of the poor.

CONCLUSION

The SMAP planning phase has illustrated the diversity of local NGOs in the area and shown that government can co-operate with selected organizations only on carefully circumscribed areas of activity. Some tensions were caused among the certain NGOs in the area competing for increased resources from participation in SMAP.

The operationalization of partnership remains a difficult objective, since NGOs resist implementing government projects if they feel that they are the mere extension of government interests and activities must therefore be jointly undertaken.

Many NGOs are seen to practise 'policy advocacy' rather than 'policy formulation', and the DA therefore encourages NGOs to concentrate more on the content of policy formulation as opposed to more generalized concerns.

THE ECOSYSTEMS RESEARCH AND DEVELOPMENT BUREAU'S (ERDB) INTEGRATED LIVESTOCK PROGRAMME

Working with people's organizations in the Philippines

Carlos C Tomboc and Gregorio D Reyes

INTRODUCTION

The Department of Environment and Natural Resources (DENR) in the Philippines is working towards the sustainable development of natural resources. Its research arm, the Ecosystems Research and Development Bureau (ERDB), conducts research on the major ecosystems. The DENR's integrated social forestry programme (ISFP) is part of the Comprehensive Agrarian Reform Programme (CARP). CARP's objective is to promote social justice and sustainable rural development through an equitable system of land-ownership and distribution. The ISFP is aimed at increasing rural income in upland areas while safeguarding the environment. The ERDB is implementing the joint CARP–ISFP research and development programme.

Since the government's extension efforts have rarely reached the poor upland farmers, ISFP seeks to extend appropriate technologies through building new links in upland communities. The ISFP is therefore seeking to work with local people's organizations (POs), such as farmers' associations, as partners in project design and implementation. In areas where effective POs do not exist, the ISFP seeks to stimulate their formation in order to receive inputs.

In this partnership, the GO has the task of technology assistance, extension and delivery of inputs and services, monitoring the overall impact on the environment, and co-ordinating linkages with other agencies. A series of 'model' income-generating projects has been drawn up – including agri-livestock, agroforestry, fuelwood, cut flowers, mangrove, aqua-silviculture, *tikog* (*Alpina officinarum*) and bamboo/rattan – which the PO selects and adapts to suit local needs and conditions.

The implementation of the project begins with a preliminary survey of the locality by the ERDB, explores potential organizational linkages and assesses the community's available resources. Analysis is also made of the community's leadership structure, felt needs, existing organizations and its potential for participation in project planning and implementation.

The PO takes charge of project implementation at field level, performed with assistance from ERDB technical personnel. Depending on the nature of problems encountered, problem-solving takes place with inputs from individual farmers, the PO and the GO staff.

For example, in the agri-livestock programme (ALP) there are three basic models (developed by the ERDB) for the delivery of inputs, which are then adapted for local needs:

- pig production
- backyard cattle fattening
- backyard goat raising.

Since the project phases out after five years, the GO works towards strengthening the PO's ability to continue the project without further technical assistance, through community organizing and networking activities. Participation from local NGOs is not encouraged, since it is felt better to establish direct linkages between the government agency and the PO. It may be considered if necessary in future.

CASE STUDY

One recent ERDB case study illustrates the process in a village selected for the ISFP. San Francisco *barangay* (village) in Luzon lies in a severely environmentally degraded upland area. It has a population of 847 belonging to the Ilocano ethnic group. Income is largely derived from farming corn, *palay* (rice), tobacco and vegetables yielding an average figure of P10,000 per year or around P833.33 per month. This is supplemented by proceeds derived from selling of light construction materials, such as bamboo. Many villagers use carpentry and masonry skills for house construction.

There were no appropriate POs found in the village. Instead, the farmers participated in the ALP through a new organization, the San Francisco Livestock Raisers Association (SFLRA), established in October 1990 in co-ordination with the area's *barangay* captain and council. It now provides the structure for facilitating the participatory exchange of needs and ideas, negotiations and the resolution of problems. It also serves as a support structure to the ALP by representing the farmers.

Farmers are encouraged to participate as a formal business partnership, with clearly defined rights and obligations. A general meeting with prospective beneficiaries was organized in co-ordination with the CENRO and *barangay* officials. This was followed by house-to-house visits with interviews for establishing household income, resources and needs. Interest in the programme was judged by attendance at meetings and willingness to shoulder counterpart expenses in the construction costs (e.g. for a goat house or pig pen). The PO, established by the members to support the objectives of the ALP, serves to keep the members active in their roles and responsibilities.

Work began in June 1990 with a communal forage area of 1.3 ha established by the co-operators. Forage grasses such as Napier (*Pennisetum purpureum*), Paragrass (*Brachiaria mutica*), Guinea Grass (*Panicum maximum*), and legumes of Desmodium, Flamengia, Acacia, Centrosema and Kakawate varieties have been planted under DENR guidance. Intensive Feed Gardens (IFGs) measuring around 15–20 sq. m each have also been established in backyards in order to supplement communal forage grass.

According to each farmer's needs, backyard pig pens and goat houses have been constructed. The project supplied fifteen bags of Pozzolan cement for foundations and flooring, matched with indigenous materials such as bamboo and *cogon* grasses for walls and roofs supplied by the farmers themselves.

A pair of does (one ready-to-breed native species and one Anglo-Nuvian) have been distributed to each of the eighteen co-operators for breeding. One piglet of Hypor or Large White breed had been given to each of the five others to be raised as breeder sows. In addition, two bucks (one Anglo-Nuvian and one Alpine) had been given to the PO to serve as communal bucks. By the end of 1990, eight kids (two males and six females) were produced out of the initial stocks. By May fifteen more were kidded (six males and nine females). This brings a total of fifteen female kids for dispersal to the second and third batch of co-operators. There is a reported 100 per cent survival record for the goats (eighteen native and eighteen graded) delivered to the area and only one of the twenty-three kids died. Meanwhile, reports show that the five piglets distributed now weigh 60–80 kilos each.

Veterinary medicines such as de-wormer, vaccines, vitamins and antibiotics along with fourteen sacks of commercial feed supplements (five hog starter mash and nine hog grower) has been provided to the co-operators.

A monthly week-long site inspection of livestock is conducted by the project co-ordinator in collaboration with the veterinarian of the local Department of Agriculture to ensure the continuing health of the livestock. The use of herbal medicine such as guava leaves for the treatment of skin infections has been encouraged by the project co-ordinator, who recently received training on livestock raising.

Technology assistance is delivered and achieved through networking with other government agencies particularly the Department of Health (DoH), the Department of Agrarian Reform (DAR) and the Department of Local Government (DLG). The Don Mariano Marcos Memorial State University in Ilocos Norte also participates through the Department of Agriculture.

Although it has been running for only a short time, results from the ALP are promising. The livestock dispersal, utilization, breeding, fattening and reproduction schemes have provided more than 100 per cent profit returns to participating farmers.

Project implementation draws its strength from the fact that the goals set are highly needs-based for both the physical environment and the people themselves. The project directly addresses specific needs either in the form of providing a supplementary source of income or promoting self-sufficiency.

The 100 per cent survival performance of the dispersed livestock illustrates the wisdom of close co-ordination and linkage with the DA service staff in areas where the project co-ordinators are not the 'experts'. Vital linkages have been established and it is proving to be among the ALP's prized assets. Previously, the government credibility in the uplands was not high, since 'experts' seemed to offer farmers very little.

The establishment of the PO facilitates participation and serves a firm organizing structure for members who have mutually bound themselves to the observance of project rules and by-laws, serving both the farmers' and the project's needs. It serves as the fulcrum for crucial decisions that have to be made on both levels. Members come together as a body to discuss issues and potential problem areas that may hamper project operation. In serving as a mechanism of exchange and interchange, it enhances participatory collaboration.

The project implementation is strengthened by close monitoring of project activities and status by two regular staff from the regional co-ordinating agency assigned to the area. A research assistant hired for the purpose of daily monitoring of project activities has improved the project implementing service.

Farmer conscientization and mobilization is also a function of these needs. The relevance of technology (due to the adverse biophysical and ecological and socio-economic condition confronting the community) has greatly enhanced project acceptability. The project is perceived as a means towards a better future. The motivation of farmer co-operators has been maintained through the appropriate delivery of services.

The concepts of shared responsibility, a perceived people-centred project orientation, and a consultative participatory approach to management all worked together to offset pockets of dissatisfaction and suspicion among the local farmers.

PROBLEMS

At this point, the major emphasis of operation rests largely upon livestock production which necessarily leaves out so much to be desired in terms of area development requirement. Other services are also needed by the farmer co-operators. As a consequence, there are ambivalent feelings among some farmer co-operators, particularly towards the Intensive Feed Gardens (IFGs) which they have each been required to set up in their respective backyards, in addition to the forage area they had communally set up in a 1.3 ha space in the project site, where forage grasses provide food for the livestock.

Varying degrees or levels of deprivation encourage resentment toward project activities that do not immediately compensate for felt needs. Thus, the IFG which was designed as a major supplement to the communal forage area to ensure the availability of forage grasses supply was often accepted out of compliance rather than conviction.

This issue illustrates the need for a more intensive educational campaign to improve farmer co-operators' perceptions and willingness to compromise and settle for a delayed reward in efforts in place of the understandable desire for more immediate compensation.

LESSONS

- The participatory approach to development (PAD) represents more than a physical manifestation of interests and involvement and is a function of inner motivation, involvement and mutually shared commitment.
- Participatory GO–PO collaboration does not necessarily involve a 'bottom-up' implementation strategy alone. In this case it is primarily a GO initiative, but with strong participation from the PO.
- Community development requires working within the context of the community's culturally instituted structure to ensure relevance and acceptability.
- The upland population clearly have potential for collective organization, self-management and other productive endeavours. However, they lack productive assets despite their self-help mechanisms. There is therefore a need to harness their potential through developmental intervention initiated by government organizations.
- The credibility of government is an important aspect of successful GO–PO collaboration. As an agency which delivers development inputs, the government organization is expected to help improve the PO's living conditions. In the case of the ALP, credibility is enhanced through prompt delivery of services and visibility and availability of co-ordinators in the area.
- The GO–PO collaboration evolves from participatory behaviour premised upon mutual agreement to move together in the same direction. Without this agreement, a wavering of commitment can easily take place, particularly when farmers' attentions are distracted by the presence of other organizations.
- A co-ordinated approach among government organizations in the implementation of their development goals is of great importance to upland areas. However, the needs of these areas are undoubtedly far more complex than can be met by the agri-livestock project alone. Other aspects may not be within the reach of the DENR.

- Upland development projects need effective monitoring and evaluation activities. This is because the upland areas are scarce in resources and other services for meeting the farmers' fluctuating seasonal needs.

THE PHILIPPINES' DEBT-FOR-NATURE SWAP PROGRAMME

Delfin J Ganapin, Jr

INTRODUCTION

The Debt-for-Nature Swap is the first major environmental programme in the Philippines to be implemented through GO–NGO collaboration. The Philippines has awoken too late to the fact that its resources have been abused. The debate on a selective or total ban on logging is carrying on at a time when productive forests have been reduced from about 16 million hectares to around 6 million, with only around 850,000 hectares of old growth or virgin forests left. The importance of 'bio-diversity' has been recognized only after 50 per cent of endemic flora has been lost. There are now increasing demands for sacrificing economic gains for ecological values at a time when there is around 20 per cent unemployment and about 60 per cent of the population exists below the poverty line.

Environmental deterioration has come from a combination of ignorance of the carrying capacity limits of ecosystems and the desire for massive profits to be had from their exploitation. However, the question has more recently become one of survival as resources become very limited in relation to the increasingly impoverished and growing population.

THE SWAP CONCEPT

Foreign debt of US$30 billion constitutes a major obstacle to development in the Philippines. Annual debt servicing alone shifts almost half of the annual budget away from other basic services. Environmental issues tend to receive low priority when short-term needs are so urgent.

Two observations inform the Debt-for-Nature concept: first, many tropical forests are found in poor and indebted countries, with little choice except to increase production and exports by clearing forests; some rich countries are now prepared to 'pay' for preservation of these 'global commons'. Second, accumulated debts cannot be fully serviced by debtor nations; private lenders had started looking for ways to minimize losses. A secondary debt market emerged for debts where substantial discounts were available on debts which are immediately paid.

Discussions on Debt-for-Nature started in 1984 when international NGOs such as the World Wildlife Fund (WWF) began a dialogue with governments. In 1987 a scheme was implemented in Bolivia and Costa Rica; the Philippines followed on 24 June 1988.

Debt-for-Nature involves an exchange or cancellation of an external debt obligation in return for environment-related action by the debtor nation. The stages of the process are as follows:

- An international NGO, such as the WWF, acquires dollar donations or grants from a donor such as USAID to buy part of the external debt on the secondary debt market.
- The NGO implements the scheme with the host country's government. In the Philippines, an agreement with the Department of Environment and Natural Resources (DENR) and the Central Bank was made. A local NGO, the Haribon Foundation (HF), was brought in to manage funds.
- The WWF finds a seller bank in the secondary debt market willing to provide the highest return on the debt to be swapped, e.g. $1 of actual payment can be equivalent to the cancellation of $2 external debt. Arrangements are then made for payment and cancellation of the equivalent external debt.
- The Central Bank is then informed of the transactions and by prior agreement releases to HF the peso equivalent of the cancelled debt.
- The programme is implemented with periodic reports to the Central Bank and other participating parties, including the donors.

Relations between the WWF, HF and the DENR were straightforward. The Undersecretary for Environment and Research of the DENR was a former president of the HF and the DENR itself is a member. The HF and WWF are more or less 'sister' organizations: the HF represent the WWF's interests in the Philippines. The HF, WWF and DENR were also previously involved in projects together, which paved the way for preparing the Swap programme: for example in 1985, the HF had a conservation strategy project funded by WWF.

The party that needed to be convinced was the Central Bank, which had no experience in funding innovative environmental projects. The criticism was also made that such a programme would have an insignificant impact on total external debt. The proposed 'swap' was for only US$2 million. Some economists also argued that the infusion of so many pesos would be inflationary and the economy needed new foreign exchange rather than the release of existing pesos. However, it was generally felt that money spent for the protection and management of bio-diversity was well spent, especially since it was a grant. This innovative method of approaching debt also coincided with growing environmental consciousness in the Philippines.

The government was well disposed towards the idea of working with NGOs on joint projects. After the 1986 revolution, many former NGO activists had entered the ranks of the bureaucracy. An NGO Desk had been established in the DENR and NGOs were represented on many of its committees (see pp. 254–8).

The DENR became official 'guarantor' to the programme. Fears that the participating NGOs may not be capable or responsible enough to handle such massive funds were dispelled by this role. The credibility of WWF and HF as responsible NGOs also helped to establish the programme.

All the agencies involved have benefited from the scheme:

- Substantial amounts were made available swiftly to both GO and NGO projects, due to the ability of NGO disbursement procedures to cut through 'red tape'.
- The profile of the WWF was raised, increasing membership and donations.
- HF and DENR also gained credibility from the programme.

The programme was organized according to a memorandum of agreement signed by the parties.

Some problems emerged in relationships between parties in field implementation at the initial stage. The project was divided into two major groups with most of the field projects under DENR and institutional strengthening and research responsibilities going to HF. This division led to suspicion.

Macro-level policy was handled by a management committee of three members, one each from HF, WWF and DENR who were old friends, and the programme therefore ran smoothly and informally. No minutes of the meetings were kept at this stage. The funding source was considered 'neutral', neither a loan nor from DENR or HF. The WWF kept a low profile in its efforts in acquiring donations. Funds resulted from a complex process involving many parties in the transactions that no one could claim outright ownership.

PROBLEMS

Finances were managed at the central level. The broad allocations are established early by the management committee while each of the programme management offices (PMOs) of the DENR and HF submit the detailed activity per activity budgets. Both the heads of the two programme management offices were Haribon members of long standing, but tension between the two also increased as the DENR head felt that the HF was handling the funds arbitrarily. A new system was set up detailing the requirements for the processing and accounting of fund releases more formally. The NGO needed to acquire some of the bureaucratic requirements of its GO partner.

The second problem of policy conflict was a more difficult problem to solve.

HF took radical views on issues outside of the programme, and campaigns for a total logging ban and a ban on loans reforestation projects. The government (through DENR) took more conservative positions on these issues.

The HF president became the spokesman of the Green Forum, a radical NGO coalition which included leftist anti-government participants. Tension between the HF president and the DENR intensified and the press misreported ill-informed HF criticisms of government policies. Influential members of the DENR, HF and WWF intervened to try to settle the issues at hand in a more rational manner. It was decided that the HF would henceforth issue press statements only on key policy issues upon agreement on the policy position by the Haribon board of directors.

The third problem of conflict of activities at the field level occurred at St Paul Subterranean River National Park project, where an HF deputy team leader was assigned to the project. The team leader was a long-standing DENR officer and a clash between 'radical' and 'conservative' outlooks occurred. The HF member felt that the DENR team leader failed to manage the project properly, while the DENR team leader felt that the deputy challenged the authority of the team leader. Further clarifications of roles and responsibilities have partly solved this.

A series of consultative conferences were held to tackle some of the problems and revealed that many participants had different interpretations of the programme. A written set of agreements was made to promote a shared vision of the programme.

A strong need which emerged was that of expanding the initial collaboration beyond the three organizations involved and embrace the host community in which the projects were being implemented and to other NGOs and GOs, both local and foreign. The exclusivity of the collaboration had caused some NGOs to accuse DENR of favouritism.

The Debt-for-Nature project continues: under the USAID-supported natural resources management programme, US$5 million has been allotted from a total of US$25 million for immediate projects of similar type and for the setting up of an endowment fund, the proceeds of which will be used by NGOs for environmental projects. This is the biggest grant of its kind in the world. The lessons from the first Debt-for-Nature Swap programme are very important as they can provide useful models for the future. The experience is helping DENR institutionalize NGO participation in key government projects and programmes and reduce mutual misconceptions about aims and objectives.

LESSONS

There are several key points arising from this experience:

- It was helpful that the parties in the collaboration had had small but successful projects prior to the bigger partnership and that key individuals were known to each other.

- Trust between the interacting parties is very important to GO–NGO collaboration. If one party feels that the other is just making use of the partnership for its own self-interest, then the collaboration starts to break down. Problems started when the NGO made public criticisms of government policies without what the government agency felt was due consultation.
- It is difficult to collaborate with NGOs that are critical of sensitive policy issues or belong to 'anti-government' coalitions. No matter how rational such criticisms are, they will still be perceived as biased and irrational.
- Mechanisms for continuing dialogue on critical issues between NGOs and GOs are crucial. Government needs to appreciate the value of constructive criticism, and NGOs need to understand the limitations of government procedures.
- Proper documentation, audits and the constant sharing of information and documents are necessary to promote trust.

NOTES

1 For example, the head of the DA's NGO Outreach Desk resigned in March 1990 to return to NGO work.

2 As armed opposition to the government has continued in the Philippines, the government has been able to use insurgency as an excuse to marginalize certain development NGOs as 'communist fronts' and control their activities. For example, the environmental NGO Haribon Foundation suffered harassment of this kind during 1991, despite its involvement with the Debt-for-Nature programme (see pp. 265–9).

3 Acronyms abound in the Philippines and some observers have also drawn attention to BONGOs, which are business-oriented NGOs formed as companies attempt a public relations exercise or tax evasion and COME'N GOs, formed on paper by entrepreneurs but which never actually exist in reality (Constantino-David 1992).

4 However, by mid-1990, the NLS had become inactive.

5 These issues were discussed by a range of NGO and government participants at the workshop on NGOs Public Sector Research and Agricultural Technology organized by ODI and IIRR held in Silang, Cavite, in the Philippines in July 1991 (Gonsalves and Miclat-Teves 1991).

6 The editors draw the reader's attention to the Philippines overview section (pp. 227–39) for a discussion of these difficulties.

7 See MBRC case study (pp. 240–43).

Chapter 7

NGO–GOVERNMENT INTERACTION IN INDONESIA AND THAILAND

INDONESIA OVERVIEW

Indonesia is a country consisting of 13,677 islands containing a diversity of cultural groups practising four formally recognized religions. The per capita GNP was US$530 in 1990, with 23 per cent contributed by agriculture, 37 per cent by industry and 17 per cent by the manufacturing sector. In 1989–90 aid receipts constituted 2.1 per cent of GDP and the total external debt was 66.4 per cent of GDP. Despite the growth of industry, 54.4 per cent of the labour force still works in agriculture. Life expectancy stands at 61. The daily calorific intake in 1988 was 2,670. Secondary school enrolment in 1985 was 41 per cent.

Since the overthrow of Sukarno in 1965 the Indonesian government has placed heavy emphasis on state security and ensuring complete loyalty to the official state philosophy of Pancasila. This has created a hostile environment to many NGOs, particularly those with Muslim, Catholic or Protestant religious origins and those working in organized development activities at the local level whose activities might have a destabilizing effect on the 'New Order' (Holloway 1989).

The organized voluntary sector in Indonesia has a history which goes back to the beginning of the twentieth century, when social welfare organizations were established in the health-care, social welfare and formal education sectors. This pattern was in response to the government's top-down approach to development which had only a limited impact on the poor in a country of enormous geographical and cultural diversity.

By the 1960s new organizations appeared with an emphasis on development rather than welfare, which aimed to

> bridge the gap between the needs of that (disadvantaged) citizenry and the goals of national government programmes. They began by responding to the diverse problems and aspirations articulated at the grassroots level, and then carving out development roles not assumed by government or business.
>
> (PACT 1989: 77)

The ideals of self-reliance began to replace the welfare orientation of many organizations. For example, the largest NGO in Indonesia, Bina Swadaya, which has its origins in the early 1960s, now trains other NGO staff and government officials in participatory approaches to agricultural development. This NGO is now generating more than 50 per cent of its own income.

During the 1970s development NGOs such as LP3ES (see case study on pp. 280–83) were established with a community focus which stressed health, small-scale industry and appropriate technology. Other NGOs developed with an advocacy orientation focusing on human rights, the environment, legal aid and consumer protection. Some NGOs were able to secure media coverage of these issues, but faced a very difficult relationship with government.

Political conditions in Indonesia since the early 1970s have increasingly made work in thc NGO sector very difficult. Political parties were compelled to form a government controlled body, student radicalism was repressed and the press was severely restrained. All organizations such as farmers and fisherfolk's associations, trade unions and women's organizations were tightly controlled under government umbrella organizations, and these contraints continue in the early 1990s. There have also been attempts by the government in recent years to replace NGO networks with government-controlled versions (Holloway 1989).

Under these conditions, the importance of NGO networks has grown in order to counter growing government interference and to provide a loose framework for informal NGO–NGO co-operation and mutual support. An example of this is the NGO Bina Desa, formed in 1976 by fourteen individuals representing organizations committed to rural human resource development. Bina Desa has trained motivators, mobilized resources and set up linkages in order to reduce the isolation felt by small rural development organizations.

There has also been co-operation between NGOs and the traditional *pesantrens*, that is Islamic educational institutions which in some cases have become entry points for NGO initiatives in local communities. The NGO LP3ES pioneered these efforts in the early 1970s, which has served with some success to expose these conservative institutions to innovative approaches to development. The process has also served to stimulate NGOs' own thinking on issues such as women's rights and the environment.

Since the early 1980s there have been signs that the government has recognized that NGO experiences may be of value to government in improving the quality of their programmes. The fall in oil prices has reduced government revenues and increased the role of aid, and therefore the role of the donors in development projects. At the local level, however, relationships with NGOs may still depend heavily on the outlook of local government officials. This has led to a situation in which there may be policy-level decisions to pursue collaboration, but local-level distrust and antagonism. Nevertheless, there are NGOs willing to attempt to work with government in order to gain for themselves a higher profile and wider access to funds and, through policy debate, to change the ways in which government is working.

THAILAND OVERVIEW

INTRODUCTION

Thailand is a country currently undergoing rapid change. Once a primarily agricultural nation, the industrialization process since the 1960s – under predominantly military rule – is expected to propel Thailand into the ranks of the newly industrialized countries alongside a number of its neighbours, with an average annual GDP growth of 10 per cent. This economic progress has been accompanied by social costs: an increasing gap between rich and poor has placed considerable pressure on the livelihoods of the 75 per cent of the Thai population in rural areas, despite the increase in rural–urban migration. Some 40 million of the population of 60 million still earn their livelihoods from agriculture or agriculture-related industries; over 10 million of these are regarded as living in poverty (Pisoun, in Sollows *et al.* 1991). While the average per capita income in the capital Bangkok is US$1,000, in the rural north-eastern part of the country, where most of the inhabitants are subsistence farmers, the average is US$235 (Ekachai 1990: 19).

The pressure on the natural resource base has also been considerable. In the south, where many of the small farmer households supplement their income with fishing, growing industrial pollution has reduced the available fish stocks. The increased use of agrochemicals has also contributed to this decline, and to numerous cases of poisoning, requiring almost 5,500 hospital admissions in 1985. The growth in demand for timber and the development of a highly profitable shrimp production business has meant that over half of Thailand's mangrove forest have been felled since the early 1970s. In the north-east, much of the tropical rainforest has been cut down in order to grow cash crops such as maize, tobacco and cassava, which has caused severe soil erosion. The government has responded in some areas by replanting fast-growing eucalyptus for export as paper pulp, which has caused conflicts over land use and forest access between forest dwellers and government agencies. Heavy logging combined with unsustainable agricultural practices have resulted in flash-flooding and also worsened soil erosion problems.

POLITICS

Although there has been a democratic movement in Thailand since the late 1920s, its progress has been erratic. National politics have been characterized by regular military interventions. Since the end of absolute monarchy in 1932, there have been over forty military coups. Conflicts over natural resources have formed an important arena for the turbulent political development of Thailand, where control of public policy secures access to land, minerals, water, forest and fisheries resources.

The 1990s have seen political events move quickly and there is cautious reason to hope that space for greater scope for participatory, NGO-based activities may be opening up. In February 1991 the elected (though highly corrupt) regime of Chatichai Choonhavan was removed from office by a coup organized by General Suchinda Kraprayoon. In April 1992 Suchinda resigned to become the unelected Prime Minister. This was seen by the urban middle classes and students as an attempt to legitimize and retain the power of the military in political life. May 1992 saw the organization of large-scale middle-class and student 'people power' pro-democracy demonstrations on the streets of Bangkok, which were brutally repressed by the army with the loss of hundreds of civilian lives, after which the government was brought down and several of the discredited generals who had taken part in the crackdown were forced from power.

After a period of caretaker government, elections in September 1992, which were the first for two decades to be conducted relatively free of military involvement, brought the Democrat Party to power with the delicate task of constructing an alliance of parties opposed to military interference in politics.

NGOs

Social welfare institutions have long been part of Thai culture and were informed by Buddhist religion and teaching. During the nineteenth century the first known voluntary organization was the Sapaynalomdaeng, which was formed by women from the elites. The organization dedicated itself to the care of soldiers injured in battle, and later evolved into the Thai Red Cross. The modern welfare sector is now well established and is co-ordinated by the National Council for Social Welfare, founded in 1948.

However, by the late 1960s the voluntary sector was beginning to concern itself with development activities. An early pioneer in this field was the Thai Rural Reconstruction Movement, affiliated to IIRR in the Philippines (see IIRR case study, pp. 244–7). The TRRM committed itself to rural development, health and literacy and also served as an important training ground for young graduates interested in active development work.

During the 1970s development NGOs continued to grow, although the continuing political conflicts ensured that most of the NGOs established remained informal, welfare-oriented groups. There has generally been little participation by the rural poor in formal politics. Nevertheless, according to local activist networks such as Asia Pacific Solidarity (Bangkok), popular grassroots movements in the form of people's organizations started to emerge in Thailand around such issues as forest protection, marine resources preservation and land rights, although the urban democratic movement has not as yet formed many links with these local initiatives. During the 1980s NGO development activities gathered momentum and it is estimated that 60 per cent of Thai development NGOs active today have originated in the period since 1984 (PACT 1989: 186).

PROSPECTS FOR THE FUTURE

There has in recent years been growing recognition of the need to promote agricultural development in ways which are both more environmentally acceptable, and which contribute to income generation among the rural poor. NGOs are increasingly recognized as having useful contributions to make in both of these areas. Contrary to what might have been expected, the military government which took power in early 1991 began to dismantle some of the restrictions that had been placed on NGO activities during recent democratic regimes. Insights into the government's vision of closer relations with NGOs is provided in Boxes 7.1 and 7.2. Box 7.1 gives extracts of a policy speech by the then Prime Minister in August 1991; Box 7.2 contains extracts of a paper by a senior representative of the Ministry of Agriculture and Co-operation presented at a national seminar in September 1991 on 'Improving relations with, and supporting the work of NGOs.'

At present more than 10,000 voluntary associations and clubs of all kinds are registered, of which only 200 have tax-exempt status as direct providers of services benefiting the public. The atmosphere of improved NGO-government relations that, as Boxes 7.1 and 7.2 indicate, was building up in 1991 was suddenly reversed by the events of May 1992.

However, as Garforth and Munro (1990) indicate, many thousands of formally registered small producer groups, with a membership of over half the total number of small farmers in the country, exist in Thailand and have had long-standing and fairly stable relations with government for such specific agriculture-related activities as input supply, marketing and savings and credit. The situation facing non-membership NGOs, whether indigenous or foreign-based, is much more volatile. All recognize that they cannot operate without being registered with, and monitored by government, and, within this tightly controlled framework, some have successfully challenged government policy on certain issues, such as the environment. But the majority prefer the politically safer strategy of supporting small farmers or grassroots organiza-

Box 7.1 Extracts from a speech by the Thai Prime Minister at the Siam Society Gala Dinner on 14 August 1991

The speech highlighted several difficulties currently faced by NGOs, including registration and screening procedures that are lengthy and treat NGOs as politically suspect; the difficulties of attracting and retaining skilled personnel, and the threat of reduced external funding as Thailand moves to 'middle-income' status.

The Prime Minister highlighted several NGO roles worthy of government support:

1 as *communicators* they have raised public awareness about e.g. the distribution of forests, and have secured a widespread ban on logging
2 as *facilitators*, NGOs mediate between government and people, and assist communities to help themselves.
3 as *providers of services* NGOs can reach some areas with certain types of services more cost-effectively than government.
4 NGOs *innovate* in the provision of services and the tackling of specific problems, and government should support and learn from this process
5 NGOs should be brought more into official *development planning* processes, to ensure that grassroots views are heard.

The Prime Minister explained that he had already emphasized in a policy statement to the National Assembly the need to support NGOs, and outlined the measures he intended to take:

1 streamlining of registration procedures, in particular the removal of security checks on personnel
2 raising of the current limits on tax deductibility for corporate contributions to charity
3 government requirements that provincial authorities co-operate with NGOs in e.g. the 6 million baht allocation for drinking water, environmental conservation and occupational training
4 government funds in certain cases to match the contributions raised by NGOs from society at large.

More widely, it was the government's intention to promote a spirit of voluntary work among commercial companies, encouraging them to sponsor particular villages or NGOs, and to provide specialist skills where these are beyond the reach of NGOs.

Other possibilities under consideration include the substitution of community service for currently compulsory military service, and the repayment of scholarships through voluntary work with NGOs during vacations.

tions in specific agricultural production issues, and collaborating with GOs mainly at a local level. Meanwhile links between the pro-democracy movement and local people's organizations remain weak.

NGOs in Thailand therefore face a difficult and unpredictable future. In common with their counterparts in Bangladesh and the Philippines, there is now some hope that the as yet fragile moves towards wider democratization and accountability will increase the political space in which NGOs can operate

Box 7.2 Extracts from a paper by the Director of Foreign Agricultural Relations, Ministry of Agriculture and Co-operation, presented at a national seminar on 10 September 1991 on 'Improving working relations with and supporting the work of NGOs'

The paper began by noting some softening of the historical lack of congruence between NGO and Ministry of Agriculture and co-operation (MAC) aspirations and objectives in agricultural development, together with an increased willingness to plan work jointly. Welcoming this trend, the director outlined the twin objectives of the new liaison committee of which he was co-ordinator:

1 to liaise with and support NGOs in rural development programmes covering agriculture, the environment and natural resources
2 to decide on the method, location and direction of joint NGO–GO activities, and to bring in other assistance as required.

As an initial step, the co-ordinator will open a dialogue with NGOs on priority areas and issues, and on modalities of collaboration. Three types of collaboration are envisaged:

1 the MAC has chief responsibility for project implementation, and is supported by NGOs
2 project implementation is shared equally
3 NGOs have primary responsibility for implementation, and GOs provide support.

Joint projects will be overseen by a joint working committee established on 9 August 1991 and chaired by an inspector representing the deputy minister.
The committee is to

1 issue regulations and guidelines on the acquisition of funding for the projects
2 designate an appropriate MAC Department to collaborate in the project
3 appoint joint advisory and monitoring sub-committees as appropriate.

Joint working parties were set up in two subject areas in March 1991:

1 In watershed conservation, a joint project is to be set up on 32 sq. km of land at Tambon Lum-sum in Kachanaburi Province. Its main purpose will be to act as a centre for raising awareness of the need for improved forest and watershed management, and for demonstrating sustainable practices. An NGO will lead the needs assessment, a GO will be responsible for the introduction of new technology, with the NGO in a supporting role, and both will be jointly responsible for awareness-raising.
2 A second joint project will promote among producers and consumers awareness of benefits of organic farming through a national campaign involving publicity, demonstrations and certification of products.

A major anticipated benefit is increased awareness of the dangers of pesticides, and a reduction in their use (see also Jonjuabsong and Hwai-Kham, pp. 289–97 in this volume).

in roles which can address pressing issues of poverty and natural resource management faced by Thailand.

INDONESIA'S INSTITUTE FOR SOCIAL AND ECONOMIC RESEARCH, EDUCATION AND INFORMATION'S (LP3ES) WORK WITH SMALL-SCALE IRRIGATION SYSTEMS

Bryan Bruns and Irchamni Soelaiman

INTRODUCTION

The Institute for Social and Economic Research, Education and Information (LP3ES) helped the Indonesian Ministry of Public Works to develop and institutionalize methods for turning over small irrigation systems to water users' associations. A series of pilot studies strengthened LP3ES's capability and explored ways to improve participation in irrigation design, construction and management. For turnover, LP3ES trained irrigation staff who worked with farmers, trained trainers, provided consultants to provincial irrigation services and collaborated in drafting regulations and manuals. Conditions for collaboration in institutional innovation included willingness to compromise, mutual trust, funding linkages and educated opportunism.

The LP3ES is an example of what Korten (1987) refers to as a 'third-generation NGO', which directs its efforts not simply at relief work or local community development but also at working with and seeking to change government activities. Formed by a group of activist intellectuals in 1971, the NGO aimed to challenge the government's centralized top-down development model. They saw a need to provide training for young people who would become key actors in an alternative model of development capable of mobilizing the participation of the people. The main aims of the NGO (LP3ES 1980) are as follows:

- To promote the advancement of economic and social sciences that will help foster the socio-cultural development of the Indonesian people through research, education and information activities.
- To provide contributions to a more integral development of human resources in Indonesia, particularly to assist the young generation in preparing themselves for the socio-economic challenges of their own future.
- To improve the knowledge and understanding of development problems in

Indonesia among the public and to promote international co-operation with national and international organizations which have common objectives.

STRENGTHENING GOVERNMENT IRRIGATION SYSTEMS

This 'turnover project' built on a series of earlier pilot studies which explored ways to improve local participation in irrigation development. The goal of transferring management of small irrigation systems back to farmers contains elements of a fourth-generation strategy to expand the role of people's organizations in development (Korten 1990) involving the state as only one actor within a broad context of many organizations striving for equitable, inclusive and sustainable development.

Irrigation is an essential resource for most of Indonesia's farmers. They depend on irrigation to grow two or three crops a year, often on one-third of a hectare of land or less per household. Farmers originally built and managed irrigation systems as local common property, but during the 1970s and 1980s the government invested in large-scale irrigation projects with little or no local participation. This continued a trend initiated by the Dutch in an attempt to attract private investment, which led to disruption of community arrangements and increased government involvement in irrigation management.

Beginning in the mid-1970s the United States Agency for International Development (USAID) funded the Sederhana (Simple) irrigation systems project to build irrigation systems in previously unirrigated areas. Over time the project included more and more sites which already had existing farmer-built irrigation systems. The project tended to design and build irrigation systems with little attention to existing arrangements and without consultation with local farmers.

Researchers from Gajah Mada University, who studied project sites, concluded that farmers took little responsibility for the operation and maintenance of irrigation systems because government-built systems did not meet their needs. In order to improve farmer participation in irrigation systems, the Ministry of Agriculture requested LP3ES to assist with the following activities:

- Research into developing social and technical aspects of irrigation development in the mid-1980s. 'Community organizers' were brought in as the key to irrigation development, living among the farmers for two to three years and acting as motivators, organizers and mediators between farmers and government agencies.[1] Formal water users' associations were established.
- LP3ES experimented with farmer participation in tertiary canal construction, with the Ministry of Public Works.
- The Ministry of Home Affairs invited LP3ES to conduct a study of water

users' associations (WUAs) to develop a WUA policy. The studies showed that only a small number of WUAs survived and that more support was required for their proper development.

The drop in oil prices in the mid-1980s created a fiscal crisis for the Indonesian government. Discussion between the National Planning Board, Public Works, the World Bank, the Ford Foundation and LP3ES developed the idea of the government 'turning over' all irrigation systems smaller than 500 hectares to farmer water user associations. These small government systems cover a total area of 1 million hectares, out of some 4 million hectares irrigated by government irrigation systems in the country.

LP3ES trained irrigation staff on social and institutional issues and they then worked directly with farmers, trained trainers, provided consultants to assist provincial irrigation service officials in institutionalizing new procedures and took part in national working groups which drafted regulations and manuals for carrying out the turnover. Other inputs in the programme were provided by an international research institute, national universities and multiple levels of government. LP3ES was the only NGO involved.

LP3ES played several roles in the turnover process. The first was training: this represented an easily understood entry role for the NGO. It soon became clear that new procedures were also required; the second role of the NGO shifted from direct implementation to the role of consultant, providing support and preparing for eventual phasing out of its role. Trainers then began to concentrate on province and section levels and spent less time in the field. LP3ES staff became provincial social and institutional consultants.

A third role undertaken by the NGO was in devising the methods for turnover and LP3ES formed part of working groups with representatives from government agencies and the Ford Foundation, preparing operating guidelines and manuals for the maintenance of irrigation systems and the formation and development of WUAs. This process involved government staff acting as partners in creating something new based on a cycle of policy to practice and back to policy, and this led to the gradual institutionalization of the new methods required for turnover.

CONCLUSION

The success of this collaboration depended upon a number of important preconditions on the part of both the NGO and the GO:

- compromise and reformism
- trust and sensitivity
- funding linkages
- coalition building
- educated opportunism.

A central condition for this type of collaboration with government is therefore a willingness to compromise, particularly on the part of the NGO. LP3ES was required to take an incremental reformist approach to development, carefully combined with advocacy and constructive criticism. The collaboration also required explicit funding linkages between the agency and the NGO, the building of mutual trust (between government agencies, the NGO and the farmers themselves) and educated opportunism in choosing which issues had the most potential for progress.

The turnover programme illustrates how an NGO can play a role in changing the ways a bureaucracy relates to farmers. In this case an NGO played a major role in training, field activities, conceptual development and institutionalization of new methods. NGOs are not just limited to working directly with local communities but instead can work at multiple levels from farmers to national policy formulation. NGOs have the advantages of flexibility, creativity, local knowledge and understanding of institutional issues. This suits them for the role of innovators collaborating to develop new approaches and consultants helping to institutionalize capabilities within agencies. Rather than replacing or substituting for government services it is therefore possible for NGOs to change the way in which government works. The innovative role played by an external funding agency in bringing together LP3ES and government departments is highlighted in Box 7.3.

Box 7.3 The Ford Foundation, LP3ES and government departments in Indonesian irrigation development

The Ford Foundation played an innovative role in encouraging collaboration between government and an NGO in restructuring the public irrigation system in Indonesia. A new flexible funding strategy was devised to allow Ford to perform a facilitating role, in which some funds were given directly to LP3ES and others to government (see the LP3ES case reported on pp. 280–83 for details of the NGO–GO collaboration).

Along with USAID, Ford brought in the NGO LP3ES to help the government and the donors to work with farmers to develop a better system. Ford saw the potential of the NGO in an intermediary role to deal with a range of different actors.

Ford's first grant was split into three and given to three university departments for them to conduct research into 'traditional' irrigation systems, co-ordinated by LP3ES. A second grant facilitated a policy dialogue between the Ministry of Agriculture, the Public Water Department and the Ministry of Home Affairs, brought into the earlier research programme. This led to an action research programme. The next stage was a World Bank grant to the government for a new irrigation project with far more stress on irrigation management issues than had been found in earlier government systems, and within the US$245 million irrigation sub-sector project, LP3ES was written into a sub-component in which the NGO acted as local consultant on training, organizational techniques and continuing research. A final Ford grant was then made to establish an Irrigation Development Studies Centre in conjunction with LP3ES.

This flexible funding strategy evolved from an identification of a potential role for LP3ES to become involved in re-shaping the public irrigation system through its proven participatory research, networking and training capacities. But the funding costs of the involvement were not clear at the outset, and could not therefore be written into a rigid proposal. Ford therefore decided to fund LP3ES directly rather than through the government. Ford also needed to persuade an initially reluctant government to accept LP3ES as possible partners at the early stages of discussions on the project. For its part, LP3ES has been very careful to build trust with public sector agencies, resisting the temptation to voice public criticisms of the project or the government. In time, the NGO's role became more conceptual rather than just acting as implementors for government, and the NGO gained in credibility from the experience.

NGO–GOVERNMENT COLLABORATION IN RICE-FISH CULTURE IN NORTH-EAST THAILAND

John Sollows, Niran Thongpan and Wattana Leelapatra

INTRODUCTION

While rice-fish culture was once a common practice in central Thailand, there was a decline in the 1970s with the introduction of the more profitable Green Revolution agricultural technologies. Many farmers found that returns were modest and uncertain, fish were readily available from other sources and heavy fertilizer and pesticide use conflicted with the culture of fish in rice fields.

However, a renewed interest in rice-fish culture began to take hold among farmers in north-east Thailand during the early 1980s, in order to compensate for rapid declines in fish stocks due to the outbreak of a hitherto unknown 'ulcerative disease', which affected many local species of wild fish.

The north-east is the poorest and largest region of the country and one with the greatest area devoted to rice production. Most farms are rainfed, but rainfall is highly unpredictable and soils are usually poor both in fertility and water-holding capacity. For poor farmers therefore any heavy investment in on-farm activity, such as intensified rice production, is of dubious wisdom, and pesticide use has therefore remained at relatively low levels. The Department of Fisheries had played an important role in the development and extension of rice-fish culture in the past but since its decline had afforded it a low priority within its programmes.

NGO roles

Instead, a number of NGOs working in the north-east were the first development agencies to notice the resurgence of rice-fish culture and made efforts to bring farmers' renewed interest to the attention of government agricultural agencies. The NGO Canadian Universities Service Overseas (CUSO) became aware of this trend through various local NGOs that it had been supporting for some time. CUSO was able to use senior-level contacts with government to communicate this information, with the result that a workshop on rice-fish culture was held at Surin Province, north-east Thailand in 1983.

The workshop was attended by farmers, staff at the local agricultural college, representatives from the Departments of Fisheries and Agriculture and NGOs working in the area, such as the Appropriate Technology Association (ATA) and the Foundation for Self-Reliance in North-east Thailand (NET).

The meeting was successful in its aim of enlightening participants on the potential importance of rice-fish culture in the north-east and in making participants more aware of each others' mandate, capabilities and objectives. This led to a commitment to the potential of rice-fish culture among all concerned parties. There were a number of direct outcomes from the workshop.

First, the Department of Agriculture (DoA) established on-farm research at its various institutes, but this led to a slow rate of adoption. The Farming Systems Research Institute (FSRI) was then established with World Bank and International Development Research Centre (IDRC) support in an effort to carry out interdisciplinary research under farmer-managed conditions. Since there was a shortage of highly qualified government specialists, CUSO supplied four fisheries scientists in order to motivate staff. The resulting studies began to demonstrate the widespread viability of rice-fish culture.

Second, and at the same time, the USAID-funded north-east rainfed agriculture development project (NERAD) with the DoA and the DoF included on-station and on-farm research into rice-fish culture in their efforts in three north-eastern provinces, which supplemented the FSRI research and led to the important finding that seed weight tended to be enhanced by rice-fish culture.

Third, the Department of Fisheries (DoF) provided technical advice to FSRI's research and development efforts. Subsequently two large DoF projects, the Thai–Australian Thung Kula Ronghai development project and the Thai–Canadian north-east fishery project, were given large rice-fish culture components. Links were established with the Asian Institute of Technology in conducting on-farm research in rice-based farming systems as part of the aquaculture outreach project in Udonthani province.

Fourth, the Department of Agricultural Extension (DoAE) included rice-fish culture as part of its extension portfolio in north-east Thailand. A training workshop in 1988 brought together participants from a number of concerned agencies and included a successful field trip to an ATA project. Contacts with relevant specialists from other agencies provided important support to the DoAE, which works in a number of diverse agricultural fields. The extension model has been simplified to reduce risks: a 0.8 ha plot is enclosed by a peripheral trench which is surrounded by a raised dike, planted with trees and crops. Modest financial help is provided to farmers in the first year but they are then encouraged to help themselves.

Fifth, ATA's work in several villages in western Khon Kaen province gained a new emphasis on rice-fish culture after a field visit by NGO staff and local farmers to a farmer in Surin, whose integrated farm had generated considerable interest. Farmers began stocking fish in their fields and results and adoption levels were good. Farmers then organized rotating work groups to raise dikes and dig ponds. One village where ATA has worked (Porn Sawan) has been used as a training site for DoF and DoAE personnel.

THE UBONRATCHATHANI PROJECT IN NORTH-EAST THAILAND

A small three-year project was begun in 1984 with funding from IDRC and the Canadian Embassy in Bangkok. Policy-makers at that time had no data with which to assess the potential of extending rice-fish culture. CUSO provided a fisheries scientist and two sites were chosen, one irrigated and one rainfed for farmer-managed, on-farm research supported by FSRI.

The project operated with highly limited facilities and budget with only one full-time member. While no permanent FSRI staff member had full-time involvement with the project, part-time contacts and good communications between the CUSO scientist and FSRI staff led to the development of expertise within the FSRI. The results indicated good returns to farmers and good rates of adoption. Although there were variations, rice yields were often enhanced and rarely reduced (Table 7.1).

The project has assisted with the adoption and practice of rice-fish culture in north-east Thailand, and illustrated its advantages to policy-makers in three departments with detailed quantitative research findings. The project also demonstrated the suitability of rice-fish culture to farmers under a range of conditions in both target areas.

The research agenda was effectively farmer-determined and led to a series of further research directions:

- on soil fertility and rice-fish culture
- twin pond nursery and stocking techniques
- low-cost hatchery development to create local self-sufficiency in seed fish.

The advantages of inter-agency co-ordination were recognized, with the DoF providing technical support to the DoA.

The Ubon project effectively demonstrated the viability of rice-fish culture in the two target areas and the reasons for this viability. Findings were corroborated by related studies elsewhere. Effects on rice yields were highly variable, but more often than not yields were enhanced. Negative effects on yields were rare. Optimal management strategies, such as those relating to stocking densities and species composition, were impossible to indicate: the 'best' system depended on each farmer's circumstances.

Table 7.1 Seasonal and annual economic balance sheets for rice-fish culture and rice monoculture for beginning farmers in the Lam Dom Noi irrigated area, Ubonratchathani Province, Thailand

Practice	*Rice-fish culture*			*Rice monoculture*		
Season	*Rainy*[a]	*Dry*[b]	*Both*[a]	*Rainy*[a]	*Dry*[b]	*Both*[a]
Areas (ha)	0.825	1.64	N/A	1.746	2.21	N/A
Production value (baht)	5,175	7,507	10,669	5,381	4,037	8,609
Inputs (baht)	6,024	2,061	7,246	965	2,329	2,598
Net benefit (baht)	-849	5,446	3,443	4,416	1,708	6,011
Net benefit/ha (baht)	-1,029	3,330	3,020	2,529	774	3,997
Time input (person-days)	132	124	222	99	113	154
Net benefit/person-days (baht)	-6.33	43.92	15.44	44.61	15.12	39.08

Source: Sollows and Thongpan (1986)
Notes: [a]Data for three farms only
[b]Data for six farms

CONCLUSION

This case study has illustrated the role of an international NGO, CUSO, in using its influence with key members of government to influence the national research agenda, and the role of a local NGO, the ATA, in bringing its experience from grassroots work with farmers into inter-agency discussion on approaches to rice-fish culture. Increased decentralization of authority in a number of government departments is assisting this process.

THE APPROPRIATE TECHNOLOGY ASSOCIATION (ATA)

Promoting rice-fish farming and biological pesticides

Lanthom Jonjuabsong and Aroon Hwai-Kham

INTRODUCTION

This section documents the activities of a Thai NGO, the Appropriate Technology Association (ATA), in three areas of agricultural technology, in all of which there have been significant interactions with government.

First, in rice-fish culture ATA provided a largely extension and training input, which complemented both the technical work of government departments (such as the Ubon Farming Systems Research and Development Unit – see pp. 285–8) and the social organizing work of other NGOs. Farmer-to-farmer visits contributed an important methodological innovation in this process, and were subsequently adopted and scaled-up by the Department of Agriculture.

Second, ATA contributed to the work of an NGO consortium in integrated agriculture, which focused on income-generating projects drawing heavily on local resources, with particular attention to the role of women in decision-making. ATA's role focused initially on training in agriculture and water resource management, with subsequent promotion and monitoring of proven technologies.

Third, in botanical pesticides, ATA played a key role in bringing together people from government, NGOs and universities who were researching, or had gained practical experience, in this area. A workshop in September 1988 led to a further programme of work drawing on the relative strengths of the various individuals and institutions concerned. While there is interest in botanicals among a number of farmers who have suffered pesticide toxicity, much work remains to be done to develop options acceptable to the majority.

BASIC PROBLEMS

While Thailand's economy is becoming increasingly industrialized, agriculture still provides employment to around 75 per cent of the working population.

Income growth in agriculture has been lower than the rate for the economy as a whole, and has been unevenly distributed. Poverty remains widespread in the north-east, for instance, where almost 70 per cent of the 14 million farmers earn less than US$610 per household per year. At the same time, agricultural land, particularly near the major cities, is being converted to industrial areas and housing estates. Land prices in these areas are rising rapidly. Environmental degradation, due largely to deforestation and some pollution, makes agriculture more vulnerable. Agricultural chemicals have become widely used to increase productivity. However, agrochemical input prices have been rising faster than those of farm products. All these factors and very limited alternatives for work in the villages make the cities, particularly Bangkok, attractive to rural youth.

Many other problems are region-specific. The central plains are fertile and high returns to farming are fairly certain: intensification of agriculture has progressed rapidly. Pesticides, in particular, are widely used. The problem of rising pesticide prices and stagnant product prices is only one of a number which have arisen. Many pesticides are highly toxic to humans if used incorrectly, and many farmers and agriculture workers are poisoned every year. Insects have rapidly developed resistance to some pesticides, thereby increasing the pressure to introduce new ones. Finally, pesticides endanger other components of the farm ecosystem, including the natural predators of these pests, thereby making crops more vulnerable to serious pest outbreaks.

In north-east Thailand, by contrast, most farms are rainfed, with infertile soils of low water-holding capacity, unpredictable rainfall and low and very uncertain unproductivity. Agricultural intensification, particularly use of pesticides, then, is very modest especially in rice farming systems. Farm family income is also low, and nutritional problems have been observed.

POTENTIAL SOLUTIONS

Farming systems which do the natural environment little or no harm and which rely on low levels of external inputs can make farming safer, more sustainable in the long term and, in some circumstances, more economically viable. In rainfed areas, water conservation is a further essential component. A number of international and Thai NGOs have been concerned to introduce elements of sustainable farming systems of this type in the hope that government policy – currently dominated by reliance on mechanization and inorganic pesticides – might be influenced. The Appropriate Technology Association, a Thai NGO established in 1982, is prominent among these.

We shall consider NGOs' work in three technologies which address these issues: rice-fish culture, highly integrated livestock-agriculture systems, and the use of botanical pesticides. Particular attention is given to the experiences of ATA in co-operating with other NGOs and government organizations in developing and disseminating these technologies.

ATA has been involved in six agriculture development projects since its inception in 1982: the farmers' strengthening project, which aimed to raise farm productivity with emphasis on soil improvement; the small-scale water resource development project, which developed drinking and agricultural water resources, in collaboration with government; the vocational enhancement project, which encouraged development of sericulture and weaving to complement rice farming; the village fresh water fishery project, which encouraged rice-fish culture to address water and nutritional needs of north-eastern villagers; the grass roots integrated development (GRID) project, which involved four agencies in efforts to upgrade villagers' living standards with an emphasis on developing self-reliance in the Thung Kula Ronghai area of the north-east; the alternative crop protection project, which identified, developed and extended botanical pesticides. We shall discuss these last three technologies on which NGOs have focused.

RICE-FISH CULTURE: THE VILLAGE FRESH WATER FISHERY PROJECT

By raising dikes around the rice fields, flooding can be controlled and water conserved for an extended period. The soil moved to raise dikes usually leaves a pond or trench behind; this allows both water and fish to be held well into the dry season. The technology is inexpensive, and provides farmers with a convenient, predictable source of fish for food or sale, well into the dry season.

Fish and rice are compatible. Rice yields are rarely adversely affected by fish, and usually increase. Fish feed on a variety of living things, including some weeds and insect pests. Much of what the fish eats, whether provided by nature or the farmer, is turned into dung, which fertilizes the field. Farmers with fish in their rice fields tend to maintain these fields more carefully, and rice, as well as fish, benefit from this.

Discussing the experiences of north-east Thai farmers with them led project workers to suspect that low technology fish culture should have considerable potential for addressing these problems. Accordingly, from 1983 to 1989, this project collaborated with two others: GRID in Roi-Et Province and the Foundation for Self-Reliance in North-east Thailand (NET) in Surin Province. Rice-fish culture was the initial technological 'wedge', and more highly integrated systems were subsequently applied, particularly in the GRID project (see pp. 293–4). The initial rice-fish farming concept was repeatedly modified over this period in the light of farmers' preferences and experiences.

At the administrative and management level, target villages were selected jointly by participants from all concerned NGOs and government agencies, in order to ensure appropriate selection. Development strategies were also considered at this level. Before activities began, local authorities were informed, in order to ensure their understanding and support. Farmer-collaborators were selected on the basis of their participation in village meetings and their interactions with project workers. Selection criteria included interest, potential for success, past experience, present activities, economic status, occupation, ownership of land, and the presence of a fish culture or trap pond.

Various agencies co-operated in these attempts. ATA supported technical education and training. The Thai Department of Fisheries provided seed fish, training in fish breeding and other technical aspects, and marketing assistance. Other government extension and development workers also acted as resource persons at these sessions. Other NGOs helped to organize villagers to build awareness and encourage farmer participation and decision-making.

ATA's field support was primarily technical. If another organization had technical personnel in a selected village, ATA would select other villages in which to provide technical services, and collaborating organizations would provide support on the social side. If the other organizations had no technical personnel in their villages, ATA would fill the gap.

ATA's technical training had several components:

- identification of existing examples of the technology and developing the operators of these examples into resource people who could share their experiences with others
- introduction of new technologies for adaptation and adoption by target farmers
- exposure to the technology through study tours, lectures, questions, discussions, and demonstrations, which was done continuously and gradually: villagers were encouraged to apply and adapt what they learned to their own circumstances
- regular, informal visits to practising villagers by project workers, to provide demonstration and advice.

Of these methods, study tours seemed the most effective, but demonstrations and informal consultations were also useful.

Within villages, technology dissemination occurred by observation and discussion of an operation with the owners, particularly if they gave advice and encouragement. Technologies often spread first to relatives, then to neighbours. Such practitioners can become resource people for farmers from other villages. In some cases, visitors can practise under the supervision of the host. In all, exposure of visitors to the host farmers and their operations can be very convincing. Some groups, on their return home, have co-operated to develop their own systems.

ATA played a key role in facilitating farmer-to-farmer visits, especially among villages and, in some cases, over distances of around 100 km. These visits essentially allow visiting farmers to see the technology at first hand, to discuss it with host farmers and, in consultation with ATA technical personnel, to modify the technologies to suit their local environment. The success of farmer-to-farmer visits has encouraged the Ministry of Agriculture to adopt it as an important component of their extension methodology. Finally, rice-fish culture can open the way to more highly integrated systems. The raised, widened dikes are usually suited to crops which do not tolerate flooding, including fruit trees. The trench or pond, by holding water into the dry season, allows garden crops and small trees to be grown during that period. Livestock manure can be a valuable fertilizer for such systems. In some cases, livestock can be raised directly over fish ponds or rice fields. Crop by-products can be used as fodder or feed for them.

INTEGRATED AGRICULTURE: THE GRASS ROOTS INTEGRATED DEVELOPMENT (GRID) PROJECT

The Thung Kula Ronghai area has the lowest agricultural productivity in north-east Thailand. Low-lying and very flat, it is highly flood-prone during the rains, and a virtual desert in the dry season. Soil fertility is poor, and some areas are saline.

This project was developed by a consortium of NGOs: ATA provided technical support; the Harry Durance Foundation supported the education programme; the graduate volunteer alumni of Thammasat University supported the project's health programme; and the Research and Development Institute of Khon Kaen University provided guidance on research, evaluation and training. The project ran from 1984 to 1989 in six villages in Roi-Et Province and took an interdisciplinary, integrated approach in its development activities. Funds, totalling some 13 million baht over six years from the Canadian International Development Agency, were provided through CUSO, a Canadian NGO.

The project was conceived as highly integrated and needs-oriented. Its components had both production-related and institution-building objectives:

- to enhance and strengthen basic health services and formal and non-formal education and improve agricultural productivity and other income-generating activities
- to enhance the capacity of a combination of villagers' groups, NGOs and GOs to identify development programmes that can respond to the basic needs of villagers in Thung Kula Ronghai area
- to enhance and strengthen the roles of women, particularly their leadership in decision-making for village development
- to enhance local self-reliance by drawing more heavily on locally available resources
- to study ways for villagers, NGOs and GOs to work together.

A project executive committee consisting of representatives of the four participating NGOs met monthly to determine operational and management policies and to advise on project activities. A provincial advisory committee, including the provincial governor, the chief district officers of the districts where the project operated, and the manager of the Thai–Australian Thung Kula Ronghai development project, advised and facilitated co ordination between GRID and related government efforts.

The project employed a village development animator (VDA) for each of the six villages. Each animator was to study and analyse the community and to support the activities of participating villagers. They were aided by five other project workers, specializing in various fields. Village development committees made decisions, planned activities, and prepared proposals, in consultation with the villagers that they represented. As villagers gradually took charge of project activities, the roles of the VDAs were gradually reduced.

ATA's role in the project changed as the project progressed. During the first half of 1985, emphasis was on various sorts of agricultural and water resource management training through seminars, meetings, study tours and idea exchange. From mid-1985 to the end of 1987 various technologies thought to be suitable were tested in the target villages. During the last two years of the project, promotion and monitoring predominated.

BOTANICAL PESTICIDES: THE ALTERNATIVE CROP PROTECTION PROJECT

Thailand produced over 35,000 tonnes of pesticide, mainly insecticide and herbicide, in 1988. Imports of pesticide have risen rapidly in recent years, from a total of some 17,000 tonnes in 1985 to almost 30,000 tonnes in 1990. Pesticide poisoning is common among farmers: a survey of 250 government hospitals and health centres in 60 provinces in 1985 revealed that almost 5,500 people were admitted for pesticide poisoning, of whom 384 died. Toxic residues, principally of organochlorine insecticides, were found in almost 90 per cent of the agricultural produce sampled by the Department of Agriculture's Division of Agricultural Toxic Materials in 1982–5.

Agriculture policies in Thailand are commonly developed in consultation with the private sector, with a view to increasing efficiency and quality, expanding markets, and improving international competitiveness. While it has had some positive results, this has led to high costs of production, environmental degradation, and dependence on uncontrollable external prices and markets.

A number of efforts have been made to identify and promote pesticides made from local plant species. Besides the attractions of availability and cost, these chemicals are usually not dangerous to humans and are less hazardous to farm ecosystems than conventional pesticides. The use of botanical pesticides, then, should lessen toxic hazards to people and the environment, and reduce dependence on externally supplied chemicals at both the national and farm levels.

In 1987 ATA began exploring the potential of botanical pesticides as an alternative to conventionally promoted practices. Experience from earlier projects had led to an increased awareness of these problems. Work was done in collaboration with CUSO which provided funding and information, and the Harry Durance Foundation, a Thai NGO which assisted in translation and implementation in the north of Thailand. ATA helped establish a network of NGOs in the north-east, and managed publicity.

Information was collected from various institutions within and beyond Thailand, and from farmers in the north, north-east and central regions. This survey led to the identification of ten pest-control methods already used by farmers, and the issuing of three publications.

From 1988 to 1990 a second phase of this project focused on trials with farmers, in order to improve and develop botanical pest-control techniques. The McKean Rehabilitation Institute acted as project holder, co-ordinated with the funding agency, and was responsible for implementation in four northern provinces. ATA looked after implementation in one central and three north-east provinces.

In 1988 information on botanical materials used and attitudes to botanical pest control were assessed from the eight provinces. Research plans were made, taking into account local conditions, and sites were selected. In September 1988 a workshop was arranged in co-operation with the Farming Systems Research Institute (FSRI) of the Department of Agriculture (DoA). Participants included workers from NGOs, government, universities and the private sector, as well as farmers. The workshop had an important awareness-creating function by bringing together different practitioners to provide information on their current work. From this the comparative strengths and weaknesses of the institutions and individuals involved in current work could be seen, gaps identified, and a future programme of work agreed.

In early 1989 pre-testing involved familiarization with farming techniques and problems, considering ways of developing technical efficiency, and studying effects on beneficial insects, livestock and humans. Pre-testing was hindered by the departure of a project worker, whose replacement was time-consuming. Field-testing began in mid-1989. In the north-east, ways of dealing with mulberry root rot took priority. In the central plains botanical control of rice pests was given attention. Once farmers were selected, experiments were planned jointly with them and implemented. Trials were monitored and problems solved in joint consultation with farmers. Results indicated that the pesticides used were harmless to farmers and beneficial to animals, and that they could be produced locally at little cost. Rice pests were effectively controlled.

In December 1989 participants at a second workshop discussed progress and problems, with emphasis on the applicability of the findings of the first workshop. In the course of these workshops, FSRI assisted in providing lodging for ATA field-workers and recommended farmer-participants. The Crop Protection Division of DoA helped in crop pest and disease diagnosis, supported with needed equipment and education, and promoted use of botanical pesticides. Various researchers helped design trials, gave technical support, analysed botanical pesticides and provided information on diseases and insect pests.

The project found that most farmers do not readily accept the idea of using botanical pesticides, preferring more conventional methods. Farmers who have suffered from pesticide toxicity, who cannot afford pesticides and who have found conventional pesticides ineffective, are the most open to trying botanicals. Groups of three or four such pioneers can help transfer the technology to other farmers.

Effectiveness of technology transfer depends on many factors. The new techniques must be effective and trials comparing them with conventional methods are important here. Inter-institutional co-operation is important since no institution by itself can effectively address the problem. Farmers who have had bad experiences with pesticides are an important link in the process. Field trials must be well monitored by dedicated workers willing to work with and learn from farmers. Technology transfer should be done well, making farmers competent practitioners who can prepare and use botanicals. Publicity is important.

Technology transfer relied on exposing visiting farmers to practitioners who act as resource people, and to farmers who have been poisoned by pesticides. Various media (videos, slides, handbooks, educational curricula) were produced. Demonstration plots were set up, training courses were held for other NGOs in their respective areas, and seminars were held with representatives of other concerned organizations.

NGO–GO CO-OPERATION: PROSPECTS AND PROBLEMS

Formally, the Royal Thai Government recognizes NGOs. All must register with the Ministry of the Interior and provincial and district authorities, employees of the ministry, must be informed of any NGOs and their activities in the areas for which they are responsible.

Practically, NGOs cannot accomplish their goals without co-operating with government. Strategies and approaches may differ and activities may be independent, but both sides should aim for complementarity and mutual support. In the alternative crop protection project, for instance, the government provided indispensable support in the form of extension and publicity, and organization of markets for organically grown products.

Operationally, the extent to which NGOs and GOs work together depends largely on how NGO workers approach local officials. Usually, however, there is co-operation between NGOs in an area and government workers who have similar responsibilities (such as in fisheries) in the same area. Both sides will modify their roles as they become more familiar with each other.

Co-operation between ATA and other NGOs or government organizations sometimes encounters problems. Principles, concepts, and working methodologies may differ. ATA, for instance, will not work in villages where conflict among villagers is a serious problem, and works with individuals, not groups. The approaches of some organizations differ from this. Co-ordination and mutual support are not always achieved as planned. Data and information are not always shared to the extent desirable.

ATA sees that the government can help in policy-making, agricultural research, data collection, technology transfer, development of production industries (e.g. botanical pesticides) and market development.

All development projects aim to help target individuals and communities help themselves to realize their potential. NGOs have certain advantages here: flexible, fast, simple approaches, clear policy guidelines and dedicated personnel. They are limited by the small scale at which they operate and, in some cases, by high turnover of personnel.

NGO activities end when goals are achieved, the project terminates, or when the government or target group is in a position to take over. They are not permanent. The government, by contrast, carries a more permanent and overall responsibility for the well-being of the population which supports it through taxes. Government agencies are usually willing to respond to NGO initiatives to work with target groups if their resources allow.

NGOs must recognize that the government has personnel at many levels in all areas; these can be mobilized if authorities deem it appropriate. They must also realize that government responsiveness is limited by budgetary and procedural factors which sometimes give government personnel an inactive image. Government workers are not free agents: their supervisors must be contacted if good co-operation is to be achieved. Government employees must give their normal work first priority; working with NGOs can sometimes lead to conflicts with normal work, and these conflicts may adversely affect both promotion and salaries of the government staff concerned.

However, the aim for both sides should be to benefit the target group, not get credit for oneself. NGO workers should be generous in giving credit to government colleagues. For smooth co-operation and clear understanding appropriate formal and continuous informal communication with the relevant government units is essential.

NOTE

1 Based on the NIA model in the Philippines. See Korten and Siy (1988) and Box 6.3 in this volume.

Chapter 8

'ROLES' IN NGO–GOVERNMENT RELATIONSHIPS: A SYNTHESIS FROM THE CASE STUDIES

NGO–GO INTERACTIONS AND ROLES

Drawing on the case studies presented in previous chapters, we identify here a number of discrete 'roles' played by NGOs and GOs in interacting with each other. Preliminary definition of some of these roles emerged from the presentation and discussion of over forty case studies by the practitioners and researchers attending the Asia workshop in Hyderabad, India, 16–20 September 1991, many of which have been presented in summary form in this volume. To some extent this represents a simplification of complex relationships, but classification of NGOs' and GOs' experiences into broad groupings of this kind is essential for analytical purposes. It is important to recognize, however, that the activities of even a single NGO are frequently diverse, and so may fall into more than one role. PRADAN in India, for example, has been active both in 'de-mystifying' technologies obtained from GOs, and in forming local groups capable of drawing on government services and interacting with the private commercial sector.

This chapter is in two parts. The first examines the types of interactions initiated by NGOs and GOs respectively, and the roles played by each. The second examines why interactions, in some cases, have failed to meet expectations. The types of interactions found between NGOs and GOs and their respective roles are summarized in Table 8.1, and the groupings of case studies under each role are presented in Table 8.2. The modes and levels corresponding to each role in Table 8.1 relate both to functions and to the institutional positions which influence functional linkages. These two types of interactions are treated separately by Merrill-Sands and Kaimowitz (1990): their linkage *relationships* referred to links between organizations and they denoted these as collaborative or competitive, and linkage *mechanisms* were related to particular functions such as planning and review processes, collaborative professional activities, resource allocation procedures and communication devices. In the present context, although some of the Merrill-Sands and Kaimowitz 'types' recur – as in, for instance, collaborative professional activities – new linkage mechanisms have been defined to reflect the fact that one side may wish to incorporate lessons from the other. The inter-institutional relationships defined here are also wider than the two possibilities – collaborative or conflictive – defined by Merrill-Sands and Kaimowitz, allowing, for instance, for the fact that one side may simply disregard the other. The final column in Table 8.1 illustrates how functional relations do not stand alone, but are influenced by inter-institutional relations.

The origins of the relationship are important because the success or failure of any interaction depends crucially on the sets of expectations formed by the various parties involved. As we discuss in the second part of this chapter, if fully collaborative relations are to succeed, a precondition is that each side should have similar expectations of the desired output, yet, much interaction has been characterized by differing assumptions and expectations.

The case studies suggest that NGOs tend to initiate links with government agencies when they come up against particular problems (such as input delivery, legal rights, etc.) or when they identify a 'gap' in government services, either in terms of inefficient provision of services or the exclusion of particular sections of the population (e.g. poor rural women in Bangladesh). If a government agency is prepared to admit its weakness in an area identified by an NGO, then there are possible grounds for interaction on a complementary basis.

From the evidence available, governments appear to have initiated interaction less frequently, and usually when they require a more efficient channel for service delivery. The drawback here is that they may not recognize the wider NGO programme, which may become 'distorted' by this type of contact.

ROLE 1: NGO–GO PARTNERSHIPS WITHIN THE RESEARCH–EXTENSION CONTINUUM

Governments devote considerable resources to agricultural technology development in research institutes, universities and field centres. Skills, equipment and facilities are located in government-sponsored centres which require investment beyond the capacity of most NGOs. Government institutes' capabilities are strengthened in many cases by links with specialist international centres. However, as we argued in Chapter 1, and as evidence from the case studies has demonstrated, the rural poor will benefit only marginally from these capabilities where government research agenda are determined by, for example, commodity or geographically based resource allocations geared towards commercial or export crops, or by scientists' own interests. Even if a substantial proportion of resources are allocated to issues broadly relevant to the rural poor, little of relevance will ensue unless specific research agenda are determined after careful consultation with clients to discover their perceptions of objectives, opportunities and constraints. Furthermore, technologies need to be tested on farmers' fields, and feedback incorporated into further phases of research.

The majority of rural poor live in difficult farming areas characterized by

Table 8.1 Respective NGO and GO roles identified from the case studies

Role	*Features*	*Mode and level*
1 NGO–GO partnerships within the research–extension continuum.	NGOs obtain and test technologies from GOs and provide feedback.	*collaborative*: each side relies on the other to contribute to agreed activities in accordance with perceived comparative advantages. In the absence of agreed inputs from one side or other, the activity cannot fully succeed. Formal agreement usually reached.
2 NGO–GO partnerships independent of the research–extension continuum.	In joint NGO–GO projects, NGOs provide social organizational and delivery components; GOs provide technical inputs.	
3 NGOs innovate; GO response varies.	NGOs innovate – whether in technical, procedural, institutional or methodological ways – in the expectation that GOs will 'scale up'.	*incorporative*: GOs may disregard NGO innovations; if they do wish to adopt them, this may be through working together in the initial stages (i.e. as NGO teaches GO) before GO incorporates lessons from the NGO into its own actions.
4 NGOs as 'networkers' among themselves and with GOs.	NGOs establish fora in which ideas are exchanged among themselves and/or between NGOs and GOs.	*informative*: NGOs provide information on activities or on technologies to each other or to GOs, sometimes leading to co-ordination among projects or activities.
5 NGOs advocate; GO response varies.	NGOs seek pro-poor administrative or legislative reform, or the full implementation of existing laws and procedures.	*conflictive*: NGOs seek to change GO practice through confrontation, lengthy negotiation or 'working from within'.

Table 8.2 NGO and government roles in agricultural technology development (ATD) identified in the case studies

Roles	*Case studies*
Role 1: NGOs as field testers	MCC crop trials (Bangladesh) PRADAN (India) LP3ES irrigation (Indonesia)
Role 2: NGOs as joint partners	Proshika livestock (Bangladesh) BRAC poultry (Bangladesh) AWS RH caterpillar (India) RKM (India)
Role 3: NGOs as innovators	FIVDB ducks (Bangladesh) BRAC irrigation (Bangladesh) CARE LIFT gardening (Bangladesh) MBRLC SALT (Philippines) MFI training, etc. (Philippines) RDRS treadle pump (Bangladesh) MCC soybeans (Bangladesh) UBON/ATA fisheries (Thailand) BAIF Myrada
Role 4: NGOs as 'networkers'	NAF forestry (Nepal) IIRR agroforestry (Philippines) Auroville reclamation (India) AKRSP extension (India) BAIF (India)
Role 5: NGOs as advocates	AKRSP wastelands (India) Proshika forestry (Bangladesh) MCC crop adoption (Bangladesh)
Government-initiated relations	KVKs (India) DENR (Debt-for-Nature) (Philippines) ERDB uplands project (Philippines) DA SMAP project (Philippines)

combinations of low and erratic rainfall, poor soils and hilly topography. It is particularly unlikely that technologies will be successfully devised for these conditions unless participatory approaches to setting research agenda and testing technologies are adopted.

NGOs' ability to build close, interactive relationships with their clients, to draw on their local knowledge and to work with them in testing new technologies and methods for managing on- and off-farm resources is evident in many of the case studies presented above. It has given rise to a widespread expectation that a division of roles in which GOs develop technology and NGOs provide field-testing, local adaptation, feedback and dissemination is a logical functional complementarity and is likely to be widely exploited (Merrill-Sands and Kaimowitz 1990; Wellard *et al.* 1990).

Ironically, only a limited number of cases emerged from the material available in which NGOs clearly drew on technologies 'on the shelf' in government research institutes. Analysis of these provides clues as to why this intuitively appealing complementarity is not more widely exploited. In the PRADAN case, the technologies drawn from government had been developed for medium- to large-scale producers, requiring access to inputs and markets and substantial capital investment. This, in turn, requires loans from the bank or government for which the rural poor generally have insufficient collateral, and in any case, represent too high a risk for them to take. The PRADAN case study demonstrates the very considerable effort needed to 'de-mystify' these technologies and scale them down to levels manageable by the rural poor. It is clear that few NGOs have the combinations of technical capability, group organizing skills and sheer perseverance to undertake work of this kind.

In the MCC case, while some success has been achieved with joint testing of a fruit-fly trap, many of the field testing programmes established jointly with GOs require a disproportionate amount of MCC staff time to bring to fruition, yet are seen by GO researchers as merely a more convenient on-farm testing service than they can obtain from government stations. On the whole, the relationship has failed to mature into one in which GO researchers work with the NGO in a deliberate effort to bring farmers' views into the research agenda. Further evidence for the one-sidedness of the relationship is found in MCC efforts to obtain official recognition for the vegetable and rice varieties which have proven highly acceptable to farmers in field trials. Official recognition is an essential step scaling up the adoption of these new varieties, and yet the National Seed Board restricts its mandate to those varieties developed by GOs, so generally refusing to recognize varieties produced by NGOs. In summary, the reasons for limited exploitation of this potential for functional complementarity are complex and, from the evidence presented, include:

- the time, effort and skill required by NGOs to de-mystify and scale down technologies developed by government (PRADAN)
- differences between the two sides in what is expected of the relationship (MCC)
- limited commitment to the relationship by GO staff, so that the burden of making the relationship work falls disproportionately heavily on the NGO (MCC)
- more widely, the fact that NGO client groups (e.g. women and the landless in Bangladesh) may differ markedly from those of government (predominantly *male farmers*), so that the very premises on which technology is developed are different.

ROLE 2: NGO–GO PARTNERSHIPS INDEPENDENT OF THE RESEARCH–EXTENSION CONTINUUM

A more commonly observed complementarity between NGOs and GOs is that in which NGOs are employed to facilitate the organization of local groups capable of using available technology, which either NGOs or GOs deliver, more widely and more efficiently than individuals could. From this relationship, government stands to gain access to a better network of distribution and a higher quality of contact with sections of the population that it is generally poor at reaching. The NGO, if the scheme goes well, stands to benefit from the greater resources available through government (especially if the government has priority access to scarce resources, such as imported vaccines).

Evidence from the case studies indicates that this relationship has worked moderately well: in Bangladesh, BRAC's experience with poultry production, and Proshika's with livestock, exhibit close parallels: each assisted in the formation of local groups, the training of local staff and, in BRAC's case, the establishment of interactive structures in which distinct but complementary roles are undertaken by individual groups. In each case, this was complemented by the acquisition of 'high technology' inputs from government (e.g. vaccines and genetically improved animals). In a different context, the youth clubs established by Ramakrishna Mission (India) served as an important conduit for assessment of local needs prior to the introduction of new technology. The Action for World Solidarity case study (India) illustrates how complex this kind of interaction can be: the technologies introduced incorporated elements of indigenous and modern knowledge drawn together over a period of years; group organization was particularly demanding since the pest-control action envisaged had to be synchronized across wide areas if it was to succeed; inputs required of government included not only scientific support but also materials and credit. Numerous obstacles had to be overcome as each of these dimensions of the collaboration was addressed, making it one of the most complex case studies, and raising questions as to whether local groups will ever be strong enough to take over the operation of this collaborative link and so allow the NGO to withdraw.

This type of partnership can work well, but suffers from two disadvantages from the NGOs' point of view. First, while it gives NGOs the opportunity to use resources provided by government to complement their own programmes, its emphasis is on the formation of groups around particular functions. This may detract from, or even be in conflict with, NGOs' efforts towards awareness creation and social organization in the wider context. Similarly, if NGOs become heavily involved in input delivery in support of functional groups, the resources they have available for these wider activities are likely to be reduced.

Second, it can be used by governments and donors as a cost-cutting exercise to fit in with wider agendas of privatization. While some of the more opportunistic NGOs may welcome service delivery contracts resulting from privatization, others may resist being pushed into this role for fear of diluting their capacity for awareness creation or innovativeness. Undoubtedly, many see the provision of certain services, including agricultural research, as properly within the mandate of government.

One of the ways in which NGOs have attempted to strengthen the long-term integrity of their package of inputs and services within their own programmes is to build a training or institutional development component into the partnership, making it clear that their involvement is for a finite period only and that their aim is to train and motivate government personnel to achieve a more effective delivery of its services in the long term without NGO assistance, and, similarly, to build up local capacity to create a 'demand-pull' to which government can respond.

ROLE 3: NGOs AS INNOVATORS

NGOs innovate – whether in technical, procedural, institutional or methodological ways – in the expectation that government will 'scale up'.[1] Almost one-third of the case studies reviewed here involve innovative approaches – whether in technology, management practices, research methods, or organizational and institutional arrangements – which offer some prospect of being scaled up by government.

It is worth reiterating that these innovations are rooted in a 'problem'- or 'issue'- oriented approach to agricultural change. By contrast with traditional, and still widely found, adherence in the public sector to commodity or discipline-based approaches to research, NGOs' prime concern in the present context is to respond to opportunities and constraints identified by the rural poor; NGOs' interest in commodity or discipline-based research extends only to what it can contribute to the issues in question. In a different dimension, NGOs' interest extends beyond trials with candidate technologies to ways in which these might fit into the wider farming system. In many cases, their interest extends beyond the farming system to the potential constraints in processing, marketing and infrastructure (whether human or physical) which need to be removed if innovations are to be introduced successfully.

In practice, most NGOs have shown reluctance to become involved in long-term programmes of trials: their interest is in what will work within their chosen geographical areas, not in the wider conditions under which a given technology may succeed, which must necessarily be the concern of GOs that have a national mandate. While NGOs have been faulted for the superficiality of some of their technology testing (Kohl 1991), the speed and flexibility with which they are able to respond to local issues through 'action research' approaches is evident.

Virtually all NGO innovativeness originates from direct experience in working with local groups. It is therefore tailored to meet such specific requirements as, for instance, landless women's need for backyard income-generating activities in Bangladesh (met by FIVDB's improved duck-rearing programme); the needs of landless people for low-risk income-generating activities (met by e.g. Proshika's and BRAC's initiatives on group ownership of irrigation pumps by landless groups); and the need for sustainable technologies and practices for managing hill farming in the Philippines (met by MBRLC's 'sloping agricultural land technology').

Many of these innovations have spread to groups similar to those for which they were originally intended, but over a wider geographical area. Thus, FIVDB's programme has produced 350,000 improved ducks, largely kept by landless women, and BAIF's artificial insemination programme accounts for some 10 per cent of the cross-bred cattle in India. In other cases, innovations have been of a more generic kind and have found application outside their original contexts. Thus, the development of participatory rapid appraisal methods by MYRADA in collaboration with other NGOs has been taken up by government departments concerned with watershed management and wasteland development in India.

Linkages with GOs – whether achieved or still at the formative stage – have been highly diverse in these contexts. In some cases, they have been at a research level. FIVDB, for instance, has reached the limits of its current research capacity, and feels that its programme would benefit from the greater genetic stability of improved breeds of duck that government research input might be able to provide. In another case, research links with GOs have provided little of value to BAIF's programmes, but fulfil a wider agenda: that of ensuring that BAIF is granted continuing status as an officially recognized non-profit research foundation, so that the significant contributions to its income from the private commercial sector remain tax-deductible. In other cases, GOs have sought to scale up NGO innovations – at times successfully, as with the technologies developed by MBRLC for Philippine hill farming, and with the farmer-to-farmer dissemination developed by ATA for rice-fish farming in north-east Thailand (see also Sollows *et al.* 1991). At other times, scaling up has been less successful owing, for instance, to GOs' adherence to civil service procedures, as when tenders put out by the Bangladesh government for local manufacturing of the RDRS treadle pump produced pumps of such poor quality that they were unworkable.

ROLE 4: NGOs AS NETWORKERS

NGOs establish fora in which ideas are exchanged among themselves and/or between NGOs and GOs. The term 'networking' is open to a wide range of interpretations. Useful definitions of research networks are provided by Plucknett *et al.* (1990) and of information exchange networks by Nelson and Farrington (forthcoming). Networking here is defined loosely as interaction among a group of institutions in order to realize anticipated benefits for themselves or for their clients.

Formal national level networks of NGOs exist in most countries. These generally have a mandate to represent NGO interests to government or to donors in one or more fields, and usually publish inventories of member NGOs and their activities. However, their performance in these tasks is uneven, and few would argue that they are capable of representing NGO interests adequately on specific (sometimes, as in agriculture, highly technical) issues at local or even national level. As the number of NGOs grows, a pressing need is emerging for organizations mandated to liaise among NGOs, and between NGOs and government in order to prevent duplication of effort and undesirable types and levels of competition among them. Some evidence is presented above from the Philippines of government efforts to liaise with NGOs in area-based development efforts in mutually complementary ways such as those described in Chapter 6 in papers on the Philippines Department of Agriculture and the Ecosystems Research and Development Bureau. In addition, a number of national umbrella organizations have been created in the Philippines which liaise actively among NGOs on specific issues. These include CODE-NGO (the Concerns of Development NGO Networks) and PHILDHRRA (Philippine partnership for the development of human resources in rural areas).

However, these high levels of interaction among NGOs appear exceptional. By contrast, the high density of NGOs in Bangladesh, for instance, has led to overlapping activities, and no successful efforts to liaise among NGOs on specific themes have been noted. Problems caused by lack of liaison among NGOs have also been noted in other regions. In eastern Bolivia, for instance, lack of co-ordination among NGOs, and inadequate liaison between 'service' NGOs and grassroots organizations is a conflictive and wasteful use of resources (Bebbington and Thiele 1993). In South Nyanza District of Kenya, district forestry officers have gained the confidence of NGOs to a degree that allows them to lead efforts towards liaison and the sharing out of tasks both among NGOs and between NGOs and GOs (Wellard and Copestake 1993).

The case studies of NGO networking discussed in this volume are more geographically localized and thematically focused than are the efforts of national-level NGO apex organizations, for example, or even of district-based attempts at liaison and, perhaps for that reason, appear to have been more successful. They include the Nepal Agroforestry Foundation, which acts as a clearing house for technologies and management practices in agroforestry. These are obtained from contacts with NGOs and GOs both in Nepal and elsewhere, assembled, classified, incorporated into the demonstration and training programmes operated by NAF, and made directly available in response to enquiries, whether from GOs or NGOs.

A similarly innovative initiative – taking the form of a single project rather than a longer-term programme – was taken by IIRR in collaboration with other NGOs and GOs in the Philippines in the assembling of an agroforestry information kit. The fact that IIRR was widely respected for its unbiased facilitating skills allowed it to bring together individuals from a wide range of institutions, and so assemble in a matter of days a range of experience which a consultant (or team of consultants) contracted in the normal way may never have been able to achieve.

Finally, in India, AKRSP, as part of an effort over several years to make training by GOs more responsive to farmers' needs, began by stimulating farmers' awareness of the technology options available to meet their needs, and then sought systematically to orient GO trainers to village conditions and to the importance of enhancing farmers' capacity to identify needs and to articulate them into training programmes. A local network of trainers from GOs and NGOs was set up to provide mutual support during the re-orientation phase.

ROLE 5: NGOs AS ADVOCATES

NGOs seek pro-poor administrative and/or legislative reform or the full implementation of existing laws. For some NGOs, generally operating at international level, advocacy is a central, and sometimes their only, function. Thus, Cromwell and Wiggins (1992), as part of a wider study on NGOs' work in local seed supply, describe how the seeds action network has campaigned on issues of genetic conservation and bio-diversity.

With others, operating at both national and international levels, the distinction between objective investigation and advocacy tends to become blurred. This applies to many of the NGOs concerned with organic farming and ecological agriculture, such as those affiliated to the International Federation of Organic Agriculture Movements.

By contrast, advocacy is not a central function in any of the NGOs reviewed here, yet, if defined loosely as the intention to persuade government to take actions favouring NGOs, their objectives or their client groups, then advocacy is evident in a number of forms at various levels of NGO–GO interaction. For instance, at a level specific to narrowly defined aspects of agriculture and to GOs working at the implementation end of government strategy, MCC, after abandoning its efforts to obtain official approval for the multiplication of imported vegetable varieties to which its client farmers had responded positively, changed strategy and began to work with and through horticulture researchers at the Bangladesh Agricultural Research Institute, in order to have varieties meeting its clients' needs submitted for clearance to the National Seed Board by BARI, in the expectation (based on MCC's earlier negative experiences) that submissions from a GO would receive a more favourable hearing than from an NGO.

More overt advocacy is evident in Proshika's efforts to promote poverty-focused social forestry actions in the Bangladesh Forest Department. Some success was achieved in acquiring medium-term (five-year) leases from local government of roadside verges in one district to allow the planting of pulse crops and trees by the landless, but attempts to persuade government to permit participatory management of multipurpose indigenous forests – since donor-funded plantings of large stands of single-purpose trees had proven particularly unsuited to their needs – fell foul of the vested interests of both the local elite, and of forestry officers, who found more scope for supplementing their incomes illicitly through new plantings than in the joint management of existing multi-purpose stands.

Undoubtedly, the most complex account of NGO advocacy among the case studies reviewed here is presented by AKRSP. Pilot projects had demonstrated the success of participatory approaches which both rehabilitated degraded forest or wasteland and generated livelihoods from the increased range and volume of forest products. The expansion of participatory approaches, however, was hampered by restrictions placed on access by villagers to large areas of land falling under the mandate of the Forest Department. Repeated efforts were necessary over several years, first by AKRSP alone, then by a consortium of NGOs, at both state and national levels before government could be persuaded to issue directives relaxing existing restrictions on access.

GOVERNMENT-INITIATED RELATIONS

The great majority of interactions reported in this book have been initiated by NGOs, which have sought to draw government resources into their programmes, to disseminate via government the technologies that they have generated, or to influence government at project, strategy or policy levels to relieve constraints faced by their clientele.

The number of initiatives taken by government is much smaller, reflecting perhaps some uncertainty over how relations with NGOs might best be taken forward, especially in countries where governments in the past have been faced with demands by NGOs which challenge policy or strategy decisions, or even their legitimacy. Evidence from Latin America (Bebbington and Thiele 1993) indicates a higher volume of government initiatives than reported there, and although wide differences in context make generalizations dangerous, it seems likely that, over the coming years, the volume of initiatives taken by GOs towards NGOs in agricultural development in Asia will increase. Widespread consultation and joint project implementation with NGOs in preventive health-care in Bangladesh (Streefland and Chowdhury 1990) and in wasteland rehabilitation in India (Poffenberger 1990) are indicative of possibilities in this direction. In the short term however, political uncertainties in Nepal make it unclear whether GOs will initiate more activity, and in Thailand, government plans for joint initiatives with NGOs set out in the second half of 1991 were undermined by the political unrest of May 1992.

Of the examples of GO-led linkage reported here, the farm science centre (KVK) initiatives in India are among the more modest. KVKs are mandated to serve as centres for demonstration and training in 'scientific farming'. Their small professional staff (twelve persons), the top-down manner in which many operate, the difficult conditions in which they are often located, and their isolation from mainstream research centres means that many are unable to make more than nominal contributions to agricultural development in their areas (World Bank 1989b). While the co-location of an increasing (but still small) number of KVKs with NGOs is a valuable expression of confidence in the NGOs, the case studies presented here suggest that only limited institutional strengthening of the NGO has taken place in two of the three cases reported. The exception is at UPASI, where the additional capacity provided to an already strong research organization financed by large tea growers has allowed greater attention to the needs of small-scale tea producers.

Other government initiatives in India include efforts over recent years to involve NGOs in watershed management and plans to hand over agricultural extension functions to them in specific geographical areas. The former have achieved some success, but the process of identifying respective roles has been a difficult one and, at times, a source of distrust. A further difficulty has been that of ensuring adequate co-ordination among government departments. The latter demonstrates increasing awareness by government of NGOs' comparative advantage in reaching the rural poor, and provides opportunities for the more opportunistic types of NGO willing to deliver services on behalf of government. Whether enough will come forward to make the scheme viable, and whether feedback systems can be designed from which government can learn, not only from this type of NGO, but also from the more reflective ones, remains to be seen.

A much wider range of GO initiatives is found in the Philippines, where consultation with NGOs in matters of economic and social development is enshrined in the 1987 Constitution, Article II of which declared that 'the state shall encourage non-governmental organizations, community-based or sectoral organizations that promote the welfare of the nation'.

In response to this macro-political initiative, several government departments began to prepare procedures for informing NGOs of their proposed activities, obtaining their views, and responding to their requirements. The structures established in the Department of Environment and Natural Resources at both national and regional levels are described in Part III, but these were predated by initiatives taken in the Department of Agriculture (DA).[2]

It is important to recall the contemporary political context: many NGOs of different types had been established in opposition to the Marcos regime between 1983 and 1987. Others had been established, or already existed, to service particular groups marginalized by the regime. The government itself had set up co-operatives, styled as NGOs but in reality little more than branches of government. An important initial problem faced by the DA was therefore to define the types of NGOs with which it could work. A second problem was to overcome NGOs' views of government as a monolithic bloc characterized by bureaucratic procedures and lack of commitment among its staff, and to develop their awareness of government procedures.

The DA's NGO Outreach Desk, set up in 1986, aimed to develop an operational model for institutionalizing the partnership between DA and NGOs. Its functions were to provide policy direction to the DA in matters affecting NGOs, and to identify collaborative activities. Numerous lessons have been learned during the life of the Outreach Desk, which have been drawn upon also by other departments such as DENR. These include the following:

- The impossibility of operating GO–NGO relations by directive: for instance, efforts to implement a 'code of conduct' for DA staff in their relations with NGOs proved unsuccessful; there may be a role for certain guidelines, but much has to grow from an atmosphere of mutual trust.
- The tensions among NGOs caused by competition for resources: this became evident in the Southern Mindanao agricultural development project, a major NGO–GO collaborative project managed by the Outreach Desk. While relations with some NGOs (e.g. MBRLC who provided much of the hill farming management technology on which the project was largely based) were cordial, relations with others – some of which had little demonstrable contribution to make to the project – were strained because of competition for public sector resources.
- The need to encourage NGOs to formulate their concerns in terms which could be incorporated into policy formulation, instead of relying on generalized advocacy.

Other GO initiatives in the Philippines occur at project level, but have strongly strategic dimensions. For instance, the Debt-for-Nature Swap programme is managed by DENR and links a reduction in the country's foreign debt with measures to reverse environmental decline. It is institutionally complex: an international NGO, the World Wildlife Fund, is instrumental in obtaining funds from a major donor (USAID) to 'buy back' part of the Philippines foreign debt in international financial markets. The Philippines Central Bank then makes available the local currency equivalent of the buy-back to a government department (DENR) which guarantees that an agreed programme of action will be carried out in collaboration with a local NGO (Haribon Foundation).

A second experience reported by the DENR falls under the Comprehensive Agrarian Reform Programme (CARP): agrarian reform is a highly political issue in the Philippines, and the gap between the Aquino government's rhetoric on land reform and what it has achieved in practice has been the subject of much criticism, particularly from NGOs such as the Philippines Rural Reconstruction Movement. Initiatives such as the integrated livestock programme implemented by the Ecosystems Research and Development

Bureau which falls under DENR can, in political terms, be seen as an effort by government to introduce income-generating activities into some of the densely populated hill areas which have benefited hitherto neither from the attention of agricultural extension services, nor from land reform. This is one of the few cases reviewed in this book in which a GO deals directly with people's organizations – i.e. membership organizations – by contrast with the NGO non-membership service organizations which form the focus of our concern.

The preliminary report presented on pp. 259–64 does not yet allow an assessment of whether the POs with which ERDB deals (and, in some cases, sets up) will develop into sustainable institutions capable of carrying forward the technologies introduced *without* the support of NGOs, but this must be an important question in the evaluation of projects of this type.

INTERACTIONS FALLING BELOW EXPECTATIONS

The case-study material provides several instances in which attempts by NGOs and GOs to work together have produced few, if any, positive results. Thus, AKRSP's attempts to make government receptive to the possibilities of joint management (by government, NGOs and the people) of forest resources has begun to produce results, but only after sustained effort over a number of years. Other examples are provided by BRAC's work on poultry: its collaboration with government has produced a more efficient system of delivering improved technology inputs, but barriers remain to the achievement of BRAC's wider goal of permanently establishing government services which are responsive to the needs of the rural poor. A further example is provided by Proshika's work in social forestry: attempts to strengthen the claims of poor people over common property resources have challenged (on the whole, unsuccessfully) existing power structures based on patronage relations between local elites and the Forestry Department.

With hindsight, it is easy enough to identify the types of interactions that have occurred, and the respective roles played by NGOs and GOs, as we did in Table 8.1. The importance of these examples in which expected benefits from interaction were not realized is that they demonstrate the difficulty of clearly identifying and agreeing upon roles and interactions in advance, particularly when NGOs and GOs perceive both advantages and disadvantages in collaboration.

Perceptions of such advantages and disadvantages are not uniform: some branches of government may see advantage in what others perceive as a threat. Similarly, perspectives vary across countries, GOs in the Philippines being generally more open than those in the other countries reviewed here. The range of perspectives among NGOs is, if anything, likely to be even wider. Even so, it is possible to summarize the range of advantages and disadvantages in collaboration perceived by each side (Table 8.3).

Not all of these views will be held by all NGOs or GOs, but experience (and discussions at the September 1991 Asia regional workshop, elements of which are summarized in Table 8.4) suggest that most GOs see improved prospects for service delivery, better information from the grassroots and so greater cost-effectiveness as advantages, and competition for donor funds, the lack of NGO accountability and NGOs' activities as highlighting the shortcomings of GO services as disadvantages of collaboration. From the NGO perspective, access to policy-making, to GO skills and facilities and to GOs' capacity for 'scaling up' are advantages of collaboration with government, against which must be set the dangers of co-optation, loss of independence and loss of credibility among NGOs' clients.

These different underlying perceptions of the advantages and disadvantages of collaboration have, in some cases, generated divergent expectations of the roles to be played by each side in collaborative projects. Although deriving from the NGOs and GOs in Bangladesh from which case-study material was obtained, there is little doubt that the divergent perceptions of roles summarized in Table 8.4 are also widespread in e.g. India and Nepal: government sees itself as the source of innovations which are then disseminated by NGOs, whereas NGOs see themselves as innovators and social mobilizers, the lessons from which are to be scaled up by government.

While Tables 8.3 to 8.5 contain highly generalized statements, our purpose has been less to argue the wide validity of such generalizations, and more to demonstrate the type and degree of differences in perception. This, in turn, indicates both the risks of failure of collaborative ventures based on such underlying differences in expectations, and the magnitude of the task of creating more open-minded perspectives on each side from which partnership arrangements having better prospects of success might emerge.

CONCLUSIONS

The case-study material presented and discussed above cannot be adequately analysed in a single dimension limited, for instance, to its agricultural characteristics. An adequate appreciation of NGOs' work, and of their potential for linking with government, requires, we have suggested, analysis in at least three dimensions, embracing the characteristics of subject matter, types of innovations and target group.

First, part of the subject matter is concerned with agriculture (e.g. MCC in Bangladesh and MBRLC in the Philippines) but large parts are concerned with innovative income-generating opportunities which may be on-farm (e.g. PRADAN'S work in India with mushroom and raw silk production), may draw on the processing of agriculture-related by-products (PRADAN's initiatives on improved processing of hides and skins) or may focus on improvements in the wider environment through tree planting and soil and water conservation (Auroville; Nepal Agroforestry Foundation; United Mission to Nepal). These are generally areas neglected by government.

Second, the types of innovations inherent in NGOs' work and in the NGO–GO interactions presented here are concerned with the development and dissemination of technologies and management practices, others relate to

Table 8.3 The benefits and disadvantages of collaboration for government agencies and NGOs – a balance sheet

From the government perspective

Benefits	*Disadvantages*
1 Better delivery facilities for government services	1 GO services shown to be inefficient by NGOs' presence and actions
2 More information available from the grassroots	2 NGO mobilization work promotes social instability
3 More interaction with 'targets'	3 Demand for government services may increase beyond the capacity to meet it
4 Enhanced cost-effectiveness	4 NGOs compete with government for donors' funds
5 Field-testing facilities for new technologies	5 Weakening of government mandate and credibility
6 Appropriate training inputs available from NGO specialists	6 The unaccountability of NGOs
7 More co-ordination of GO–NGO activities possible; more control of NGOs in general	

From the NGO perspective

Benefits	*Disadvantages*
1 Improved access to policy formulation	1 Co-optation by government and greater bureaucratic controls
2 Access to specialist research facilities and expertise	2 The NGO grows to assume a more bureaucratic character
3 Opportunity to improve government services 'from within' by training	3 Loss of NGO autonomy and independence
4 Access to new technologies 'from above'	4 Relegation to mere delivery activities to the detriment of NGO's wider programme
5 Opportunities for passing on technologies and models for replication or 'scaling up'	5 Government takes credit for NGO achievements
	6 'Substitution' by the NGO for government services perpetuates government inefficiency
	7 Loss of credibility among clients and a tendency to maintain existing social and political conditions

more generic issues such as innovations in the methodologies of diagnosis (MYRADA), of training (AKRSP), of research (Ramakrishna Mission; MCC) and of dissemination (ATA).

Third, NGOs' concern extends beyond farmers to cater for other – often more needy – sectors of the rural poor. Thus, backyard poultry production for

Table 8.4 Views from the Asia Regional Workshop on constraints to successful NGO–GO collaboration

1 Lack of understanding of each others' goals

2 Inability of government to identify the types of NGO that might become reliable working partners

3 Restrictive government procedures

4 Problems of attitude (distrust, etc.) on both sides

5 Lack of clear government policy and guidelines on NGOs

6 Poor communications among NGOs and between NGOs and GOs

7 Sharp contrasts between the 'top-down' working methods of government, and the participatory approaches of NGOs

8 Lack of existing linkages among institutions

9 Poor understanding of relative strengths and weaknesses on both sides

10 Lack of NGO accountability to their clients, or to the public at large, for the ways in which resources are used

Once a partnership has been established, problems may then be caused by three sets of factors:

1 Incompatible working methods (e.g. BRAC's attempt to work with the government banking system, which was slow and bureaucratic; RDRS's attempt to work through the government's tendering system, which halted progress)

2 Conflicts over vested interests in the rural power structure (e.g. Proshika's forestry programme)

3 Contradictory objectives of collaboration, for example, BRAC's interest in establishing sustainable government services in its poultry project contrasted with government's desire for more efficient input delivery in the short-term; other examples include the numerous points at which AKRSP's views on policy reform in the forestry sector were denied a hearing by government

the landless is an important component of the livestock work of BRAC and Proshika. Both have also made 'lumpy' technologies accessible to landless irrigators, and FIVDB has focused on women in its duck-rearing programme.

Analysis of their respective roles supports a view that interaction between NGOs and GOs is both multifaceted and dynamic. Some cases of interaction are clearly collaborative, a successful outcome depending on adequate fulfilment of prearranged obligations by the respective partners; other types we have termed 'incorporative' where NGOs offer lessons from innovative experience for incorporation, and so scaling up by GOs' programmes, although GOs may, of course, choose simply to disregard what is on offer. At another level, NGOs simply establish fora for information exchange on activities both among themselves and between themselves and GOs to generate complementarity, or simply to avoid duplication of effort. Other interactions largely derive from attempts by one side (usually NGOs) to influence the other, and may be conflictive.

Table 8.5 Conflicting perspectives on roles in agricultural technology development

GOs' perspectives	*NGOs' perspectives*
The government agency undertakes technological innovation due to its superior R & D resource base and access to wider facilities (e.g. HYV seeds, pedal threshers in Bangladesh) and expects NGOs to utilize the results at local level.	The government agency replicates or 'scales up' technologies provided by locally networked NGOs, since government has little expertise in participatory innovation at the grassroots level. It may work with NGOs providing further R & D support. NGOs may become trainers on larger-scale government projects.
The NGO disseminates technologies 'handed down' by governments since it is unable to manage its own R & D due to its comparatively low resource base. The NGO may find that it is forced to specialize in input delivery.	The NGO is active in small-scale community-centred technological innovation, often in the social organizational context (e.g. BRAC's irrigation groups) or in technical development (e.g. RDRS's treadle pump). It then looks for support for further R & D or wider distribution from other NGOs or government agencies.

Productive outcomes may, in principle, just as well be generated by conflictive as by collaborative interaction – though, in practice, issues over which there is conflict often require long periods of searching for productive solutions, as the AKRSP experience demonstrates – and solutions may ultimately prove elusive (Proshika).

Whatever the mode and level of interaction, it is important to note, first, that the types and levels of interactions that may be achieved are determined not only by the perceived advantages and disadvantages of particular opportunities for interaction, but also by the roles that each side sees as 'appropriate' for the other. Both sets of factors are conditioned by wider strategies on each side towards rural development, and by the historical, political and economic context in which state–NGO relations have evolved. Differences between countries are particularly important in this respect: thus, while in Bangladesh government agencies may argue that NGOs should disseminate technologies which they – government – develop, in the Philippines there has been a willingness on the part of government to recognize NGO achievements in technology development (e.g. MBRLC/SMAP) and build the technology into government programmes.[3]

Second, many NGOs interact with government at more than one level. AKRSP, for instance, has worked in a collaborative mode with government staff to set up participatory approaches to training, but in a (mildly) conflictive mode with higher echelons of government to influence policies towards joint management of forest reserves. However, there are limits to the extent that multiple levels of interaction can be pursued: IIRR's reputation as a neutral broker which allowed it to bring together NGO and GO professionals to develop an agroforestry information kit may have been seriously impaired if, for instance, it had placed itself in opposition to government over land reform. NGOs that are vocal in their advocacy of land reform see little prospect of functional collaboration with government, a view which government reciprocates. Some specialization of roles may therefore be anticipated between those NGOs whose areas of concern allow the possibility of collaborative, incorporative or informative links with government, and those whose more overt political activism rules it out.

Third, 'roles' evolve over time. Many NGOs would regard as unacceptable a role in which they substitute for services that government should normally be expected to provide. However, many begin by performing a 'substituting' role in the expectation that government may learn from the approaches that they are adopting and eventually take over. This philosophy, for instance, strongly underlies PRADAN's work and leads to a function which is both catalytic and complementary to that of government. NGOs' evolving relationship with grassroots organizations provides an important additional dynamic: many aim to strengthen local organizations so that, when NGOs withdraw, these can take over many of the functions previously performed by the NGO, and can generate a 'demand-pull' on government to ensure adequate continuing provision of the services required. The difficulties of making progress in this dynamic context should not be underestimated. For instance, it took BRAC five years to persuade the government to replicate its poultry model. Nor is it clear that grassroots organizations will have the necessary leverage to make changes at the middle and higher levels of government service. In its sericulture programme, for instance, BRAC had sufficient influence at district level to have disciplinary action taken as a sanction against poor performance by local level government officers. It is not easy to imagine that grassroots organizations will have the power to do likewise.

Finally, external funding agencies have played an important role in many of the NGO initiatives reported here (PRADAN is a notable exception). In some cases, this has been part of a wider and explicit objective to facilitate productive NGO–GO partnerships (see p. 284 on the Ford Foundation's work with LP3ES in Indonesia). On other occasions, donor funding for NGOs has been stimulated by a sense of frustration with public sector organizations and by a view that NGOs provide an 'alternative'. Such a view is innocuous where it implies an alternative approach which government might eventually adopt. It is problematic where it is conceived in the sense of NGOs substituting for government services: problematic both because it is potentially unsustainable, and because it places increasing responsibility for what should be government's role into the hands of organizations which are unelected and, therefore potentially unaccountable for their actions. The problems and prospects of the roles that external funding agencies might play is, however, a focus of the policy discussion in the final chapter, and so is not further elaborated here.

NOTES

1 A more detailed discussion of the term 'scaling up' and its implications is provided in Chapter 9.
2 A practical outcome of the Asia regional workshop were the visits by joint groups of senior NGO and GO officials from India and Bangladesh to Filipino NGO and GO contacts made at the workshop in order to see at first hand the structure and functioning of NGO Desks in the DENR.
3 Nevertheless, the credit for the MBRLC's SALT technology has occasionally been claimed by government personnel as their own. However, the NGO has been reluctant to enter into a dispute about this and prefers to regard such attempts to claim credit as proof of the technology's success.

Chapter 9

POLICY IMPLICATIONS: RECONCILING DIVERSITY WITH GENERALIZATION

DIVERSITY AND GENERALIZATION

A central question in this concluding chapter is – given the diversity of the case-study material reviewed in Chapters 3 to 7 – to what extent generalizations might be made, whether concerning the nature of NGOs and GOs, the types of interactions that might usefully be promoted, or the preconditions for successful collaboration. To address this question, we first review the types of diversity evident in the case studies. We then argue that generalization is possible in certain areas, and conclude by drawing out the implications for policy formulation among GOs, funding agencies and NGOs themselves.

DIVERSITY

In Chapter 1 we presented in Figure 1.2 a scheme for classifying NGOs, and indicated from the numerous types those that were of particular interest in this study. First, considerable variation still remains in the type of NGO reviewed here – diversity in the extent to which they are autonomous from north-based funding or management influences, in the extent of their concern with the formation of grassroots groups intended eventually to be autonomous, and in the type and extent of participatory approaches that they promote.

Second, we have argued throughout that NGOs' actions and their relations with the state will be determined to a large extent by the evolution of the political, economic and cultural context in which they operate. In these contexts lies a source of diversity: as our conclusions to Chapter 2 indicate, there are wide differences between the environments of political repression (characteristic of Indonesia today, and of the Philippines under Marcos), those environments in which governments are non-democratic and non-antagonistic to NGOs, but bureaucratic (Thailand; pre-1990 Bangladesh) and those in which governments are broadly democratic but exhibit varying capacities to tolerate and respond to the demands of NGOs (India, the Philippines and, recently, Bangladesh and Nepal). These differences necessarily imply differing NGO strategies and actions: NGOs operating under repressive regimes are likely to work in 'safe' localized issues which allow them to work without compromising their principles, but at the same time without drawing political attention to themselves. At the opposite end of the spectrum are the politically pluralist societies in which government facilitates (Philippines) or at least tolerates (India; Bangladesh) NGO interaction with the state and where NGOs engage both in joint actions and in policy dialogue (providing that they are large enough in the latter cases to overcome the transaction costs of obtaining access to the public sector) principally because they sense a possibility of influencing the state's policies or actions towards the needs of their own clients. Needless to say, conditions vary within countries over time, as well as between countries, and, in politically unstable environments, NGOs tend to develop strategies which fulfil their broad objectives and philosophies in liberal periods, without exposing them to reprisals under illiberal regimes. At the time of writing, elections in the Philippines have returned a new government, while in Thailand recent repression of the democratic movement by the military may have altered the terms on which NGOs will work with governments.

Third, governments' capacity to deliver rural services varies both within and between countries, and has implications for the extent to which 'gaps' occur which NGOs may seek to fill, often in the hope of eventually attracting government services into the area. The government's physical and institutional infrastructure for rural service delivery is particularly strong in India, and, by contrast, weak in the outer islands of the Philippines. But even in India, which has the added advantage of a comprehensive strategy for rural poverty alleviation, the measures implemented by government had only limited positive impact on livelihoods among the rural poor.

A fourth type of diversity is found in the actions undertaken by NGOs. We are less concerned here with the types of innovative activity undertaken by NGOs (whether in technology, for example, or in method of diagnosis, investigation or dissemination) or the level (whether through interaction with farmers or, at the other extreme, with central government) – although considerable diversity exists among these – but more with the political implications of such actions. A potentially useful distinction is that between radical and non-radical NGOs (Clark 1991), a distinction which often cuts across the traditional divisions between left and right. 'Radical NGOs' regard all development efforts to be in some sense political, since they must involve the transfer of power to those previously denied access to it. 'Non-radical NGOs' are content to attempt to work in what they regard as an apolitical context.

Korten (1990) also distinguishes between the humanitarian assistance provided by some NGOs and 'serious development assistance to the poor' provided by others, which demands attention to political and economic empowerment.

The case studies presented here illustrate the point that few if any development interventions undertaken by NGOs are politically neutral. However, different activities undertaken by NGOs in relation to government agencies engender different levels of engagement with political realities, and it is possible to conceive of a continuum of NGO activities in terms of their increasing political sensitivity, with 'non-radical' NGOs at one end and 'radical' NGOs at the other (Table 9.1).

Table 9.1 NGO activities in order of increasing political sensitivity

1 Delivery of inputs and services
2 Developing new technologies and methods
3 Developing new social innovations which bear on technological change
4 Policy level lobbying (e.g. for land reform)
5 Grassroots organizing (e.g. for legal rights, land tenure, higher wages)

Non-radical NGOs have for obvious reasons found more favour with government in partnerships generally found within categories 1 and 2 in Table 9.1.

The difficulties facing NGOs in levels 3 to 5 of this continuum serve to emphasize the importance of tackling the more fundamental issues in relation to government. As Clark points out, development does not take place on the basis of 'projects', which will

> remain irrelevant to the majority of the needy unless used as beacons to light up pathways for others – notably the state – to pursue. Popular participation on a significant scale will only come about through reforms in official structures, not through multiplying NGO projects.
>
> (Clark 1991: 75)

If NGOs are to make their policies relevant to wider sections of the poor then government agencies must benefit from their influence. The task then becomes one of improving the efficiency of government in the services it seeks to provide (e.g. through training: see pp.131–5), as well as increasing its accountability to a wider section of the population (e.g. though seeking the application of laws relating to access to common property resources, such as Proshika's social forestry programme in Bangladesh). While the first is a relatively straightforward task, the second is far more challenging.

The four types of diversity outlined above give rise to a fifth, that is in the types of *NGO–GO interaction*. The types of roles played by NGOs and GOs in interactions between them were identified in Chapter 8, and were classified into modes ranging from the collaborative to the conflictive.

WHAT GENERALIZATIONS ARE POSSIBLE?

The types of diversity outlined above highlight the importance of *context* in influencing NGO actions and NGO–government interaction. The more important is context, the greater the risks that experiences simply transferred from one location to another will not be successfully replicated. Attempts to generalize from the experiences reported in Chapters 3 to 7 should therefore be approached with extreme caution. Nevertheless, we now argue that certain general statements can be made about the institutional characteristics of NGOs, and the strengths and weakness these imply in relation to agricultural technology development and dissemination.

First, institutional characteristics of NGOs and related strengths and weaknesses: NGOs vary considerably in size. From the case studies reviewed here, they range from Mag-uugmad with some 20 staff to BRAC, which has over 6,000. Even the largest, however, is small compared with government departments, which commonly employ tens of thousands of staff. NGOs' small size, and (often) their commitment to decentralized decision-taking, result in flexible and non-hierarchical modes of operation. These qualities facilitate rapid response to the needs of chosen client groups.

Second, NGO approaches to agricultural technology development and dissemination tend to be more strongly issue-oriented than are those of government. In defining the issues, they tend to probe beyond the farm gate (since their mandate is not restricted to agricultural research) and seek to relieve several relevant constraints simultaneously, as the AWS and PRADAN case studies demonstrate. NGOs tend to conduct research at levels appropriate to the issues identified. In practice, this means that their work tends to be of a strongly applied type, producing results relevant to local issues, perhaps at the expense of identifying, for instance, how widely applicable the results might be. NGOs' concern is at least as much with action as with research. They rarely have available the amounts and continuity of funding necessary for long-term investigation. They – and the agencies which fund them – have a keen interest in seeing the results of experimentation applied. This can produce commendably quick results, but also runs the risk of implementing inadequately tested technologies.[1]

Third, an important contextual characteristic of NGOs' work is that it is focused on defined client groups. These may generally be described as the 'rural poor', but are differentiated into resource-poor farmers, rural women and the landless, for example. Five areas in which NGOs have demonstrated particular strengths in their work with these groups were identified in Chapter 2:

- technologies and management practices adapted to difficult areas
- technologies to meet the needs of the rural landless
- technologies to meet the needs of women
- approaches that 'de-mystify' complex technologies and made them suitable for neglected groups
- approaches helping local groups to form which then carry forward the technology in a sustainable fashion, linking in with input suppliers and markets.

Fourth, an important characteristic common to all of the above is that NGOs' small size and clearly identified target groups facilitate approaches which are fundamentally more participatory than those of government. This has resulted in particular methodological innovations, including diagnostic methods based on 'participatory rapid appraisal' (e.g. MYRADA case study) and farmer-to-farmer dissemination strategies (e.g. Thailand rice-fish farming case study), but its wider importance lies in the fact that change developed by these means tends to be consistent with the aspirations, opportunities and constraints of local communities and so more likely to achieve sustained success than that introduced by more conventional methods. The participatory approaches adopted by many NGOs are empowering in that they aim to establish local membership organizations capable of taking forward members' interests – both through action and advocacy – in a sustainable fashion, and so taking over many of the functions initiated by the NGO.

Fifth, smallness of size has a number of disadvantages: although there are exceptions (e.g. BAIF and MCC) it generally prevents NGOs from undertaking the more basic or strategic types of research, and gives them little interest in identifying from their work what might be generalizable beyond the confines of their immediate target group. Government research and extension services have a major concern – and comparative advantage – in both of these areas.

Sixth, smallness of size and concern with local issues bring further disadvantages: NGOs rarely address the wider structural and policy factors which ultimately influence the environment in which they operate.[2] Their capacity for research of all kinds – whether on local issues or wider policies – is resource-constrained: while NGOs are attracting increasing numbers of qualified graduates, individual NGOs are unlikely to have the specific skills, analytical facilities or access to wider information systems necessary to address research issues in depth. NGOs' capacity for documenting and disseminating the results of their work tends to be severely constrained, thus depriving others of potentially useful results. Furthermore, they have a poor record of setting up the types of co-ordinating mechanisms that might help to overcome shortcomings of this kind, and might serve as a basis for the pooling of skills or facilities.

Finally, it has been widely observed (e.g. Clark 1991) that a gap exists between reality and the claims that NGOs have made for themselves. There is, for example, a lack of NGO accountability either to governments or to the rural poor. NGOs are particularly exposed to charges of inadequate accountability when operating in democratically governed countries. Second, while espousing participatory approaches to change in matters affecting their clientele, NGOs rarely permit participation by the poor in influencing the size, structure and objectives of their own organization.

THE SCOPE FOR CLOSER NGO–GO LINKS

The wide range of NGO–GO interactions noted in Chapter 8, and the fact that both the wider political context and the immediate NGO and GO institutional context in which they occur are dynamic, mean that future interactions will not be a simple extrapolation of those currently observed. We speculate that, first, although not as severely as in Africa or in Latin America, governments in Asia will continue to experience some financial stringency which will limit their capacity to deliver a full range of services to the rural poor. In some countries, this capacity will further be limited by corruption and inefficiency. The trend for donors to look to NGOs as democratizing influences on the state as a whole, and on its institutions, appears set to continue as, therefore, will funds for NGO–GO collaborative projects.

Second, a large number of 'opportunistic' NGOs willing to act in a service delivery role in order to take advantage of this increased funding is therefore likely to emerge over the next decade. While not as innovative, nor as committed to social mobilization as many of the NGOs reviewed in this book, these nonetheless have the capacity to deliver services efficiently and generate pressures on government via feedback from the rural poor.

Third, as part of the same trend, donors may follow specific examples identified in Africa (Wellard and Copestake 1993) and Latin America (Bebbington and Thiele 1993) in which funds have been granted to NGOs to allow them to commission investigations by GO scientists into specific issues that have arisen in the course of their work.

Fourth, some donors have channelled substantial sums into broad-based rural development projects in which an NGO has played a pivotal role in 'brokering' interaction among other agencies – both GO and NGO – in both project design and implementation. An example is provided by the Indo-German Watershed Development Programme, based in Ahmednagar, India (Box 4.1) co-ordinated by a church-based NGO, the Ahmednagar Social Centre. In other settings – e.g. the Southern Mindanao agriculture project – donor funding has been channelled through government to create a project in which NGOs have mutually agreed roles. In the majority of countries reviewed here, the risk of inadequate consultation by GOs if they were to take the lead role in brokerage or in project design is severe, and so wider experimentation with NGO-led initiatives is to be expected.

Finally, as mutual trust grows between NGOs and GOs in projects focusing on agricultural innovation, the likelihood is enhanced of representation by NGOs and farmers' associations on committees governing future research priorities, thus directly stimulating response to local needs.

UNRESOLVED ISSUES

Service delivery

The success of certain NGOs in delivering services to the poor has prompted many governments and donors to take a simplistic view of the potential that NGOs offer for delivering services to the rural poor. But it is not in the interests of any of the relevant actors (the NGOs, the poor or governments) for NGOs to replace the state in its role of channelling external resources into the countryside. This is likely to lead to a continuation of existing 'top-down' strategies which run counter to the participatory approaches of most NGOs and fail to address the need for improving the responsiveness of governments to people.

Donor pressures towards structural reform and privatization in many countries underlie the increased interest in NGOs as 'service deliverers'. Larger numbers of service contracts for NGOs are likely to change the overall composition of the NGO community by encouraging the establishment of NGOs specializing in this field which may, in practice, be semi-commercial in outlook. As experience from South America shows, to take on more contracts means, for some NGOs, that necessary time for reflection and innovation is reduced (Aguirre and Namdar-Irani 1992).

Korten (1990) rightly points out that a wholesale conceptualization of NGOs as service deliverers may serve to compromise the integrity of the wide-ranging programmes followed by many of the better NGOs: 'the distinctive role of the voluntary sector is not to serve as a cheap contractor to implement government-defined programmes' (Korten 1990: 207).

Promoting institutional sustainability

Many NGOs seek to create or strengthen local membership organizations capable of carrying forward the projects and programmes that have been initiated in a sustainable fashion, so allowing the NGO to withdraw after a limited period. Several instances were observed among the case studies of where this had proven possible (including PRADAN, ATA, IIRR). However, a major difficulty in some cases (e.g. BRAC poultry) lay in securing a continued input of the government services that had been provided.

Some NGOs (e.g. BRAC) have used a combination of training government personnel in the requisite skills, and pressures on the mid-level government hierarchy to ensure that commitments are honoured. Whether these strategies will succeed over the long term remains to be seen.

Innovation and scaling up

Many NGOs have developed technologies (hardware, systems of delivery and forms of organization for receiving and deploying technologies to enhance incomes among the rural poor) which have later been taken up by government. We have seen that some technologies have been developed by NGOs, while others have been adapted or replicated from other sources. As Clark points out:

> In such cases NGOs often play a catalytic or seeding role – demonstrating the efficacy of a new idea, publicising it, perhaps persuading those with access to greater power and budgets to take notice, and then encouraging the widespread adoption by others of the idea
>
> (Clark 1991: 59)

However, two points should be made about NGO innovations. The first, already mentioned, is that while innovativeness in participatory and group organizing methods is widespread, few NGOs have the requisite skills or facilities to generate *technical* innovations with a potential for wide adoption. The second is that there are limits to the types of innovations most governments will tolerate. For example, issues of forest land access (Bangladesh: Proshika) and forest land security of tenure (Philippines: MFI) have brought NGOs into conflict with governments whose interests remain linked to those of local elites.

An obvious question raised by 'successful' NGO-based activities is whether they can be replicated on a larger scale by government or other NGOs. Questions about the role of government are important even in those countries where the density of NGOs is high. In Bangladesh, for example, even if the combined efforts of all the NGOs working in the country are taken *together*, it is unlikely that more than 20 per cent of the rural poor will be reached. This would assume that these efforts are in some way compatible or co-ordinated, which of course most are not.

These questions leads back to one of the starting-points of this research, namely that some kind of relationship between government and NGOs is needed if poverty issues are to be addressed on a comprehensive, country-wide basis.

We have argued in this book that the relationships between them cannot be viewed on a purely functional basis. The relationship is conditioned by a range of social, political and historical factors and a functional conceptualization does potential damage to both parties:

- The status of government as inefficient, unimaginative and unmotivated becomes a self-perpetuating reality as increased resources flow to the more dynamic NGOs which may take over many of the functions of government, though on an unaccountable, piecemeal basis.
- The 'inherent' advantages of the NGOs themselves are gradually worn away by increased funding, professionalization, bureaucratization and the shifting of objectives away from social 'mobilization' towards service delivery and income generation.

A recent conference on 'scaling up' (Hulme and Edwards 1992) distinguishes between strategies in which an NGO can increase its size in order to expand its operations, and those in which it can transfer strategies to or have a 'catalytic' effect on other agencies.

Although to reach a wider section of the population (particularly in service and input delivery roles) may be in the interests of all parties, it may also bring significant problems: 'What scaling up puts at risk are precisely the features and strengths for which NGOs are valued' (Paul and Israel 1991: 13), because

- NGOs' ability to react quickly and adapt to local requirements is undermined as the scale of operations is increased.
- More distance may be introduced between the NGO and the people with whom it works.
- Expansion of the NGO reduces the commitment of its staff as the operation becomes more professionalized.

Paul *et al.* (1991) argue that there are ways around the scaling up dilemma. One is to support a multiplicity of small, local NGOs in order that they retain their comparative advantages and create more pluralism. Another is to support NGO–NGO linkages. A third path is for NGOs to act as 'nurseries' for ideas, emphasizing the R & D functions of the NGO which can then be taken up and adapted by other organizations and agencies. During these processes the NGO can play a crucial training role. A good example of this from the Asian case studies is that of MBRLC, which developed the SALT technology, which has now been taken up in the government's SMAP programme.

Inter-agency communications

It is clear from the case studies that while complementarities have often been acknowledged between NGOs and GOs, far less thought has gone into the practical aspects of creating linkages. Many of the documented collaborations have been on an *ad hoc* basis, formed from local, spontaneous action or from higher-level contacts achieved around the efforts of key personalities.

In general, there has been less attention paid to the question of building trust and informal alliances, the sharing of decision-making or the political questions of process management (Bebbington and Farrington 1992). The formal mechanisms for creating mutually agreed partnerships rarely exist.

Only in the Philippines have there been attempts to establish the institutional framework for linking NGOs and GOs in collaborative activity. The NGO Desks, often staffed by ex-NGO personnel themselves, provide an exploratory model for achieving linkages in which some other Asian governments have expressed a keen interest. The initiatives which have resulted from these desks are still in their early stages, and it is difficult to assess their impact. Political changes in the Philippines will no doubt be crucial in determining how these fledgling institutions fare.

NGO–NGO relations remain an area in which far more could be done by NGOs towards presenting their aims, ideas and needs to both the public and the government more clearly. The Philippines, perhaps more than any other country, has seen the development of NGO coalitions and alliances around a range of issues such as health and land reform, and can offer potentially illuminating experience on inter-agency communication and collaboration to others. The more typical condition of national-level NGO umbrella organizations is, however, typified by ADAB in Bangladesh. While playing an important co-ordinating role and providing a forum for discussion on development issues, ADAB has suffered from frequent internal arguments (e.g. over what kinds of NGOs it should represent; domination of small NGOs by larger ones) and a continuing lack of direction. Much clearly remains to be learned and implemented in this field, and, given the diversity of NGO interests, progress will inevitably be slow.

IMPLICATIONS

The implications of the findings of this book are now analysed in respect of:

- theory and practice of agricultural innovation
- the contribution that NGOs (and NGO–GO links) primarily seeking rural livelihood enhancement through agricultural change can make to wider processes of pluralism and democratization
- the future level and type of interactions among institutions.

AGRICULTURAL TECHNOLOGY DEVELOPMENT

The case studies presented in this volume add weight to growing criticism of the performance of hierarchical public sector research and extension services in generating and disseminating adoptable technologies for the rural poor. Although of localized scale, NGOs' 'action research' approaches have been based on a sound, participatory assessment of needs and opportunities. They represent processes in which the rural poor have been extensively involved, so that any product they generate – whether a technology or management practice – has a stronger likelihood of adoption.

Overall, the evidence is strongly supportive of the notion of 'multiple sources' of technology generation (Biggs 1989b). While scope for closer interaction between NGOs and GOs exists, it is likely to be:

- functionally complex: simple conceptualizations in which GOs conduct research and NGOs disseminate are not borne out by the evidence
- of multiple modes and levels: *collaborative* interaction is only one of four types identified, is likely to be the most difficult to achieve, and will not in all circumstances be the most productive. While successful interaction at local level may be the easiest to achieve, in some cases it may not succeed unless effective higher-level links are in place
- ultimately, conditioned in its scope, mode and level by the wider political and economic context of NGO–state relations.

NGOs, DEMOCRATIZATION AND THE STATE

Since the mid-1980s an awareness of the activities of NGOs has transformed our thinking about the relationship between the state and the wider community. For example, Robertson's (1984) survey makes almost no reference at all to 'private voluntary organizations' in its comprehensive discussion of development planning and participation.

The control of productive assets is a crucial issue which continues to expose the conflicts of power and interest in most communities, and the ways in which wider government structures are drawn into these. Korten argues that 'The current need is to achieve true economic democracy based on meaningful participation in the ownership and control of productive assets for reasons of equity, productivity and environmental responsibility' (Korten 1990: 174).

Robertson argues that the state has the capacity to mobilize and allocate economic resources which it may do in the interests of the majority of its citizens or more usually, in the interests of only a few: 'planning is a means to organize economic growth and distribute its benefits, but in itself it offers no guarantees that either task will be done equitably or efficiently' (Robertson 1984: 89). He goes on to point out that projects, co-operatives and committees represent the arenas in which 'the complex encounters between state and people take place' (Robertson 1984: 141). To these other forms of organizations we need, in the 1990s, to add that of the NGO.

In this discussion it is of course necessary to dispense with what Korten (1990) calls the 'benevolent state myth' in which it is assumed that unlike other types of institutions, which represent particular sectoral interest groups, the state acts in the best interests of everyone. Most aid flows from north to south have been premised on this assumption. It is this logic which maintains that governments should decide the most economically and socially productive allocations of development resources. As Korten correctly points out, one of the achievements of NGOs has been to question the right of governments to act on behalf of people in determining the development agenda. The case studies presented here show that interaction between NGOs and government occurs on many issues and that partnerships of several kinds have been forged.

Contradictions in different actors' conceptions of public/private sector roles therefore persist. For example, there is still widespread adherence to 'top-down' models of technological change in many public sector agencies, even when there is a rhetorical acknowledgement of the potential of NGOs in developing appropriate technologies. The rhetoric tends to translate into narrow conceptions of NGO roles. Many NGOs for their part continue to see the state as monolithic, although, as the evidence demonstrates, some departments and individuals are more progressive than others.

The issue is therefore not one of whether or not governments and NGOs might work together, but how. As Clark (1991: 75) argues, NGOs can 'oppose the state, complement it or reform it but they cannot ignore it'. The question then becomes one of examining the *types* of relationships that are possible under particular circumstances and their implications for the rural poor.

IMPLICATIONS FOR INSTITUTIONS AND INTER-INSTITUTIONAL RELATIONS

Implications for NGOs

NGOs have traditionally defined themselves in relation to the state. As the state is transformed by the wider agendas of political change, democratization and a reduction in resource flows to governments, NGOs are required to rethink their role (Bebbington and Farrington 1993). Some will be content to become sub-contractors for services, receiving a greater share of public resources, while others will take on new roles influencing policy. Political circumstances are changing in many Asian countries, but it would be a mistake to conceive of this change as a linear movement towards greater freedom and democracy. Events in Thailand in May 1992 and the 1992 election result in the Philippines illustrate the fragility of the structural conditions under which NGOs are operating.

Individual NGOs clearly need to assess the advantages and disadvantages of varying modes of interaction with GOs in terms of their individual development philosophies. The possibilities interweave into a rich fabric: an NGO that is content to fulfil a service delivery role under contract may fear that it is compromising its principles only with the more extreme of the governments noted here. An NGO seeking to develop innovative ways of responding to the needs of the rural poor may find that service delivery contracts reduce its space for reflection and action in this principal role, but it may engage government in a number of other modes, ranging from the collaborative to the conflictive. Those pressing for social reform in a highly conflictive mode may find it difficult to engage with government at any other level.

On a wider scale, NGOs have much to learn from each other. This book represents one of what remain a limited number of initiatives to encourage the documentation and exchange of experiences. While a number of inter-NGO networks and coalitions are being created on specific thematic and geographical bases in the Philippines, for example, such efforts have barely begun in other countries, yet offer the potential of catalysing new ideas and preventing the duplication of effort. National-level umbrella organizations of the type that exist in practically every country have generally proven too remote to address highly location-specific issues of agricultural technology. As the densities of NGOs operational at field level increase, the need for institutionalized means of communication (and, where appropriate, co-ordination) among NGOs and between NGOs and government will become all the more urgent.

Implications for governments

A number of the approaches used by NGOs which might, with advantage, be adopted by GOs emerge from the above discussion:

- a focus on the rural poor which incorporates participatory approaches to diagnosis, investigation and dissemination; approaches which embrace food systems, not merely farming systems; approaches which treat on- and off-farm resource use in a holistic fashion
- a scaling-down and de-mystifying of what were initially complex technologies in order to make them more accessible to the rural poor
- the re-orientation of GO staff towards practical 'learn by doing' training techniques which address farmers' problems directly.

Perhaps NGOs' flexibility, work ethic and ability to respond rapidly to issues as they arise offer the most powerful lessons – lessons which go to the heart of how GOs are organized and managed *as institutions*. Farrington and Bebbington (1991) discuss several factors which limit GOs' institutional capacity to conduct systems-based agricultural technology development and dissemination to meet farmers' needs:

- excessive centralization of decision-making authority, the consequences of which for inflexible and inefficient management are also discussed by Merrill-Sands and Kaimowitz (1990)
- inappropriate reward systems, generally based on scientific publications, which are likely to bias agricultural technology development effort away from farmers' needs and unlikely to offer adequate regards to field-oriented agricultural technology developments
- inflexible programming and budgeting procedures, which leave no 'unallocated' resources which might be used to respond to needs (whether of NGOs or of farmers directly) as they arise.

Careful observation of the working practices of NGOs would allow at least some steps towards alleviation of these public sector constraints, though it should be emphasized that the larger size of the public sector and its (at least nominally) stricter accounting and reporting procedures are unlikely ever to allow it the flexibility that NGOs enjoy.

A more profound change would be to take lessons from NGOs' institutional structure and incorporate them into GOs. This would mean decentralizing authority within GOs, and increasing the flexibility and adaptiveness of local offices (c.f. Sollis 1991)[3] It would also involve structuring local offices of GO programmes along the lines of NGOs' small relatively informal field offices, while retaining the co-ordinating mechanisms made possible by the presence of the overlying institutional structure of the public sector.

The likelihood, of course, is that this will rarely happen, partly because of GO resistance, and partly because representative organizations of the rural poor do not exist everywhere: in the short term, the most that can be hoped for is that NGOs will be invited into these programmes as surrogate representatives of the concerns of the rural poor. Ultimately though, and sooner rather than later, it should be the poor themselves who are there, with NGOs advising as experienced specialists.

Implications for funding agencies

Funding agencies can usefully support several of the initiatives indicated for governments in the previous section. They might, for instance, support decentralization policies, or sponsor collaborative NGO–GO projects which have the potential to lead to longer-term influences by one side or the other.

With the wide interest in structural adjustment programmes, much of the interest of conventional bilateral and multilateral lending agencies has been in NGOs as more efficient agencies than government for delivering services to the rural poor. Some donors see the promotion of better contacts between NGOs and GOs as a way of increasing the efficiency of the latter (Bebbington 1991).

Those funding agencies (including the major international foundations such as the Ford Foundation) which have a tradition of funding NGOs directly have exerted an influence which has ranged from facilitation in which NGOs are supported in pursuit of their own agendas (probably the majority of NGO funding until recently) through constructive influence in which the donor contributes and supports an innovation (for example, Ford's funding strategy with LP3ES in Indonesia) to distortion (for example, the pressure placed by CIDA in Bangladesh to bring NGOs largely against their judgement – and in the end unsuccessfully – into a training role in the government's co-operative programme for the poor).

Non-radical NGOs' activities have tended to attract more sympathetic interest and attention than radical NGOs, and this has often translated directly into higher funding from the international community. But it is not always possible to attach clear labels. The LP3ES irrigation case in Indonesia, stimulated and supported by an innovative Ford Foundation system of flexible funding combines both radical and non-radical roles.

However, as Clark (1991) points out, the funding proposals of some NGOs may be determined by the types of activities that donors are known to be willing to fund, and this has produced tension between radical and non-radical NGOs to the extent that the former fear that scarce funds will be absorbed and the development agenda co-opted.

Overall, it is clear that no single donor strategy will be adequate for the multiplicity of NGO types and for actual or potential NGO–GO interactions. Donor support for service contracts is likely to increase, but care needs to be taken to avoid drawing the more catalytic NGOs excessively into such contracts, and so undermining their innovative capacity. Other donors may wish to encourage GOs to increase their awareness of NGOs' needs and their responsiveness to NGO initiatives by supporting joint NGO–GO projects and programmes. A third category of initiatives – more difficult for donors to administer because of their open-endedness – are the innovative 'process'-type approaches supported (and, to some extent, conceived) by the Ford Foundation, examples of which are given in Chapter 2 and in Box 7.3.

FUTURE PROSPECTS

NGOs are highly diverse in size, access to funds and philosophy, and so in the range of activities they pursue. Not surprisingly, the types of interactions between NGOs and governments reported in this book are also highly diverse.

None the less, it has been possible to draw some general lessons concerning NGOs' strengths and weaknesses, and those exhibited by GOs involved in the development and dissemination of agriculture-related technologics and management practices. The evidence supports the hypothesis of a complementarity of functions between NGOs and GOs: to a degree, each side can support and learn from the other. However, interaction between the two sides is likely to be dynamic and complex. Full, formal collaboration has proven the most difficult mode of interaction to achieve, and can be found in only a minority of the case studies reviewed.

For the future, much will depend on the broad characteristics of governments that come to power within the region. The building up of mutual trust and confidence, often over several years, is generally a precondition to more formal types of collaboration, and is unlikely to occur where wider relations between NGOs and the state are hostile.

Administrative reform is also of prime importance, especially in the South Asian countries: a precondition for effective collaboration is that government staff share the concern of NGOs to address issues of direct relevance to the rural poor. Yet reward systems remain constructed around criteria such as the numbers of journal articles published by scientists, and so remain largely irrelevant to the needs of the poor.

Many NGOs will undoubtedly continue doing the things that they already do well: working with farmers face to face, forming groups, developing participatory technologies, lobbying governments, networking between governments and other NGOs (locally, nationally and internationally) and identifying sections of the population which have been marginalized by official efforts.

Development policy in the 1990s is moving increasingly towards an emphasis on securing political rights and representation for all sections of the community, and opening governments to increased accountability to the poor in service provision, legal rights and policy formulation. The facilitation of such changes requires creative thinking among all the actors involved (NGOs, farmers' associations, government agencies, donors) to find ways of ensuring that the interests of disadvantaged groups are represented in and addressed by governments. NGOs, if they are allowed to by governments and donors – and if they themselves can rise to the challenge by recognizing their own strengths and weaknesses – will have an important role to play in this process.

There is therefore a need to form more generalized plans for the future which may, in the first instance, focus at the macro-level relationship between NGOs and governments, but will ultimately have implications for their interactions at sectoral level. Part of the responsibility for the current high profile of NGOs lies with donors, who have seen NGOs as better 'value for money' and in some cases more 'poverty focused' in their activities. But the challenge for the 1990s for the donors, government and the NGOs themselves is to assign clear roles in accordance with the goal of sustainable, equitable development.

In each country, the circumstances of these negotiations will vary. In Indonesia, the political climate is primarily cool towards NGOs, and the challenge for many is to work locally without antagonizing government, but laying the foundations for change and participation for the future, as many NGOs in the Philippines under Marcos attempted. In Bangladesh, the recently restored constitutional democracy creates a new room for manoeuvre and different sets of rules apply. Here, roles must be found for NGOs in which long-term institutional change is brought to government, accountability is improved and there is more participation by disadvantaged groups in the 'development process'.

Despite the diverse contexts in which they operate, NGOs are beginning to share concern in a common theme: that of enhancing the responsiveness of government services to the needs of the rural poor. It is principally towards this area that the growing volume of NGO–GO interaction will be directed over the coming decade.

NOTES

1 See Kohl (1991) for a well-documented example of inadequate testing.
2 Exceptions in the present context include Proshika and AKRSP.
3 Sollis (1991: 19) argues: 'The key to converting a disabling state into an enabling one in terms of social service provision is identified in decentralization policy.'

REFERENCES

Abad, F. (1991) 'The future of agrarian reform and a renewed call for NGO participation', *Kaisahan Occasional Paper no. 91-2*, Quezon City.
Abed, F.H. (1991) 'Extension services of NGOs: the approach of BRAC', Paper to National Seminar on GO-NGO Collaboration in Agricultural Research and Extension, Dhaka, 4 August.
Abedin, Z. (1991) Collaboration between NGOs and public sector agricultural research in Bangladesh', Paper presented at the Asia regional workshop on NGOs, Natural Resources Management and Links with the Public Sector, Hyderabad, India, 16–20 September.
ActionAid-Nepal (1988) 'ActionAid-Nepal: sector policy papers', Kathmandu: ActionAid.
—— (1990) 'ActionAid-Nepal: annual progress report July 1989 – June 1990', Kathmandu: ActionAid.
Adhikary, P.K. and Suelzer, R. (1985) 'Self-sustaining development programme to strengthen self-help potentials of self-organizing groups', unpublished paper, Tansen.
Affal, K.E. (1988) 'NGOs as vehicle for socio-economic development: a study of international NGOs operating in Nepal', unpublished report, Department of Economics, University of Lancaster.
Aguirre, F. and Namdar-Irani, M. (1992) 'Complementarities and tensions in NGO–state relations in agricultural development: the trajectory of AGRARIA (Chile)', Agricultural Research and Extension Network, *Network Paper no. 32*, London: ODI.
Ali, R. (1986) 'Revegetating barren land: the Aurovillian experience', Tiger Paper, RAPA, FAO, XIII (1).
ANGOC (1984) *An Overview of Philippines NGOs*, Manila: ANGOC..
Ashby, J.A. (1987) 'Farmer participation in on-farm variety trials', Agricultural Research and Extension Network, *Discussion Paper no. 22*, London: ODI.
Auroville Afforestation (1989) 'A brief overview', *Auroville Today* 8, July.
Auroville (1990) 'Auroville', mimeo, Auroville Fund.
Ayers, A. (1992) 'Conflicts or complementaries? The State and NGOs in the Colonization Zones of San Julian and Berlin, Eastern Bolivia', Agricultural Research and Extension Network, Network paper no. 37, London: ODI.
Bairacharya, D. (1983) 'Deforestation and food/fuel context: historic political perspectives from Nepal', *Mountain Research and Development* 3(3): 227–40.
Barrett, A. (1991) 'NGOs and the state in Bangladesh: a case study of the Land Reform Programme', Paper to the Development Studies Association Conference, Swansea, 11–13 September.

Bear, M. and Rahman, M.A. (1989) *Mid-Term Review for LIFT in Gaibandha*, Bangladesh: CARE.

Bebbington, A.J. (1989) 'Institutional options and multiple sources of agricultural innovation: evidence from an Ecuadorean case study', Agricultural Research and Extension Network, *Network Paper no. 11*, London: ODI.

—— (1991) 'Sharecropping agricultural development: the potential for GSO–government cooperation', *Grassroots Development* 15(2): 20–30.

Bebbington, A.J. and Farrington, J. (1992) 'The scope for NGO–government interactions in agricultural technology development: an international overview', Agricultural Research and Extension Network, *Network Paper no. 33*, London: ODI.

—— (1993) 'Governments, NGOs and agricultural development: perspectives on changing inter-organisational relationships', *Journal of Development Studies* 29(2) 199–219.

Bebbington, A.J. and Thiele, G. (1993) *Non-Governmental Organizations and the State in Latin America: Rethinking Roles in Sustainable Agricultural Development*, London: Routledge.

Biggs, S.D. (1989a) 'Resource-poor farmer participation in research: a synthesis of experiences from nine national agricultural research systems', *OFCOR Comparative Study Paper no. 3*, The Hague: International Service for National Agricultural Research.

—— (1989b) 'A multiple source of innovation model of agricultural research and technology promotion', Agricultural Research and Extension Network, *Network Paper no. 6*, London: ODI.

—— (1989c) 'The role of management information systems in agricultural research, policy, planning and management in the Indian Council of Agricultural Research, mimeo, University of East Anglia, Norwich.

Campbell, J.G. (1979) *Community Involvement in Conservation: Social and Organisational Aspects of Proposed Resource Conservation and Utilisation Project in Nepal*, Kathmandu: APROSC.

Carew-Reid, J. and Oli, K.P. (1991) 'The international union for the conservation of nature and NGOs in environmental planning in Nepal', Paper presented at the Asia regional workshop on NGOs, Natural Resources Management and Links with the Public Sector, Hyderabad, India, 16–20 September.

Carroll, T. (1992) *Intermediate NGOs: Characteristics of Strong Performers*, West Hartford, Conn.: Kumarian Press.

Chambers, R., Pacey, A. and Thrupp, L-A. (eds) (1989) *Farmer First: Farmer Innovation and Agricultural Research*, London: Intermediate Technology Publications.

Clark, J. (1991) *Democratising Development: The Role of Voluntary Organisations*, London: Earthscan.

Clay, E.J. (1988) *Food Strategy in India*, Report for ODA, London: Relief and Development Institute.

Constantino-David, K. (1992) 'The caucus of development NGO networks: the Philippines experience in scaling up NGO impact'; in D. Hulme and M.D. Edwards (eds) *Making a Difference? NGOs and Development in a Changing World*, London: Earthscan.

Coulter, J. and Farrington, J. (1988) *India – Renewable Natural Resources Sector: Review of ODA-Supported Research and Development*, Report for ODA, mimeo, London.

Cromwell, E. and Wiggins, S. (1992) *Sowing beyond the State: NGOs and Seed Supply in Developing Countries*, London: ODI.

Dean, L. (1990) *Local Initiatives for Farmers' Training (LIFT) Gaibandha July 1986–June 1990: Final Report*, Bangladesh: CARE.

de Boer, J.A. (1988) 'Sustainable approaches to hillside agricultural development', Paper presented at USAID/Nepal seminar, Winrock International/Agricultural Research and Production Project, Kathmandu, Nepal, 16 December.

Denholm, J. and Rayachhetry, M.B. (1990) 'Major agro-forestry activities of non-governmental NGOs operating in Nepal: a survey', HMG Ministry of Agriculture – Winrock International, Kathmandu, Nepal, September.

Deore, P.A. (1990) 'Manual on cross-bred cows', Pune: BAIF.

Ekachai, S. (1990) *Behind the Smile: Voices of Thailand*, Bangkok: Thai Development Support Committee.

Faaland, J. and Parkinson, J.R. (1986) *The Political Economy of Development*, London: Francis Pinter.

Farnworth, E.G. (1991) 'The Inter-American Bank's interactions with non-governmental environmental organisations', Paper presented at the Third Consultative Meeting on the Environment, Caracas, 17–19 June.

Farrington, J. and Bebbington, A.J. (1991) 'Institutionalisation of farming systems development – are there lessons from NGO–Government links?', Paper for the FAO Expert Consultation on the Institutionalisation of Farming Systems Development, Rome, 15–17 October.

Farrington, J. and Bebbington, A.J. (1993) *Reluctant Partners? Non-Governmental Organizations, the State and Sustainable Agricultural Development*, London: Routledge.

Farrington, J. and Biggs, S.D. (1990) 'NGOs, agricultural technology and the rural poor', *Food Policy*, December: 479–92.

Farrington, J. and Martin, A. (1987) 'Farmers' participation in agricultural research: a review of concepts and practices', Agricultural Administration (Research and Extension) Network, *Discussion Paper no. 19*, republished 1988 as *Occasional Paper no. 9*, London: ODI.

Farrington, J. and Mathema, S.B. (1990) 'Managing agricultural research for fragile environments: Amazon and Himalayan case studies', *Occasional Paper no. 11*, London: ODI.

Feder, E. (1983) *Perverse Development*, Manila: Foundation for Nationalist Studies.

Fowler, A. (1990) 'Doing it better? Where and how NGOs have a comparative advantage in facilitating development', *AERDD Bulletin* 28, February.

—— (1991) 'The role of NGOs in changing state–society relations', *Development Policy Review* 9(1): 53–84.

Freire, P. (1972) *Pedagogy of the Oppressed*, Harmondsworth: Penguin.

Friedman, J. (1992) *Empowerment: The Politics of Alternative Development*, Oxford: Basil Blackwell.

Garforth, C.J. and Munro, M. (1990) 'Rural people's organisations and agricultural development in the Upper North of Thailand', Department of Agricultural Extension and Rural Development, University of Reading.

Garilao, E. (1987) 'Indigenous NGOs as strategic institutions: managing the relationship with government and resource agencies', *World Development* 15, Supplement: 113–20.

Gilbert, E. (1990) 'Non-governmental organisations and agricultural research: the experience of the Gambia', Agricultural Research and Extension Network, *Network Paper no. 12*, London: ODI.

Gilmour, D.A. and Fisher, R.J. (1990) *Villagers, Forests and Foresters*, Kathmandu: Sahayogi Press.

GoI (Government of India) (1988) 'Report of the dairy production recording and genetic evaluation team, January 20 – February 20 1988', Indo-US Subcommission on

Agriculture, New Delhi.

Gonsalves, J. and Miclat-Teves, A.G. (eds) (1991) 'GO–NGO collaboration in the area of agriculture and natural resources management in the Philippines', Proceedings of workshop held by International Institute of Rural Reconstruction (IIRR) and Overseas Development Institute (ODI), Silang, Cavite, 18–20 July.

Gronow, J. and Shrestha, N.K. (1990) 'From policing to participation: reorientation of forest department field staff in Nepal', HMGN Ministry of Agriculture - Winrock International, Kathmandu, Nepal, September.

Hashemi, S.M. (1989) 'NGOs in Bangladesh: development alternative or alternative rhetoric?', mimeo, Dhaka.

Hassan, S.T., Satish, S. and Sitapati Rao, C. (1990) 'Watershed management and irrigation utilisation in chronically drought prone areas', New Delhi: Planning Commission, GOI, and Administrative Staff College of India.

Hassanullah, M. (1991) 'NGOs and public sector agricultural research and extension in Bangladesh', Paper presented at the Asia regional workshop on NGOs, Natural Resources Management and Links with the Public Sector, Hyderabad, India, 16–20 September.

—— (ed.) (forthcoming) Papers from the Bangladesh national workshop on NGO–government links in agricultural technology development (provisional title), *Bangladesh Journal of Extension Studies*.

Haverkort, B., van der Kamp, J. and Waters-Bayer, A. (eds) (1991) *Joining Farmers' Experiments: Experiences in Participatory Technology Development*, London: Intermediate Technology Publications.

Hazlewood, P.T. (1987), *Expanding the Role of Non-Governmental Organizations in National Forestry Programme*, New York: World Resource Institute.

Healey, J. and Robinson, M. (1992) *Democracy, Governance and Economic Policy: Sub-Saharan Africa in Comparative Perspective*, London: ODI.

Henderson, P.A. and Singh, R. (1990) 'NGO–government links in seed production: case studies from The Gambia and Ethiopia', Agricultural Research and Extension Network, *Network Paper no. 14*, London: ODI.

Holloway, R. (ed.) (1989) *Doing Development: Government, NGOs and the Rural Poor in Asia*, London: Earthscan/CUSO.

Hossain, M. (1990) 'Bangladesh economic performance and prospects', *Briefing Paper*, November, London: ODI.

Hossain, M. and Jones, S. (1983) 'Production, poverty and the co-operative ideal: contradictions in Bangladesh rural development policy', in D.A.M. Lea and D. P. Chaudhri (eds) *Rural Development and the State*, London: Methuen.

Howell, J. (ed.) (1988) 'Training and visit extension in practice', *Occasional Paper no. 8*, London: ODI.

Hulme, D. (1990) 'Can the Grameen Bank be replicated? Recent experiments in Malaysia, Malawi and Sri Lanka', *Development Policy Review*, 8(3): 287–300.

Hulme, D. and Edwards, M.D. (eds) (1992) *Making a Difference? NGOs and Development in a Changing World*, London: Earthscan.

Huq, M.F. and Sabri, A.A. (1991) 'Livestock development programme for the poor: experience of Proshika', Paper presented to BRAC-organized workshop on livestock, Dhaka.

ICAR (1988) *Report of the ICAR Review Committee*, New Delhi: ICAR.

IDS (1982) 'Mid-term review in the areas of women's income generation, and NGOs programmes', Kathmandu: IDS.

—— (1985) 'Non-governmental institutions and processes for development in Nepal', Kathmandu: IDS.

IUCN, UNEP and WWF (1990) 'Empowering communities: primary environmental care in "Caring for the world: a strategy for sustainability" ', Second draft, Gland, Switzerland: IUCN.

Johnson, H.D. (1989) 'Planting a forest', *San Francisco Examiner*, November.

Kabeer, N. (1989) 'Monitoring poverty as if gender mattered', *Discussion Paper no. 255*, Institute of Development Studies, University of Sussex.

—— (1991) 'Gender, production and well-being: rethinking the household economy', *Discussion Paper no. 288*, Institute of Development Studies, University of Sussex.

Kaimowitz, D. (1990) *Making the Link: Agricultural Research and Technology Transfer in Developing Countries*, Boulder, Colo.: Westview Press in co-operation with ISNAR.

Khan, A.R. (1979) 'The Comilla model and the integrated rural development programme of Bangladesh: an experiment in cooperative capitalism', *World Development* 7: 397–422.

Kohl, B. (1991) 'Protected horticultural systems in the Bolivian Andes: a case study of NGOs and inappropriate technology', Agricultural Research and Extension Network, *Network Paper no. 29*, London: ODI.

Korten, D.C. (1987) 'Third generation NGO strategies: a key to people-centred development', *World Development* 15, Supplement: 145–59.

—— (1990) *Getting to the 21st Century: Voluntary Action and the Global Agenda*, West Hartford, Conn.: Kumarian Press.

Korten, F. and Siy, R.F. (1988) *Transforming a Bureaucracy*, West Hartford, Conn.: Kumarian Press.

Kramsjo, B. and Wood, G.D. (1992) *Breaking the Chains: Case Studies of Proshika Groups in Bangladesh*, London: Intermediate Technology Publications.

Krinks, P. (1983) 'Rectifying inequality or favouring the few? Image and reality in Philippine development', in D.A.M. Lea and D.P. Chaudhri (eds) *Rural Development and the State*, London: Methuen.

Kumar, S. (1991) 'Anantapur experiment in PRA training', in J. Mascarenhas, P. Shah, S. Joseph, *et al.* (eds) *Participatory Rural Appraisal: Proceedings of the February 1991 Bangalore PRA Trainers' Workshop, RRA Notes no. 13*, August, London: IIED, and Bangalore: MYRADA.

Lea, D.A.M. and Chaudhri, D.P. (eds) (1983) *Rural Development and the State*, London: Methuen.

Lehmann, A.D. (1990) *Democracy and Development in Latin America: Economics, Politics and Religion in the Postwar Period*, Cambridge: Polity Press.

Lewis, D.J. (1991) *Technologies and Transactions: A Study of the Interaction between New Technology and Agrarian Structure in Bangladesh*, Dhaka: Centre for Social Studies.

—— (1993) 'Catalysts for change', NGOs, government and agricultural technology in Bangladesh', *Journal of Social Studies*, Dhaka.

Lewis, D.J. with McGregor, J.A., Glaser, M., White, S.C. and Wood, G.D. (1993) *Going it Alone: A Review of Rural Female-Headed Households in Bangladesh*, Centre for Development Studies, University of Bath, occasional paper.

Liamzon, C.M. and Salinas, A.M. (1989) 'Strategic assessment of NGOs in agrarian reform and rural development', in A.B. Quizon and R.U. Reyes (eds) *A Strategic Assessment of Non-Government Organisations in the Philippines*, Asian Non-Government Organisations Coalition for Agrarian Reform and Rural Development, Manila: ANGOC.

Lobo, C. and Kochendörfer-Lucius, G. (1992) ' "The rain decided to help us". Pimpalgaon Wagha: an experience in Maharashtra State, India', Report prepared for GTZ, mimeo, Ahmednagar/Stuttgart.

LP3ES (1980) *Description of Objectives and Activities*, Jakarta: LP3ES.

LSP and ATC (1990) *Gramaseva - Annual Reports 1980-1990*, Narendrapur: Ramakrishna Mission.

McGregor, J.A. (1988) 'Credit and the rural poor: the changing policy environment in Bangladesh', *Public Administration and Development* 8: 467-82.

—— (1989) 'Towards a better understanding of credit in rural Bangladesh', *Journal of International Development* 1: 467-86.

Mandal, B. and Das, C.S. (1990) 'Farming systems research and extension (FSR/E) as a tool for technology transfer', *Farming Systems Research Newsletter* 4(1) March.

Maniates, M.F. (1990) 'Organizational designs for achieving sustainability: the opportunities, limitations, and dangers of state-local collaboration for common property management', mimeo, Kathmandu.

Mascarenhas, J., Shah, P., Joseph, S., Jayakaran, R., Devavaram, J., Ramachandran, V., Fernandez, A., Chambers, R. and Pretty, J. (eds) (1991) *Participatory Rural Appraisal: Proceedings of the February 1991 Bangalore PRA Trainers' Workshop, RRA Notes no. 13*, August, London: IIED, and Bangalore: MYRADA.

Merrill-Sands, D. and Kaimowitz, D. (1990) *The Technology Triangle: Linking Farmers, Technology Transfer Agents and Agricultural Researchers*, The Hague: ISNAR.

Ministry of Finance (1991a) *Economic Survey: Fiscal Year 1990-1991*, Nepal: HMGN Ministry of Finance.

—— (1991b) *Budget Speech: Fiscal Year 1991-1992*, Nepal: HMGN Ministry of Finance.

Morgan, M. (1990) 'Stretching the development dollar: the potential for scaling up', *Grassroots Development* 14(1): 2-12.

Mukhopadhyaya, S. (1988) 'Factors influencing agricultural research and technology: a case study of India', in I. Ahmed and V. Ruttan (eds) *Generation and Diffusion of Agricultural Innovations: Role of Institutional Factors*, Aldershot: Gower.

Musyoka, J., Charles, R. and Kaluli, J. (1991) 'Inter-agency collaboration in the development of agricultural technologies at national and district levels in Kenya', Agricultural Research and Extension Network, *Network Paper no. 23*, London: ODI.

Nebelung, M. (1987) 'Against the net? Mobilisation of assetless agricultural labourers as a development strategy: the case of Bangladesh. A note on some research findings', *Bangladesh Sociological Review* 1(2): 113-19.

Nelson, J. and Farrington, J. (forthcoming) *Information Exchange Networking for Agricultural Development: A Review of Concepts and Practices*, Wageningen: Technical Centre for Agricultural and Rural Co-operation.

New Era (1989) 'Cooperation with non-governmental organizations in agriculture/rural development projects in Nepal', Kathmandu: New Era.

Oram, P. and Bindlish, V. (1981) *Resource Allocations to National Agricultural Research: Trends in the 1970s: A Review of Third World Systems*, Washington, DC: IFPRI, and The Hague: ISNAR.

Orr, A., Nazrul Eskim, A.S.M. and Barnes, G. (1991) *The Treadle Pump: Manual Irrigation for Small Farmers in Bangladesh*, Dhaka: PACT.

Osborn, T. (1990) 'Multi-institutional approaches to participatory technology development: a case study from Senegal', Agricultural Research and Extension Network, *Network Paper no. 13*, London: ODI.

PACT (1987) 'The non-governmental organization sector of Nepal: NGO functions and capacities, needs and potentials for strategic gains in development', April, Kathmandu: USAID.

—— (1989) *Asian Linkages: NGO Collaboration in the 1990s - A Five Country Survey*, New York: PACT.

Pathak, B.R. (1990) 'Baseline survey report of Mahankal, Thakani and Bhotechaur Village Panchayat for farming system', Kathmandu: ActionAid-Nepal.

Paul, S. and Israel, A. (eds) (1991) 'Non-governmental organisations and the World Bank: an overview', in *Non Governmental Organisations and the World Bank: Cooperation for Development*, Washington, DC: World Bank.

Plucknett, D.L., Smith, N.J.H. and Ozgediz, S. (1990) *Networking in International Agricultural Research*, Ithaca, NY, and London: Cornell University Press.

Poffenberger, M. (ed.) (1990) *Forest Management Partnerships: Regenerating India's Forests*, Executive summary of the workshop on Sustainable Forestry, New Delhi, 10–12 September, Delhi: Ford Foundation and Indian Environmental Society.

Pray, C.E. and Echeverria, R. (1989) 'Private sector agricultural research and technology transfer links in developing countries', *Linkages Theme Paper no. 3*, The Hague: ISNAR.

Putzel, J. and Cunnington, J. (1989) *Gaining Ground: Agrarian Reform in the Philippines*, London: War on Want.

Quizon, A.B. (1989) 'A survey of government policies and programmes on NGOs in the Philippines', in A.B. Quizon and R.U. Reyes (eds) *A Strategic Assessment of Non-Government Organisations in the Philippines*, Asian Non-Government Organisations Coalition for Agrarian Reform and Rural Development, Manila: ANGOC.

Quizon, A.B and Reyes, R.U. (eds) (1989) *A Strategic Assessment of Non-Government Organisations in the Philippines*, Asian Non-Government Organisations Coalition for Agrarian Reform and Rural Development, Manila: ANGOC.

Ramakrishna Mission Loka Siksha Parishad (1990) *Farming Systems Research, Annual Report, 1989–90*, Narendrapur: Ramakrishna Mission.

Raman, K.V., Anwer, M.M. and Gaddagimath, R.B. (eds) (1988) *Agricultural Research Systems and Management in the 21st Century*, Hyderabad: NAARM Alumni Association.

Randhawa, N.S. (1988) 'Expected technological advances to meet the needs of the 21st century: focus on agriculture', in K.V. Raman, M.M. Anwer and R.B. Gaddagimath (eds) *Agricultural Research Systems and Management in the 21st Century*, Hyderabad: NAARM Alumni Association.

Rhoades, R.E. and Booth, R.H. (1982) 'Farmer-back-to-farmer: a model for generating acceptable agricultural technology', *Agricultural Administration* 2: 127–37.

Richards, P. (1985) *Indigenous Agricultural Revolution: Ecology and Food Production in West Africa*, London: Hutchinson.

Riddell, R. and Robinson, M. (forthcoming) *Working with the Poor: NGOs and Rural Poverty Alleviation* (provisional title), London: ODI.

Robertson, A.F. (1984) *People and the State: An Anthropology of Planned Development*, Cambridge: Cambridge University Press.

Robinson, M.A. (1991) 'Evaluating the impact of NGOs in rural poverty alleviation – India country study', *Working Paper no. 49*, London: ODI.

Samarasinghe, S.W.R. de A. (1991) 'Lessons of Bangladesh elections', *Economic and Political Weekly*, 14 September.

Sanyal, B. (1991) 'Antagonistic co-operation: a case study of non-governmental organisation, government and donors' relationships in income generating projects in Bangladesh', *World Development* 19(10): 1,367–79.

SAP-Nepal (1988) 'Strengthening Nepalese non-governmental organizations: human resource development needs assessment', mimeo, Kathmandu.

Satish, S. and Farrington, J. (1990) 'A research-based NGO in India: the Bharatiya Agro-Industries Foundation's cross-bred dairy programme', Agricultural Research and Extension Network, *Network Paper no. 18*, London: ODI.

Satish, S. and Siva Mohan, M.V.K.A. (1987) *Mid-Course Evaluation of the World Bank*

Aided Maheswaram Watershed Development Project, Andhra Pradesh, India, Hyderabad: Administrative Staff College of India.

SCF-US/Nepal (1990a) 'Semi-annual report October 1990–March 1991', Kathmandu: SCF-US/Nepal.

—— (1990b) 'Phaseover and sustainable community development', Kathmandu: SCF, US/Nepal.

Schwartzentruber, C. (1985) 'A review of MCC's horticulture programme: 1980–1985', *MCC Special Report*, Dhaka: MCC.

Seth, S.L. and Axinn, G.H. (1991) 'Institutionalising a farming systems approach: a case from India', Paper presented at the 11th International Farming Systems Research Symposium, Michigan State University, October.

Shiva, V. (1991) *The Violence of the Green Revolution: Third World Agriculture, Ecology and Politics*, London: Zed Books.

Shrestha, I. (1990) 'Women in development in Nepal: an analytical perspective', Kathmandu: IDS.

Shrestha, N.K. (1991) 'An overview of NGOs and the agricultural activities initiated by them in Nepal', Paper presented to the Asian regional workshop on NGOs, Natural Resources Management and Links with the Public Sector, Hyderabad, India, 16–20 September.

Skar, H.O. (1987) *Norwegian Aid and the Third System: Non-Governmental Organizations in the Development Aid Process – Future Directions*, Oslo: Norsk Utenrikspolitisk Institutt.

Sollis, P. (1991) 'Multilateral agencies and NGOs in the context of policy reform', Paper presented to the Conference on Changing US and Multilateral Policy toward Central America, Washington, DC, 10–12 June.

Sollows, J.D. and Thongpan, N. (1986) 'Comparative economics of rice-fish culture and rice monoculture in an irrigated area of Ubon Province, north-east Thailand', in *Proceedings, First Asian Fisheries Forum*, Manila, Philippines: ICLARM.

Sollows, J., Jonjuabsong, L. and Hwai-Kham, A. (1991) 'NGO-government interaction in rice-fish farming and other aspects of sustainable agricultural development in Thailand', Agricultural Research and Extension Network, *Network Paper no. 28*, London: ODI.

SSNCC (1991) 'International non-governmental organizations in Nepal', Kathmandu: SSNCC.

Steinberg, D.J. (1986) *The Philippines: A Singular and Plural Place*, Boulder, Colo.: Westview Press.

Streefland, P. and Chowdhury, M. (1990) 'The long-term role of national non-government development organisations in primary health care: lessons from Bangladesh', *Health Policy and Planning* 5(3): 261–6.

Tadem, E.C. (1990) 'Agrarian reform implementation in the Philippines: disabling a centrepiece programme', mimeo, Manila.

Tamang, D. and Hill, R. (eds) (1990) *NGOs in Nepal: Change and Challenge*, Kathmandu: SEARCH.

Tandon, R. (1989) *NGO-Government Relations: A Source of Life or A Kiss of Death*, New Delhi: Society for Participatory Research in Asia.

Thiele, G., Davies, P. and Farrington, J. (1988) 'Strength in diversity: innovation in agricultural technology development in eastern Bolivia', Agricultural Research and Extension Network, *Network Paper no. 1*, London: ODI.

UMN (1991) *UMN 35 Years in Nepal*, Kathmandu: United Mission to Nepal.

UNDP (1988) *Agriculture Sector Review – Bangladesh*, Dhaka: UNDP.

Unia, P. (1991) 'Social action group strategies in the Indian sub-continent', *Development in Practice, 1991* 1(2): 84–96.

Villegas, G. (1990) *GO-NGO Collaboration: The Philippine Experience*, Manila: Ang Makatao.
Wellard, K. and Copestake, J.G. (eds) (1993) *Non-Governmental Organizations and the State in Africa: Rethinking Roles in Sustainable Agricultural Development*, London: Routledge.
Wellard, K., Farrington, J. and Davies, P. (1990) 'The state, voluntary agencies and agricultural technology in marginal areas', Agricultural Research and Extension Network, *Network Paper no. 15*, London: ODI.
White, S.C. (1991a) 'NGOs and the state in Bangladesh: resolving the contradiction?', Paper to the Development Studies Association Conference, Swansea, 11–13 September.
—— (1991b) 'Evaluating the impact of NGOs in rural poverty alleviation: Bangladesh Country Study', *ODI Working Paper no. 50*, London: ODI.
Williams, A. (1990) 'A growing role for NGOs in development', *Finance and Development* December: 31–3.
Wood, G.D. (1981) 'Rural class formation in Bangladesh', *Bulletin of Concerned Asian Scholars* 13: 2–18.
—— (1985) *Labelling in Development Policy*, London: Sage.
Wood, G.D. and Palmer-Jones, R. (1990) *The Water Sellers*, London: Intermediate Technology Publications, and West Hartford, Conn.: Kumarian Press.
World Bank (1988) *The Poverty Challenge in the Philippines*, Washington, DC: World Bank.
—— (1989a) *Environment and Natural Resource Management Study*, Washington, DC: World Bank.
—— (1989b) *India: Agricultural Research: Prologue, Performance and Prospects*, (draft), Delhi: World Bank Office.
—— (1991a) *How the World Bank Works with Non-Governmental Organisations*, Washington, DC: World Bank Publications.
—— (1991b) *World Development Report 1991: The Challenge of Development*, Oxford: Oxford University Press.

INDEX